QUATERNARY GEOLOGY AND ENVIRONMENT OF CHINA

Quaternary Research
Association of China

China Ocean Press Beijing
Springer-Verlag Berlin Heidelberg New York Tokyo

Editor in Chief
Professor Dr. Liu Tung-sheng

Members of the Editing Group
Sun Yi-yin
Lu Rui-lan
Zhao Xi-tao
You Yu-zhu
Wei Lan-ying

Published by: China Ocean Press
                        Fuxingmenwai Street, Beijing, China

Distribution by:
Domestic—   China Ocean Press
Foreign—   Springer-Verlag Berlin Heidelberg
                        New York Tokyo

Springer-Verlag Berlin Heidelberg New York Tokyo is the Exclusive Distributor
for all countries outside the People's Republic of China

ISBN 3—540—13148—5 Berlin Heidelberg
ISBN 0—387—13148—5 New York Heidelberg

PREFACE

China is a country with a vast territory and a variety of natural environments and geological conditions. Consequently, the Quaternary deposits are complete in sequence and various in sedimentary types and bear abundant fauna and flora fossils. It was on this richly endowed land that the ancient men created the splendid material culture, which furnishes favourable conditions for the Quaternary research. For the recent tens of years, the production, education departments and scientific institutions in China have contributed great efforts to the Quaternary research and various branches involved of China in the fields of production and basic theory and considerable progress has been made. For the purpose of greeting the 11th INQUA Congress in upcoming August, 1982 and summarizing the results in the Quaternary research in recent years, a symposium was held in February, 1982 at Beijing by Quaternary Research Association of China and part of the papers submitted to that symposium are collected in this contribution, all together 42 papers with the whole texts and 18 abstracts. The contents of the papers can be divided into 6 parts. The first part: Quaternary stratigraphy, 8 whole texts and 5 abstracts; the second part: the lithology and genesis of Quaternary deposits, 10 whole texts and 2 abstracts; the third part: Quaternary fauna and flora, 6 whole texts and 3 abstracts; the fourth part: Quaternary paleogeography, 13 whole texts and 5 abstracts; the fifth part: the prehistoric man and his material culture, 1 whole text and 1 abstract; the sixth part: the natural resources of Quaternary deposits and their utilization, 4 whole texts and 2 abstracts. The papers collected here are by no means sufficient to represent all the latest results achieved in recent years, but, they involve all the fields of the Quaternary research and reflect on one aspect of the Quaternary research of China.

We are heartily indebted to the relevant departments and colleagues for their enthusiastic helps and vigorous supports during the symposium and the process of editing,

i

examining and publishing the papers of this contribution.
Owing to the pressing of time, there are bound to be over-
sights and omissions during editing, we hope the readers
could give us their valuable comments without hesitation.

Liu Tung-sheng
May 3, 1982
Beijing

CONTENTS

iii

II. Lithology and Genesis of Quaternary Deposits

III. Quaternary Fauna and Flora

IV. Quaternary Paleogeography

vi

PLEISTOCENE STRATIGRAPHY AND PLIO/PLEISTOCENE BOUNDARY IN CHINA

LIU TUNG–SHENG
(Institute of Geology, Academia Sinica)

AND DING MENG–LIN
(Institute of Geology, National Bureau of Seismology)

Abstract—Continental Pleistocene sediments of East China can be grouped into four main different depositional stratigraphical types: the intermontane basin, the loess, the cave deposits, and the maritime plain deposits. According to the biostratigraphical, lithostratigraphical and magnetostratigraphical studies, the four main deposit types all favor to draw the Pliocene/Pleistocene boundary at a place in coincidence with the boundary of Matuyama/Gauss Epoch. This boundary corresponds with the boundaries lying between the upper and lower Nihewan series (middle and early Villafranchian), Yuanmou series (in narrow sense) and Sagou series, and also between the Wucheng Loess and the underlying Red Clay.

A considerable amount of studies on the Quaternary stratigraphy and the boundary between the Pleistocene and the Pliocene in China have been made. The Quaternary deposits in China were classified into cave, gravel, fluviolacustrine and earth-like four accumulation types by Young[1] at the 18th International Geological Congress in 1948.

Since the founding of New China in 1949, most geologists adopted the proposal suggested by the 18th International Geological Congress, that in continental strata, Villafranchian was taken as the lower Pleistocene of the Quaternary [2,3]. During the 70s on account of magnetostratigraphic studies, some authors suggested to place the lower boundary of the Quaternary at the boundaries of various polarity events in the magnetostratigraphic time scale [4,7]. In recent years, lithostratigraphic, biostratigraphic and magnetostratigraphic studies were made in a great amount of cores from the North China maritime plain, which provides a basis for the revision of the classification and correlation of the Quaternary stratigraphy as well as the study of the boundary between the Pleistocene and the Pliocene in China. On the basis of the latest available magnetostratigraphic and other chronostratigraphic data obtained from the northern part of the North China maritime plain along with the research results from other districts, the authors tend to put forward in this paper some suggestions on the stratigraphy of various types of Pleistocene deposits and to discuss the boundary between the Pliocene and Pleistocene in China.

1

I. REPRESENTATIVE SECTIONS FROM VARIOUS TYPES OF QUATERNARY DEPOSITS

1. Alluvial–Lacustrine Deposits in Intermontane Basins

The intermontane basins in China were mainly formed in Late Tertiary and influenced by the Himalayan Movement, such as the Fenwei Graben between Shanxi and Shaanxi Provinces and the Yuanmou Basin of Yunnan Province. Lacustrine of fluviolacustrine sediments were deposited successively from Middle-Late Miocene to Quaternary. Generally, the thickness of the sediments is 500–600 m, sometimes the greatest thickness exceeds 1,000 m.

(1) Yangyuan Basin

It is located at the northwestern part of Hebei Province. Sediments in this basin include lacustrine deposits of an age from Pliocene to Pleistocene and the name Nihewan series was established from a village located here [8]. At the margin of the basin, where Nihewan series covers a denudated surface of the Pliocene *Hipparion* Red Clay, the exposed Nihewan series is 100–150 m in thickness. Vertebrate fossils found from the upper part (yellow Nihewan) of Nihewan series amounts to 22 genera and 37 species. The characteristics of this fauna are that it includes both remnants of Pliocene species, such as *Proboscidipparion sinensis, Nestoritherium* sp., and forerunners of Pleistocene species, such as *Equus sanmeniensis, Eucladoceros boulei, Elaphurus bifurcatus,* etc.

It is considered in China that Nihewan series at Yangyuan is a type section of the Lower Pleistocene, and can be correlated with the Villafranchian of Europe [3,9].

In recent years, a new fossil horizon in the lower part of the Nihewan series (green Nihewan) is found, which contains forms older than those previously found in the upper part (yellow Nihewan), probably of Middle or Late Pliocene to Early Pleistocene in age, such as *Lynx* sp., *Zygolophodon* sp., *Hipparion* sp., *Paracamelus* sp., *Antilospira* sp. [9,10].

In 1979 Cheng Gou-liang and others [11] made a magnetostratigraphical study of the Nihewan series at Hongya and Haujiatai profiles. According to these authors the Matuyama/Gauss boundary is situated at the boundary between the upper and lower Nihewan series.

2

(2) Yuanmou Basin

It is located at the southern bank of Jinsha River in the northeastern part of Yunnan Province. In the basin, Pliocene and Pleistocene fluviolacustrine sediments are 600 m in thickness, consist mainly of sand and gravels, sand and clay with lignite beds. Since 1938 on account of the finding of teeth of *Equus yunnanensis,* Yuanmou formation was taken as the type section of the Lower Pleistocene in South China, and correlated with the Nihewan series in North China. In 1961, Zhou Ming-zhen studied the *Enhydriodon* cf. *falconeri* found in the Shagou lignite bed at the lower part of the Yuanmou Basin and held that the horizon bearing *Enhydriodon* could be correlated with the Dhok Pathan bed of the Siwaliks and the age of the Shagou lignite bed turned to be Late Pliocene. Later, this Pliocene strata was subdivided from the Yuanmou formation (in broad sense) and named Shagou formation by You Yu-zhu et al.. The Shagou formation comprises mainly *Hipparion* sp., *Dicerochinus* sp., *Serridetinus* sp., *Stegolophodon banguaensis, Stegodon yuanmouensis, Chilotherium, Enhydridon* cf. *falconeri,* etc. Most of them are fossils of Late Pliocene [12]. You et al. restricted the use of the name Yuanmou formation (in narrow sense) to Early Pleistocene. The fossils found in the Yuanmou formation (in narrow sense) overlying the Shagou formation consist *Canis yuanmouensis,. Vulpes* cf *chikusanesis, Cynailurus* sp., *Felis tigris, F. pardus, Hyaena licenti, Rhinoceros* cf *sinensis, Equus yunnanensis Stegodon yuanmouensis, S. zhaotongensis, Sus scrofa, Muntiacus* cf. *bohlini,* etc. Though some Tertiary relics still existed, but *Equus yunnanensis* was already quite flourishing at this time. They gave a name to the Middle Pleistocene fossil bearing beds, Shangnabang formation to the upper part of the Yuanmou formation. The fossils are characteristic to those of the Middle Pleistocene in South China, such as *Stegodon orientalis* [12].

Qian Fan and others subdivided the Yuanmou formation (in broad sense) into four parts totally 28 layers from bottom upwards [13]. The paleomagnetic results published by Li Pu in 1976 [6], and Chen Guo-liang et al. in 1977 [7] were quite consistent. They all held that sediments above the upper portion of the second part was of Matuyama Epoch and the lower portion of the second part was of Gauss period. They all claimed the lower boundary of the Quaternary should be placed near by the lower boundary of the Mammoth event of the Gauss normal polarity epoch.

As for the boundary between Shagou formation and Yuanmou formation (in narrow sense) (the upper and lower portions of the second part of the Yuanmou formation of Qian), according to the paleomagnetic studies by Li and Cheng, it lies right at the boundary between Matuyama and Gauss Epochs. This result together with the biostratigraphic evidence in Yuanmou Basin, all favors that the boundary between the Pleistocene and the Pliocene lies at the beginning of the Matuyama reverse polarity epoch, which is around 2.43 million years.

3

2. Loess Deposits

Loess is the most widespread Quaternary deposit in North China. It occupies about 530,000 km², with various thickness. The thickest reaches more than 150 meters. It is a succession of continental deposits from Early Pleistocene up to recent. The author subdivided the loess into Lower Pleistocene Wucheng Loess, Middle Pleistocene Lishih Loess and Upper Pleistocene Malan Loess and Holocene Loess [14].

The stratigraphy of the loess can be described according to their depositional characteristics into:

(1) Plateau loess

Loess plateau situated at the center of the loess mantle in China, where loess accumulated continuously to a great thickness, and deposited on an uneven erosional surface, covering various rock units but generally on the Lower Pliocene *Hipparion* Red Clay.

Fossils found from Wucheng Loess include *Nycterutes sinensis, Hypolagus brachypus, Hipparion (Proboscidipparion) sinensis, Hipparion* sp., *Sus lydeckkeri, Gazella* sp., etc.

The faunal characterestics are that there are some relics from the Pliocene, such as the *Probosidipparion,* and also the Pleistocene elements, *Equus sanmeniensis*. In general, the nature of the Lower Pleistocene loess fauna is similar to that of the Nihewan fauna (upper Nihewan series).

Recently the author and Dr. Fritze Heller made a paleomagnetic study from a core drilled by the 2nd Hydrogeological Brigade of the Shaanxi Geological Bureau in 1980, near Heimukou, Lochuan. More than 250 samples were measured (135 meters of loess), at the paleomagnetic laboratory of the Institute of Geophysics of the Erdegebignossie Technisches Hochschule, at Zurich, Switzerland, under the supervision of Prof. W. Lowrie.

The preliminary results show that the Brunhes/Matuyama boundary is situated at a depth of 62 m or so and the Jaramillo event was recorded at a depth of 70 m or so. Further downward at a depth of about 110 m in the Wucheng Loess there is a group of normal polarity samples which could be no others but the Olduvai event within the Matuyama reverse polarity epoch. Below the Olduvai there are more than 20 meters of Wucheng Loess, which contacts directly with Pliocene Red Clay.

From the magnetostratigraphic study of the loess from Lochuan, it shows that the loess is a deposit laid prior to the Olduvai event of the Matuyama Epoch. Therefore, in plateau loess successions the lower boundary of the Quaternary seems to be lower, down to the lower limit of the Olduvai event, though not yet quite definite, probably it situates at the boundary of matuyama and Gauss Epochs.

4

(2) Valley loess

Within the loess plateau, loess usually covers terraces developed along large river valleys (Luohe River, Huanghe (Yellow) River and Weihe River). On the river terraces, Wucheng Loess usually is absent. Lishih Loess and Malan Loess cover on the fluvial-lacustrine deposits sometimes forming the benches of the terraces or constituting a part of the terrace deposits.

The lacustrine deposit underneath the Middle Pleistocene Lishih Loess is of Early Pleistocene in age, which can be differentiated into a younger member and an older one.

The younger lacustrine member is discovered from Sanmenxia, Henan Province, and the upper part of Youhe formation, Weinan, Shaanxi Province [15].

Mammalian fossils in this member are similar to those from the upper part of Nihewan, which is of late Early Pleistocene.

The lower lacustrine member includes Leijiahe group, Youhe group, and the mammalian fossils studied by Huang, W. P. and Xue, X. X. [15,16] show a characteristic transitional from the Pliocene *Hipparion* fauna to the Nihewan fauna. There are archaic elements such as *Hipparion, Cervarites, Annacus, Pentalophodon, Qurliqnoria, Sinocastor; Presphneus, Oioceros, Sinotherium.* Within the same fauna there are Early Pleistocene elements such as *Equus* cf. *sanmeniensis, Leptobos* sp., *Elephas (Archidiscodon) youheensis, Crocuta (Hyaena) licenti, Nyctereutes sinensis, Ochotonoides* sp., *Sus* sp., etc.

This fauna, according to the authors, if having been studied extensively, might be comparable with the early Villafranchian fauna in Italy. Xue has pointed out that *Elephas youheensis* (sp. nov.) could be compared with Etouaires element. So its stratigraphical position is not yet very definite. It might be the earliest Pleistocene, or the uppermost Pliocene in continental Quaternary stratigraphy of China.

3. Cave Deposits

(1) South China

It is the so called "*Stegodon-Ailuropoda* fauna" including cave fauna of Early Pleistocene to Late Pleistocene. Recently, Huang [17], has subdivided this fauna into: (i) Early Pleistocene *Gigantopithecus* fauna, found in caves situated at a height about 100 meters above the rivers. *Gigantopithecus blacki*, a characteristic form of this fauna, is more advanced than *Gigantopithecus bilaspurensis* from the Middle Pliocene Dhok Pathan Series [18]. (ii) Middle Pleistocene Bijiashan fauna, it is the typical *Stegodon-Ailuropoda* fauna. Caves with this fauna are situated at a height about 60 meters above the rivers. (iii) Late Pleistocene Liujiang Manfaunna. Caves with Liujiang Man (*Homo sapiens*) generally are situated at about 25 meters above Liujiang River.

(2) North China: Cave deposit of Zhoukoudian

Zhoukoudian caves are situated at the foothills of Yanshan Mountains, south-west of Beijing. It is the famous place where "Peking Man" (*Homo erectus Pekinensis*) was first discovered. The bottom of the Peking Man Cave is situated about 5 meters above the Zhoukoudian River. From top to bottom, 17 layers have been excavated. It is named Zhoukoudian group from 1 to 13 layers. Fossil mammals mainly come from layers 1–10. This Zhoukoudian fauna resembles the Hoxinian of the British Isles and Steinheim of Germany, and Trinil of Indonesia [19]. Paleomagnetic study of layers 1–13 shows that they belong to the Brunhes normal epoch. Layer 14 to layer 17, mainly a kind of tranquil water deposit which might be formed at a stage when the caves were situated underground, samples from these layers are of negative measurements and could be explained that they were deposited at the Matuyama reverse epoch and named Longgushan (Dragon Bone Hill) group [20]. There is no evidence of Lower Pleistocene/Pliocene contact in Zhoukoudian caves.

4. North China Maritime Plain

This plain began to develop at the end of the Mesozoic age. Cenozoic sediments reach a depth of about thousands of meters. Quaternary deposits consist mainly of fluvial-lacustrine deposits interbedded with marine layers, with a thickness reaching about 300–500 meters. In recent years, magnetostigraphic studies have been undertaken on the cores recovered from deep drillings by various authors. In 1977 Li and others studied a series of cores at Jizhong depression. The hole is 650 meters deep. Brunhes, Matuyama, and Gauss polarity epochs were recorded. They were at the opinion that the lower boundary of the Pleistocene in this core might be situated at the Olduvai event of the Matuyama reverse polarity epoch, which is at a depth of 361 meters below the ground surface. The boundary of the Middle and Lower Pleistocene is at the boundary between the beginning of the Brunhes polarity and the end of the Matuyama Epoch, at 140 meters deep [4].

In 1978 Zhang et al. studied the Ca. 13 core, the depth of the hole is 600 meters. Brunhes, Matuyama, Gauss and Gilbert polarity epoches were recorded. As for the lower boundary of the Pleistocene, these authors put it at the Mammoth event of the Gauss Epoch. Its depth is 410 meters [21].

1979, An Zhi-sheng and others studied cores from a deep drilling hole at the Beijing depression which is 801.9 meters deep. In this hole Jurassic rocks were encountered at 793.7 meters beneath the ground surface. Brunhes, Matuyama and Gauss polarity epochs were recorded. Within a marine layer at a depth.

6

about 420 meters below they discovered *Hyalinea baltica, Globigerina bulloides* and other foraminifers. The total number of them amounts to 28 species. Since *Hyalinea baltica* is a form of the Calabrian deposits in Europe, these authors suggested that the boundary of Pliocene/Pleistocene should be set at the base of the *Hyalinea-Globigerina* horizon and in turn it coincides with the boundary of Matuyama and Gauss. Its depth is 468 meters [5].

II. THE PLIO/PLEISTOCENE BOUNDARY

On account of the above-mentioned stratigraphical characteristics of the four main depositional types of Quaternary sediments in China, the Pliocene/Pleistocene boundary can be summarized and discussed as below.

1. The Nihewan Series (Sanmenian Series) which contains a Villafranchian fauna, has long been taken as the Lower Pleistocene of China, Recently the Nihewan series has been subdivided into two parts, the lower Nihewan and the upper Nihewan. A restudy of the fossils listed by P. Peviteau and P. Teilhard shows that they are of advanced types and could be correlated with those of middle Villafranchian. From the fact that those fossils described by Piveteau were come from the upper Nihewan series, and that those fossils newly discovered from the lower Nihewan beds (the lower part of the Sanmen group, or the Youhe group) show characteristics archaic to those of the Nihewan fauna, the Youhe fauna can be compared with the early Villafranchian fauna of Europe. Therefore, it is feasible to draw the boundary of Pliocene/Pleistocene at the boundary between the lower and upper Nihewan Series.

2. The opinions favorable to draw the Pliocene/Pleistocene boundary at the base of the Mammoth event of the Gauss normal polarity epoch or even still lower, which makes the time span of the Quaternary up to 3.0–3.5 million years (BP) are not in accordance with the biostratigraphical studies. This would include the lower Nihewan series, (the lower part of the Sanmen group and the Youhe group and the Sagou group) into the Lower Pleistocene. Since vertebrate fossils in these beds are forms corresponding to those of the Early Villafranchian of Europe, it is probable this fauna is a representative of the Late Pliocene. However, in China, more extensive study of the Villafranchian fauna are needed. Especially important is that the strata which contain Early Villarfranchian fauna, are mostly deposits succeeded by those containing Middle or Late Villafranchian fauna uninterruptedly.

3. Some of the magnetostratigraphical studies of the Nihewan series and the loess profiles, have put the Pliocene/Pleistocene boundary at the base of the Olduvai event. It seems to be a little higher than the lithostragraphical boundary, because at Lochuan Loess profile, beneath the Olduvai event there still exist 20 and more meters of Wucheng Loess deposits, which are clearly demarcated from the underneath Pliocene Red Clay. Therefore, from the view of lithostratigraphy, the boundary seems to be appropriately situated at

7

the base of the loess, which might turns out to be the boundary between the Matuyama and the Gauss Epochs. Furthermore, from the biostragraphical study of the type locality of the Wucheng Loess (Lower Pleistocene) at Liushu-gau, Wucheng, Shanxi Province, where Nihewan fauna were found at the basal part of the loess profile, the occurrence of the Nihewan fauna at the basal part of the Wuchen Loess also favors the Early Pleistocene boundary being situated at the base of the loess.

4. Biostratigraphical, lithostratigraphical and magnetostratigraphical studies of the intermontaine basins deposits, the loess, the cave deposits and the mari-time plain deposits, all favor to draw the Pliocene/Pleistocene boundary at a place coinciding with the boundary of Matuyama/Gauss Epoch, which is of 2.43 million years (BP) in age. This boundary corresponds with the boundary lying between the upper and lower Nihewan series, (upper and lower Sanmenian), Yuanmou series (in narrow sense) and the Shagou series, the boundary between the Wucheng Loess and the underlying Red Clay. The lithological characteris-tics of all those four types of sediments are quite obvious and fauna containing in them are also distinguishable.

REFERENCES

[1] Young, C. C. 1948, The Pliocene-Pleistocene boundary of China, Rept. of 18th Ses-sion, Part IX, Intern. Geol. Congress.

[2] Young, C. C. 1949, The Pliocene-Pleistocene boundary. Science. (in Chinese) Vol. 31, No. 11.

[3] Liu Tung-sheng et al., 1964, Problems of Quaternary geology, Science Press.

[4] Li Hua-mei, Wang Jun-da et al., 1977, Paleomagnetic investigation on drilling cores in the Hengshui area, North China Plain, Geochemica, No. 3.

[5] An Zhi-sheng, Wei Lan-ying et al., 1979, Magnetostratigraphy of the Core S-5 and the transgression in the Beijing area during the Early Matuyama Epoch, Geochemica, No. 4.

[6] Li Pu, Qian Fang et al., 1976, The Preliminary study of Yuanmou Man fossil dated by paleomagnetic method, Scientia Sinica, No. 6.

[7] Cheng Guo-liang, Li Su-ling et al., 1977, Discussion of the age of *Home erectus yuan-mouensis* and the event of Early Matuyama, Scientia Geologica Sinica, No. 1.

[8] Teilhard de Chardin, P. et Piveteau, J., 1930, Les mammifères fossiles de Nihewan (Chine). Annales de Paléontologie. Tome XI.

[9] Group of Nihewan Cenozoic, 1974, Observation on the Later Cenozoic of Nihewan, Vertebrata PalAsitica, Vol. XII, No. 2.

[10] Tang Ying-jun, You Yu-zhu, Li Yi, 1981, Some new fossil localities of Early Pleis-tocene from Yangyuan and Yuxian Basins, northern Hopei, Vertebrata PalAsitica, Vol. XIX, No. 3.

[11] Cheng Guo-liang, Lin Jin-lu et al., 1978, A Parliminary paleomagnetic survey of the Nihewan beds, Scientia Geologica Sinica, No. 3.

[12] You Yu-zhu, Liu Hou-yi et al., 1978, Later Cenozoic strata in Baguo Basin, Yuanmou, Yunnan. Collected Papers of Stratigraphy and Paleontology, Chinese Academy of Geological Sciences, No. 7, Geology Press.

[13] Qian Fang, Pu Qing-yu et al., 1977, The Quaternary Glacial Epoch and stratigraphic correlation of sediments in Yuanmou Basin, Yunnan. Collected Papers of the Quater-

nary Glacial Geology in China. Geology Press

[14] Liu Tung-sheng et al., 1965, Loess deposits of China, Science Press.

[15] Xue Xiang-xu, 1981, An Early Pleistocene Mammalian fauna and its stratigraphy of the Youhe River. Weinan, Shaanxi. Vertebrata PalAsiatoca, Vol. XIX, No. 1.

[16] Huang Wan-bo, Cheng Shao-hua, et al., Discovery and significance of the Early Pleistocene mammalian fauna from Lingtai, Gansu Province. (in press)

[17] Huang Wan-bo, 1979, On the age of the cave-fauna of South China, Vertebrata Pal-Asitica, Vol. XVII, No. 4.

[18] Woo Ju-kang, 1962, The mandibles and dentition of *Gigantopithecus*. Palaeontologia Sinica, New Series D, No. 11.

[19] Hu Chang-kang, Study of mammalian faunas from Loc. I Choukoudien. (in press)

[20] Yang Zi-gen, Mu Yun-zhi, 1981, A new note on Late Cenozoic geology in Zhoukoudian Area. Kexue Tongbao Vol. 26, No. 17.

[21] Zhang Hong-cai, et al., 1978, Paleomagnetic study of two sediment cores from the northern coastal region of China. Oceanologia et Limnologia Sinica, Vol. 9, No. 2.

THE LOWER BOUNDARY OF THE CONTINENTAL QUATERNARY IN SOME AREAS OF CHINA BASED ON PALYNOLOGICAL DATA

Song Zhi-chen, Liu Jing-ling, and Tang Ling-yu

(Nanjing Institute of Geology and Palaeontology, Academia Sinica)

Abstract—The present paper discusses the lower boundary of the continental Quaternary in some areas of China based on the climatostratigraphy. The writers are of the view that the first climatic deterioration which resulted in the changes of the composition of Neogene flora marks the beginning of Quaternary. This phase is equivalent to that below the first mammal fauna of the Villafranchian, and its age can also be referred to 2.5—3.1 million years.

The lower boundary of the Quaternary, or the Pliocene-Pleistocene boundary, has been a subject of dispute for a long time in the geological circle. Following the recommendation adopted by the 18th International Geological Congress, the Pliocene-Pleistocene boundary should be drawn on the basis of the changes in marine faunas, at the horizon of the first indication of climatic deterioration in the Italian Neogene succession, and the marine Calabrian stage and its terrestrial equivalents — the Villafranchian stage, were placed to Lower Pleistocene as its basal member.

Unfortunately, it was discovered from later investigation that the marine Calabrian stage is not completely corresponding to the terrestrial Villafranchian stage in geological age, assuming that the base of the Calabrian is at or near the zone boundary N21-22 of planktonic foraminifers, almost equivalent to the Gilsa normal event of Matuyama reversed epoch, about 1.8 million years in chronology. The Villafranchian stage is noted for a series of mammal fauna [1]. Combining with the studies of fossil vertebrates of some adjacent localities, the section of Villafranchian is characterized by the presence of three mammal fauna zones: the lowest one is equivalent to the fauna of Perrier-Etouaires locality, with an age of 3.1 to 2.5 million years by K-Ar method. The fauna of Perrier-Roca-Neyra locality is dated at 2.5 million years; and the youngest fauna which contains *Mimomys pliocaenicus* is dated by means of two overlying lavas of 1.8 and 1.92 million years respectively. Therefore, the marine Calabrian stage may be correlated to the upper part of the Villafranchian in chronology [1,2].

The so-called macrofloras of the upper Pliocene Reuver clay and Tegelen clay from Netherlands are not the same in terms of detailed investigation of palynology. In the former, the elements relating to the present-day plants of East Asia and North America (such as *Sequoia, Taxodium, Sciadopitys, Nyssa* and *Liquidambar*) are rich in species, amounting to 50% of the total number of species. And in the latter, only such East Asian and North American elements as *Tsuga, Pterocarya, Carya* and *Eucommia* have been discovered, with exotic species only constituting about 15% of the total species of Tegelen flora; or otherwise there occur a lot of new living elements of holarctic distribution which are hardly to be discovered in the Reuver flora. Zagwijn [2] pointed out that the disappearance of warmth-loving elements in the Tegelen flora was caused by the change of cold climate. Hence, between the floras of Reuver and Tegelen, there is a cooling climatic phase. This phase of cold climate has been named as the Praetiglian. In the light of the recommendation of the 18th International Geological Congress, the Pliocene-Pleistocene boundary should be placed at the horizon of the first indication of climatic deterioration in the Neogene succession, and from the viewpoints of paleobotany and palynology, the Pliocene-Pleistocene boundary in western Europe may be defined at the base of the Praetiglian cold stage, i.e. between the Reuverian and Praetiglian.

Based on the referable data, the climatic fluctuations began to happen at the middle period of Neogene, which resulted, however, insignificantly in the composition of flora. Some palynologists believe that the climatic deterioration which strongly influenced the composition of flora would be a lower temperature than that of the present day in each area. Therefore, this would be the same case with the so-called first climatic deterioration of Neogene and so was the climatic deterioration in the periods between Reuverian and Praetiglian. Similar cases have been discovered in other parts of Europe and below the first mammal fauna of the Villafranchian stage as well.

Of course, an important geological boundary may be defined on the basis of sufficient paleontological evidence as well as materials of sediments, geological texture and chronology, and the lower boundary of the Quaternary would also be defined in the same way. In the meantime, one of the Quaternary characteristics is of course the arrival of glaciation (ice age). Although the glaciers appeared only in some areas, their influence was widespread in vast areas, due to the strongly climatic fluctuations, being cool in advance and warm in retreat of ice sheets. Some scientists have attached greater importance of glaciation to the defining of lower boundary and series of the Quaternary. Since glaciation and climatic fluctuations as well as changes of floras and faunas are supplemental with each other and they are the reflections of different sides on one factor, it is reasonable to take climatic changes as an important reference to the defining of the lower boundary of the Quaternary, known scientifically as climatostratigraphy.

It is the aim of the present paper to tentatively discuss the lower boundary of the continental Quaternary of China, based on the palynological data and the viewpoint of climatostratigraphy. In the writers' opinion, the first climatic

11

deterioration which led to a change in the composition of Neogene flora is the beginning of the Quaternary. This phase is equivalent to the sedimentary phase below the first mammal fauna of the Villafranchian, being likewise 2.5–3.1 million years in age. The lower boundary of the continental Quaternary in some areas of China may be briefly stated as follows:

I. THE NIHEWAN FORMATION OF HEBEI PROVINCE

1. The Nihewan formation in Yangyuan County of Hebei Province is well known for its rich fossil mammals. The age of the formation was assigned as a stratotype of the early Pleistocene in North China by most geologists and paleontologists many years ago. Based on the lithological and paleontological evidences, this formation may be generally divided into two parts. The upper part consists of grayish-yellow or grayish-white sand, gravel, clayey sand, silty-clay and clay; while the lower one is composed mainly of grayish-white or grayish-green sand, sandy-gravel and siltstone beds. The formation lies unconformably or disconformably on the *Hipparion* Red Clay layer.

Attaining more than 40 m in thickness, the upper part of the Hutouliang section of Nihewan formation is mainly composed of sand and gravel deposits, while the lower consists of fine-grained beds. In the upper part, only a few pollen grains have been found, whereas in the lower, there is plenty of spores and pollen[3]. According to the characteristics of palynological assemblage, these zones may be recognized in ascending order as follows:

Zone I: The pollen of conifer is absolutely dominant (up to more than 95%), among which the pollen grains of *Pinus, Picea* and *Abies* are especially predominant. The pollen of broad-leaved trees and herbs are present sporadically.

Zone II: This zone is characterized by the predominance of herbs, amounting frequently to more than 70% of the assemblage. They are mainly the pollen grains of Artemisia, Chenopodiaceae, Cruciferae, etc., with small amounts of grains of broadleaved trees, such as *Betula, Quercus, Ulmus* and *Tilia* (sometimes up to more than 10%).

Zone III: This zone is very similar to Zone I, but the percentage of the pollen of cupressaceae and herbs is higher than that of Zone I. Similar assemblages have also been found in the sections of Hongya, Chengdiang. The above-mentioned variation of palynological zones clearly reflects the evidence of climatic fluctuations. So far as our knowledge goes, it is the first real colder phase of Zone I in this region since the Neogene time. It is presumed that the climate then was cooler than that of the present. Therefore, we may take this climate cooling phase as a mark for defining the Plio-Pleistocene boundary in the region. In other words, the lower boundary of the Quaternary should be placed at a level below Zone I, corresponding generally to the boundary between the *Hipparion* Red Clay layer of Pliocene and the Nihewan formation. It is recorded that the paleomagnetic age of the Hongya section

12

of Nihewan has been determined to be 2.6 m.y.

2. The Yuxian Basin: The palynological zone IV (211.8–126.6 m) from Borehole No. Yu–142 indicates the rich presence of coniferous pollen grains, including *Pinus* (8—63%), *Picea* (10—68%), *Abies* (1—13%) in association with a few *Ephedra, Tsuga, Podocarpus, Cedrus* and *Ketelceria*. The broad-leaved trees are represented by pollen grains of *Ulmus, Betula, Quercus, Tilia*, etc., herbaceous plants by *Artemisia* and Chenopodiaceae, while spores of ferns consist chiefly of Polypodiaceae, reaching up to the highest percentage of 17%. The aspect of the pollen assemblage is comparable to that of Zone I of Hutouliang. Thus the lower boundary of Quaternary should be drawn at a level below Zone IV in the Yuxian Basin.

II. THE YOUHE FORMATION OF SHAANXI PROVINCE

The Youhe formation of Early Pleistocene in Shaanxi Province, more than 200 m in thickness, may be generally divided into two parts, the lower consists of grayishwhite or grayish-yellow fine sand, grayish-green clay and subclay; while the upper contains mainly brown-yellow or brown-red subclay with fine sandy beds. The lower sequence which is an equivalent of the Nihewan formation, overlies unconformably the Lantian formation of Pliocene. Three sections in this area, considered as equivalents of the lower part of Youhe formation, have been studied palynologically by the Institute of Botany, Academia Sinica and others[4]. Based on the characteristics of the palynological assemblage, four zones may be recognized in ascending order as follows:

Zone I: The dominant elements are woody plant pollen grains (57—96%), in which the coniferous pollen grains amount to 61% (39—84%), while the broad-leaved tree pollen grains are 39% (16—61%). The coniferous pollen grains are mainly *Abies, Picea, Tsuga, Pinus* and Cupressaceae, whereas the broad-leaved tree pollen grains are *Ulmus, Celtis, Betula, Carpinus, Ostrya Quercus, Tilia* and Rosaceae;

Zone II: Herbaceous plants are chiefly composed of common herbaceous plants in northern temperate zone: the broad-leaved tree pollen grains are *Ulmus, Celtis, Carpinus, Quercus, Tilia,* Rosaceae, *Juglans, Pterocarya* together with a few *Carya, Liquidambar,* etc.;

Zone III: The broad-leaved tree pollen grains are composed chiefly of *Salix, Juglans, Ulmus, Corylus, Carpinus, Quercus, Celtis*, etc., and

Zone IV: The pollen of woody plants is absolutely dominant, including *Pinus, Abies, Tsuga, Ulmus, Celtis, Juglans, Quercus, Tilia, Podocarpus, Larix, Cedrus,* etc., along with only a small amount of herb pollen grains.

From what has mentioned above, it can be seen that in the first palynological zone may be found plenty of pollen grains of conifers, among which the richer presence of pollen grains of *Abies* (more than 20% in the total of woody plant grains) is of special importance, as it is a good indicator of the cooling climate.

13

Moreover, a few typical Tertiary tropical and subtropical elements have been discovered in this assemblage. Judging from these features, we can assume that the temperature during that time was much cooler than that of Neogene which was in the subtropical climate. Thus, the first zone may belong to the close of the first glacial period or the beginning of the first interglacial period in North China. It seems, therefore, certain that the boundary of Plio-Pleistocene should be placed at a level below the Youhe formation which coincides with the lower boundary of the Nihewan formation.

III. SHANXI PROVINCE

1. The Zhangcun formation in the Yushe Basin. The Zhangcun formation, 150–300 m thick, consists of grayish-green, grayish-black, grayish-white, yellow-green sands, sandy clays, silt soils and violet, brown clays or sandy clays alternated with grayish yellow and gray sand layer. The base gravel and sand gravel beds are in an angular unconformity with the underlying Renjianao formation. In this formation abundant fossils were found such as vertebrates, abdominals, insects, birds, plants and microfossils. Among these fossil mammals are *Rhinoceros orientalis*, *Myospalax* sp., *Ochotona* sp., while there are plants such as *Quercus*, *Acer*, *Ulmus*, *Populus*, *Salix*, *Carpinus*, and *Fagus* along with a few branches and fruits of conifers.

The pollen assemblage: Numerically, it is dominated by woody plants (60–94%), such as Cupressaceae (?5–57%), *Larix* (2–26%), *Salix* (5–36%), Ulmaceae (5–21%). The elements of herbs including mainly *Artemisia* (5–19%) and Chenopodiaceae (2–30%), are much less than those of woody plants (15%). It is interesting to note that the percentages of cold-loving conifers such as *Pinus* (13–14%), *Abies* (7–9%) and *Picea* (12–23%) markedly increase in the upper part of the Zhangcun formation, while the elements of the broad-leaved plants, such as Ulmaceae (2–4%), Oleaceae (3%), *Betula* (3%), *Ulmus* (5–13%), *Celtis* (3%), *Salix* (1%) and *Quercus* (4%) decrease. Thus, according to the principle of climatostratigraphy, the Plio-Pleistocene boundary at the Yushe Basin may be defined at the base of Zhangcun formation.

2. The Datong Basin. On the basis of the drilling data from Shuoxian, Yingxian, Shanyinxian and Datong City, the Quaternary of Datong Basin, more than 600 m thick, consists principally of grayish, yellow, grayish-green silty clay and may be divided into the following 4 palynological assemblage zones in ascending order:

Zone I: Zone of *Ulmus*, *Pinus*, *Tsuga* and Chenopodiaceae. NAP frequency is less than AP frequency. Pollen grains of Ulmaceae, *Tsuga* and Chanopodiaceae are especially rich;

Zone II: Being characterized by the poverty of pollen:

Zone III: Zone of *Picea*, *Abies*, *Pinus* and *Artemisia*, with low frequencies

14

of *Tsuga,* and

Zone IV: Zone of *Pinus, Betula,* Chenopodiaceae and Polypodiaceae. The quantity of AP is comparable to that of NAP. The elements of broadleaved trees (mainly *Betula*) of this zone are more abundant than those of all other zones.

Judging from the plentiful coniferous pollen grains, Zone III from the Datong Basin is indicated by a remarkable cooling phase. So far as the writers are aware, this major cooling phase is the first remarkable climatic deterioration since the Neogene time in this region and may be taken as a mark of the beginning of Quaternary in the Datong Basin. In other words, the lower boundary of the Quaternary in this basin is at the base of the sporo-pollen assemblage of Zone III.

IV. HENAN PROVINCE

The sections of the Sanmen formation from Sanmenxia area in Henan and the fauna contained in them have been researched in detail. This formation, 20–30 m thick, is composed chiefly of coarse sandstone, sandy conglomerate and marlite; in the upper part being silt soil and reddish soil[5]. It is unconformable with the underlying *Hipparion* Red Clay layer, and may be correlated with the Nihewan formation in Hebei and the Villafranchian in Europe, belonging to early Pleistocene in age. A lot of palynological fossils, such as *Pinus, Picea, Juniperu, Betula, Ulmus, Carpinus, Typha, Artemisia* Gramineae and Liliaceae have been found in this formation from Sanmenxia Area.

V. YUNNAN PROVINCE

1. The Songhua Basin, Kunming. The Neogene and the Quaternary deposits in the Songhua Basin situated in the northeastern part of the Kunming Basin may be roughly divided into two members: the lower member represented mainly by lignite, calcareous shale alternated with sandy clay rocks, about 200 m in thickness, and the upper member composed mostly of sandy clay rock, sandy conglomerate and clay rock, with a thickness of 170 m.

Palynologically, four developmental stages of Late Tertiary and Early Quaternary vegetation were recognized in the Songhua Basin[6]. In the first stage, the temperate plants (*Pinus, Quercus* and *Castanea*) were plentiful in pollen assemblage along with some warm-temperate to subtropical elements *Podocarpus, Carya, Ilex, Aralia,* etc., reflecting a rather warm climate. In the second stage, the percentages of Pinaceae and Fagaceae were predominant again, and some subtropical elements (*Platycarya, Liquidambar, Pittosporum, Antidesma* and *Sympolocus*) were distributed, showing that the climate might be both wetter and warmer. In the third stage, the amount of Pinaceae and Fagaceae decreased, while some subtropical Gymnosperms like *Podocarpus, Dacrydium, Keteleeria* and *Sterculia* increased, marking a still warmer climate. In the fourth stage, because the pollen grains of some conifers (*Abies* 15%, *Pinus* 8% and *Cupressus*), the broad-leaved trees (*Alnus, Betula* and *Tilia*) and herbs increased, the assemblage showed some changes significantly, indicating that the climate became cooler and drier. It was the first climatic deterioration in this region and the lower boundary of the Quaternary might be assigned to the base of this stage. In other words, the lower boundary of the Quaternary should be defined between the sandy and gravel beds and the lignite layers in the Songhua Basin.

2. In the neighbourhood of Kunming City, the lower boundary of the Quaternary was designated at the base of the sporo-pollen assemblage II. This is because the pollen grains of *Picea* and *Abies* amount to 31.6% and *Betula* to 28% of the total number of this assemblage, showing a worse deterioration in climate.

3. The Zhaotong Area. The strata of the Neogene-Quaternary are continuous deposits in this area. According to the available data of borehole, the rock unit is about 460 m thick, and may be divided into three members and three palynological assemblages as follows:

The upper assemblage is characterized by the presence of *Pinus* (sometimes accounting for more than 10%), *Picea, Abies* and *Keteleeria,* a decreased number of *Polypodiceasporites* and a dominance of *Tricolpollenites* in angiosperm along with a few elements of *Liquidambar, Rhododendron* and Caryophyllaceae. The elements of Betulaceae (*Betula* and *Alnus*) are on the obvious increase, but the hydrophylic organic elements have not yet been found.

In the middle assemblage (200–404 m thick, equivalent to the second member and the upper part of the first member), there are a good many of *Polypodiceasporites.* The pollen of Pinaceae is sparsely present, while that of angiosperms is represented by *Tricopopollenites* of mainly *Quercus,* etc. Meanwhile, *Myrica, Celtis, Corylus, Carya, Juglans* and hydrophylic organic elements are common.

The lower assemblage (equivalent to the lower part of the third member, below 404 m) is characterized by the rarity of pollen and spores.

The upper assemblage may be of Quaternary and the middle assemblage of Pliocene respectively. As a result, the lower boundary of the Quaternary should be defined upward on lignite layers in the Zhaotong Area and might be correlated to the fourth stage in the Songhua Basin from the palynological viewpoint.

VI. SHANGHAI AND THE NEIGHBOURING AREAS

The loosely deposited Quaternary lies unconformably on the Tertiary or much older sedimentary rocks in Shanghai and its neighbouring areas, composed chiefly of fine sand, clay, sandy clay and gravel beds. The pollen analytic

16

results[7] clearly indicate the presence of climatic fluctuations reflected by the alternative appearance of the dominant conifer grains and broad-leaved tree grains. For example, on the sporo-pollen diagram of Zone I of Borehole No. 1 in Shanghai there are many spores of *Ceratopteris* with a few conifer pollen grains: but on the diagram of the Qinlinggang section of the Shimin Mt. area in Zhejiang, there are rich conifer pollen grains (sometimes up to more than 90% of the total number of pollen and spores), including *Pinus, Picea* and *Keteleeria* associated with a few *Larix, Abies, Cedrus, Tsuga* and Cupressaceae, etc. The presence of a large number of conifers indicates a cold climate at that time in this region. This cooling phase would be older in age than that of Zone I of Borehole No. 1 in Shanghai. It becomes apparent that the Plio-Pleistocene boundary might be drawn at a horizon below Zone I in Shanghai and its neighbouring areas.

In a word, it is a complex problem to define the lower boundary of the Quaternary. Only relying on sufficient available data can this problem be readily solved. But all the facts mentioned above have shown that the climatic deterioration relating to the glaciation at the outset of Quaternary should be taken into fuller account, and in addition to other respects, such climatic factor may be considered as the characteristic symbol to define the lower boundary of the continental Quaternary. This symbol can be evidenced more or less in the relative deposits of North China, East China and some areas of Southwest China (i.e. Yunnan Province), but in South China this symbol has not yet been discovered. On account of widespread influences of glacial climate, the writers tend to believe that such climatic fluctuations are likely to be also found in the relative deposits in South China, though different in degrees.

REFERENCES

[1] Zagwijn, W. H., 1957, Vegetation, climate and time-correlations in the early Pleistocene of Europe. Geol. en Mijhn., n. Ser., 19 (7), 233—244.
[2] Zagwijn, W. H., 1974, The Pliocene-Pleistocene boundary in southwestern Europe. Boreas (3), 94.
[3] Liu Jing-ling, 1980, Pollen analysis and geological age of the Nihewan formation. Monthly, Jour. Sin. Soc. 25 (3—4), 584—587.
[4] Cenozoic Palynol. Gr. Inst. Bot., Acad. Sin. and Inst. Geol. and Min., 1966, Study on Cenozoic palaeontology from Lantian Area, Shenxi. in: Inst. Verteb. Paleont. and Paleoanthrop., Acad. Sin. "Symp. Cenozoic Lantian Field Meeting, Shenxi", 157—182, Sci. Press, Peking.
[5] Song Zhi-chen, 1958, Plant fossils and sporo-pollen complex of the Sanmen Series. Quart. Sin., 1 (1), 118—130.
[6] Li Wen-yi, Wu Hsi-fang, 1978, A palynological investigation on the Late Tertiary and Early Quaternary and its significance in the paleogeographical study in Central Yunnan. Acta Geogr. Sin., 33 (2), 142—155.
[7] Liu Jing-ling, Ye Ping-yi, 1977, Studies on the Quaternary sporo-pollen assemblages from Shanghai and Zhejiang with reference to its stratigraphic and paleoclimatic significance. Acta Paleont. Sin., 16 (1), 1—10.

THE LOWER BOUNDARY OF THE QUATERNARY IN THE HIMALAYAN REGION IN CHINA

WANG FU–BAO

(Department of Geography, Nanjing University)

AND LI BING–YUAN

(Institute of Geography, Academia Sinica)

Abstract—We take the lower boundary of Matuyama reversed epoch as the beginning of Quaternary, about 2.4 m.y. from today.

To establish and subdivide the Late Cenozoic successions in the Himalayan Region is one of the essential tasks in the study of the age, process, amplitude and cause of the uplift of the Qinghai-Xizang Plateau. Having worked for many years in the area, the authors have accumulated plentiful data and therefore, attempted to have a tentative discussion on the lower boundary of Quaternary era here.

I. THE STRATIGRAPHY AND AGES OF THE LATE CENOZOIC

In the Himalayan Region a number of Pliocene Basins are scattered, the largest of which is Zada Basin, on the northern slope of the Himalayas, with an area of about 7,000 km². In all basins, without exception, the Pliocene and the Pleistocene series as thick as several hundreds to a thousand metres are exposed. The Pliocene mainly includes lake facies mudstone, siltstone and sandstone, with an abundance of fossils, called Jilong group and Zada group; the Pleistocene is coarse gravel deposits, with very little fossils, and the early Pleistocene is called Gongba conglomerate. We shall illustrate these further with the Jilong group exposed in Gyirong Basin.

Gyirong Basin is an intermountain basin on the northwest of Mt. Xixabangma (8,012 m); the section is in the eastern part of Gyirong County:

The Gongba conglomerate of the early Pleistocene

(18) Greyish-yellow coarse gravel bed, N 35° E/7°	>30 m
(18) Yellow gravel bed	40 m

———————————— **Unconformity** ————————————

Jilong group of the Pliocene series

(16) Greyish-yellow and dark grey mudstone, sandstone and siltstone, con-

18

taining fossils: *Qinghaicypris subpentanoda* Huang, *Euoypris* sp., *Candona* sp., *Leuco cytherella sinensis* Huang, *Candoniella plana* Huang.

The sporo-pollen consists of *Cedrus* (25.8—60.1%), *Picea, Pinus, Quercus, Carya, Cathayensis* and Chenopdiaceae, Caryophylacea, Polypodiaceae, etc.

45 m

(15) The upper part is greenish-grey and brownish-yellow fine sandstone with a thin layer of grey-black mudstone.   15 m

The middle part is greyish-yellow sand and gravels with ferrous sandstone.
                                                                3 m

The lower part is grey fine sandstone and ferrous sandstone.   1.5 m

(14) Brown-yellow ferrous sandstone.   2 m

(13) Greyish yellow fine sandstone with mudstone and brownish red car-
bonaceous fine sandstone, containing *Gangetia* ex. gr. *rissoides* Odhmer; *Candoniella cylindrea* Huang; *Picea, Pinus, Quercus, Ketelleeria, Parciperites pavissacus, Ephedra, Pteris,* etc.   27 m

(12) Brown-yellow ferrous conglomerate.   5.2 m

(11) Grey siltstone, containing *Gangetia* ex gr. *rissoides* Odhmer.   1 m

(10) Grey gravel bed, bearing gravels composed of sandstone and slate.   1.7 m

(9) Grey fine sandstone with ferrous sandstone, containing the sporo-pollen, which includes *Hippophae, Castanea, Cedrus, Pinus,* Chenopdiaceae, *Onychium* and *Riccia, Fossombronia,* etc.

(8) Gravel bed.   4.3 m

(7) Grey to brownish yellow ferrous sandstone, containing *Cyclocyris vulgaris* Huang, *Candona angusticandta* Huang, *Leucocythere dorsocuberosa,* etc. 2.5 m

(6) Grey-black gravel bed.   1 m

(5) Grey, greyish yellow and greyish purple fine sandstone, containing the sporo-pollen including *Pinus, Cedrus, Onychium, Riccia,* etc.   5 m

(4) Grey-black mudstone alternated with siltstone, containing the sporo-pollen mainly *Riccia* and *Fossombronia,* with a small number of *Cedrus* and *Pinus.*   9.6 m

(3) Grey siltstone.   1.7 m

(2) Grey gravel, S35°W/6°   3 m

———————————————— Unconformity ————————————————

(1) Jurassic beds

The Orma Basin, to the south of Gyirong Basin, contains, at a horizon corresponding with the 12th layers just mentioned above, fossils of the *Hipparion* fauna, namely, *Hipparion gyirongensis* Ji, Xu et Huang, *Chilotherium xizangensis* Ji, Xu et Huang, *Metacervulus capreolinus* Teilhard et Trassart, *Palaeotragus microdon* Koken, *Gazella gaudryi* Schlosser, *Ochotona gyirongensis* Ji, Xi et Huang, *Hyaena* sp.[1]. These fossils show that layers 2—16 of the section in Gyirong belong to Pliocene, while the upper gravel bed early Pleistocene.

The section at the southern slope of Yagru Xongla, east of Gyirong Basin, can be divided into three parts: the upper is the glacio-fluvial deposits of 80 m thick; the middle is siltstone, with a thickness of 80 m, containing *Leucocy-*

19

therella sp., *Condona* sp., *Condoniella* sp.; the lower part is lake facies fine sandstone, sandstone and mudstone, with fossils: *Hipparion* sp.; *Pisidum* sp.; *Sphaerium* cf. *corneum* (Lin), *S.* sp., *Unio tschiliensis* Sturany, *U.* cf. *douglasia* Grif et Pidg; *Adelinlla regularis* Yu, *Staja xizangensis* Yu; *Ilyocyris gibba* (komdohr), *Limnocythere dobiosa* Li; *Quercus, Cedrus, Pinus, Picea, Podocapus,* etc. *Cedrus* sometimes amounts as high as 70.9%[2] of xylophyta pollen.

The contacts of above-mentioned three parts are unconformable. In the middle part, the palaeomagnetic data show it belongs to Matuyama Epoch, while in the lower part, it is referred to Gauss normal epoch. Therefore, the lower lacustrine deposits in the section at the southern slope of Yagru Xongla belong to Pliocene, the middle to Early Pleistocene and the upper to, perhaps, Middle Pleistocene.

II. DISCUSSION ON THE LOWER BOUNDARY OF THE QUATERNARY

As early as the beginning of 1960s, the authors placed the Gongba conglomerate as the lowest layer of Quaternary, because features of the tectonic movement and the natural environment before and after the formation of Gongba conglomerate display distinct changes.

1. The Intense Uplift of the Himalayas—the Beginning of Quaternary

The 1st episode of the Himalayan movement beginning in Late Eocene (55—42 m.y.) made possible the emergence of the whole Qinghai-Xizang Plateau from the sea. Then the 2nd episode of the Himalayan movement starting from Miocene Epoch (33—27 m.y.) added height to the Himalayas and as a result formed, on both sides, a depression area, the southern one being called Siwalik Depression Basin and the northern one, Gandise-Yarlung Zangbo Basin. And the 3rd episode of the Himalayan movement starting from Late Miocene (22—10 m.y.), the Gandise-Yarlung Zangbo Depression Basin uplifted and new mountains formed. But in Pliocene, the Himalayan Area was occupied by a number of large lakes, the sediments of which contain rich tropical and subtropical sporo-pollen and *Hipparion* fauna. The altitude at that time was only about 1,000 m[3]. The intense uplift of the Himalayas began in Late Pliocene. The tectonic type of the uplift seems different from those of the previous Himalayan episodes. It was block-faulted uplift, which we call "plateau movement", and so we refer the beginning of the uplift to the beginning of Quaternary.

2. Gongba Conglomerate—Lower-boundary Bed of the Quaternary

In the process of the intense uplift of the plateau, on the margins of the plateau

20

and piedmonts of the mountains very thick molasse appeared, as is seen along the northern slope of the Himalayas.

Gongba conglomerate was the piedmont diluvial, alluvial and shore deposits formed along with the disappearance of the Pliocene lakes. It lies unconformably on top of Pliocene lacustrine deposits with a thickness of 200—over 400 m. From the lithology of Gongba conglomerate and a few fossils, we learn that, during the time of the intense uplift of the plateau, the weather was dry and meanwhile there were cyclical changes of cold and warm weather. The spores and pollen obtained from boring between Kunlun Mts. and Tanggula Range show a division of three cycles[4] (from bottom to top): *Pinus, Picea, Betula, Ulmus*—*Chenopodiaceae, Ephedra* to *Picea, Pinus*—*Chenopiciaceae* to *Betul, Quercus, Picea*—no sporo-pollen or *Chenopodiaceae*. It is obvious that the tectonic type and natural environment during the formation of Gongba conglomerate are strikingly different. So we deem that the Gongba conglomerate was the lowest layer of the Quaternary in the area.

3. The Lower Boundary of the Quaternary

According to the palaeomagnetic data in the section at the southern slope of Yagru Xongla, the lower lacustrine deposit belongs to Gauss normal epoch and the middle Gongba conglomerate belongs to Matuyama reversed epoch. The former was formed when the tectonic movement was comparatively stable and the climate was mild, and the age is referred to Pliocene. The water system and natural environment were distinctively different when the latter was deposited. The similar condition is also found in the Himalayan Area and deposits of all the basins. This is the reason why we put the lower boundary of Matuyama reversed epoch at the beginning of Quaternary, about 2.4 m.y. from today.

The determination of the age is also confirmed by data obtained from other parts of the plateau. The Quguo Lake at the southern slope of Tanggula Range disappeared about 2.24 m.y. ago; the lithology of the deposits and fossils in the ancient lake in Kunlun Mts. pass display clear distinction before and after 2.56 m.y.; some of the lakes in Qaidam Basin took shape less than 2.32 m.y. ago. In conclusion, we think it appropriate to put the lower boundary of Matuyama reversed epoch at the beginning of Quaternary as far as Qinghai-Xizang Plateau is concerned.

REFERENCES

[1] Ji Hong-xiang et al., 1980, The *Hipparion* Fauna from Gyirong Basin, Xizang. The Series of the Scientific Expedition to Qinghai-Xizang Plateau, Palaeontology of Xizang (Book I), Science Press.

[2] Huang Ci-xuan et al., 1980, Analysis of the spores-pollen assemblages and the age of

the deposits from Dati palaeo-lake basin at Yagru Xongla, South Xizang. The Series of the Scientific Expedition to Qinghai-Xizang Plateau (Palaeontology of Xizang) (Book I) Science Press

[3] Li Ji-jun et al., 1979, A discussion on the period, amplitude and type of the uplift of the Qinghai-Xizang Plateau. Scientia Sinica. vol. 22, no. 1
[4] Kong Zhao-chen et al., 1981, Neogene-Quaternary palynoflora from Kunlun Mts. to the Tanggula Range and the uplift of the Qinghai-Xizang Plateau. Studies on the period, amplitude and type of the uplift of the Qinghai-Xizang Plateau. Science Press.

Table 1 The correlation of the lower boundary of the Quaternary in the Himalayan region in China

Age	Pagri	Gyirong	Yagru Xongla	Bulung	Zada
Q_3 Q_2	Moraine	Terrace gravels	Yagru Xongla gravels 80 m	Terrace gravels	Terrace gravels
	————U_2————	————U_2————	————U_2————	————U_2————	————U_2
Q_1	Gongba conglomerate 200—300m	Gongba conglomerate 100—200 m	Gongba conglomerate 80—90 m	Gongba conglomerate >280 m	Gongba conglomerate 100—200 m
	————U_1————	————U_1————	————U_1————	————U_1————	————U_3
N_2	Lacustrine deposits >30 m	Jilong group >300 m	Dati formation >400 m	Bulong conglomerate >300 m	Zada group >800 m

CORRELATION OF QUATERNARY STRATA BETWEEN INLAND AND COASTAL AREAS IN NORTH CHINA

HU LAN–YING, HUANG BAO–REN,
(Nanjing Institute of Geology and Palaeontology, Academia Sinica)

LI HUA–MEI
(Institute of Geochemistry, Academia Sinica)

AND YANG LIU–FA
(Nanjing Institute of Geography, Academia Sinica)

Abstract—According to the recent investigation of the writers, the Quaternary inter-beds of marine and continental facies in northern coastal area of Bohai Sea contain three marine beds. Among them the lower marine bed is a mixture of marine and continental facies, and can be correlated with Middle Pleistocene lake formation such as Erlangjian formation in Qinghai Lake area and a part of Nihewan formation at Hutouliang section in North China, also with Middle Pleistocene Holsteian glacial deposits in Europe. The middle marine bed may be of Upper Pleistocene and the upper marine bed belongs to the Holocene.

There are a variety of Quaternary continental strata in the inland of North China, for example, loess, paleosol, moraine, fluvial and lake deposits, alluvial deposits and flood deposits. These strata yield rich plant and animal fossils. Anthropolithic and lithic wares have also been found in some places. In the past years, much work was done on these strata which make it known that thick Quaternary interbeds of marine and continental sediments were deposited in coastal area of North China. These beds have great significance for the study of the changes of sea and land and the vicissitude of the coastal lines of ancient sea. Based on the micro-fossils, the paper aims at the correlation on the Quaternary strata between the inland and the coastal areas of North China.

Numerous continental brackish water ostracods such as *Limnocythere dubiosa* Daday, *L. sancti-patricii* Brady et Robertson and *L. binoda* Huang occurred in the Pleistocene in Gansen district, Qaidam Basin[1]. This ostracod assem-blage is similar to that of the Middle Pleistocene Holsteian interglacial deposits in Europe[3]. It was also found from the Middle Pleistocene Erlangjiang formation (lake facies) in the lake terrace on the southern bank of Qinghai Lake[4]. In 1980 this assemblage was discovered from upper Middle Pleistocene Hutouliang interglacial lake deposits in the lower and middle valley of Sanggan

23

River, Hebei[2].

Recently, the writers have studied the core samples from a rather standard formation in Bohai Coastal Area, the drilling well is about 600 m in depth. The age of the lowermost core, according to the paleomagnetic data, is about 3 million years. In the light of the microfossils the geological ages of the cores in ascending order range from Pliocene to Holocene. The major cores of this drilling well yield numerous marsh, pool and lake microfossils, while some upper cores contain shallow sea fossils. According to the discovered ostracods and foraminifers, the cores from 0 to 77 m deep contain three marine beds.

Based on the ostracod fossils, there are three continental beds under the lower marine bed. The lower continental bed yields ostracod *Leucocythere gong-hensis* Huang which has been found from the Pliocene Qugou formation in Gong-he Basin, Qinghai. Therefore its age may belong to Pliocene. The middle continental bed contains *Ilyocypris kaifengensis* Lee which has been found from the Lower Pleistocene Nihewan interglacial deposits in Hebei. So it may be of Early Pleistocene age. The upper continental bed at the depth of 77—260 m bears ostracods *Ilyocypris bradyi* Sars and *Candoniella albicans* (Brady) which belong to early Middle Pleistocene.

Among the interbeds of marine and continental sediments at drilling depth of 0—77 m, the first marine bed (in ascending order) is at the depth of 75m—77m, according to the paleomagnetic measurement, it was deposited about 0.28—0.29 millions years ago. It yields brackish water lake ostracods *Limnocythere dubiosa* Daday, *L. sancti-patricii* Brady et Robertson and *L. binoda* Huang. On the basis of the ostracod fossils, this bed can be correlated with the continental Middle Pleistocene Erlangjiang formation in Qinghai Lake area and with the Middle Pleistocene interglacial Hutouliang formation in Hebei as well. This first marine bed also yields neritic ostracods *Echinocythereis bradyi* Ishizaki and *Aurila* sp. etc. and foraminifers *Pseudononion minutum* Zheng and *Ammonia confertitesta* Zheng. Due to the ecology of living forms of these fossil genera, it may be said that the sea water transgressed into the continental brackish lake. According to the continental brackish water lake ostracod fossils, the age of the lower marine bed may be late Middle Pleistocene.

Above the lower marine bed, there is a continental bed, the age of which, by the paleomagnetic measurement, is about 0.15 to 0.28 millions years old. This continental bed occurred at the depth of 46 to 75 m and yields the ostracods of marsh and pool facies such as *Ilyocypris gibba* (Ramdohr), *I. biplicata* (Koch) and *Candoniella albicans* (Brady) and it also contains charophyta and gastropods. Its age may be of early Upper Pleistocene.

The second marine bed was deposited about 0.11 to 0.15 millions years ago. It is at the depth of 38 to 40 m and bears plentiful neritic ostracods, such as *Neomonoceratina dongtaiensis* Yang et Chen, *Sinocytheridea latiovata* Hou et Chen, *Leguminocythereis hodgii* (Brady), *Echinocythereis bradyi* Ishizaki, *Sinocythere sinensis* Hou and *Xiphichilus sinensis* Yang et Hou and foraminifers

24

Elphidium advenum (Cushman), *Massilina pratti* Cushman et Ellisor, *Ammonia confertitesta* Zheng and *Asterorotalia inflata* (Millet). It may belong to middle Upper Pleistocene.

The continental bed between the third and second marine beds at the depth of 16 m to 39 m was formed about 0.01 to 0.11 million years age and yields pool and marsh ostracods *Ilyocypris bradyi* Sars and gastropods. It may be of late Upper Pleistocene.

The third marine bed according to paleomagnetic measurement began to deposit about ten thousand years ago. It is at the depth of 0 to 16 m, yielding marine ostracods *Loxoconcha* sp., *Sinocytheridae latiovata* Hou et Chen, *Leguminocythereis hodgii* (Brady), *Neomonoceration dongtaiensis* Yang et Chen and *Albileberis sinensis* Hou and foraminifers *Elphidium limpidum* Ho, Hu et Wang, *Stomoloculina multangula* Ho, Hu et Wang, *Pseudononionella variabilis* Zheng and *Rosalina floridana* (Cushman). The age of this bed is of Holocene. The continental bed is not seen above the third marine bed.

All of the above mentioned three transgressions belong to neritic facies. Among them the second one attains the greatest depth, probably not deeper than 50 m. Based on these fossil ostracods and foraminifers which are similar to those of the shallow sea of Bohai and Huanghai (Yellow) Seas, the three transgressions came from the ancient sea in the southeast with the salinity less than 34‰.

From the correlation between the interbeds of marine and continental sediments in the coastal area of Bohai Sea and the continental beds in inland of North China, we have learned that the early transgression occurred after the formation of Lishi Loess or before the deposition of Malan Loess, or during the period when Qinghai Lake was fully developed and the late age of Pleistocene Nihewan Lake in Hebei or in the period when Peking Man existed. The second transgression could happen in the time of the relics of Xujiayao Man.

From the relation between transgresion and interglacial, this Early transgression is equivalent to the Hutouliang section in Hebei which can be correlated to the Middle Pleistocene Holsteian interglacial in Europe. The early transgression may be correlated to Dagu—Lushan interglacial. The second transgression may be equivalent to Lushan—Dali interglacial and the third transgression occurred after the Dali glacial.

REFERENCES

[1] Huang Bao-ren, 1964, Ostracoda from Gansen district, Qaidam Basin. Acta Palaeont Sinica. 12 (2),
[2] Huang Bao-ren, 1980, A preliminary study of Pleistocene Ostracoda from middle and lower Sanggan River Valley. Kexue Tongbao, 25 (3).
[3] Lüttig, G., 1955, Die Ostrakoden des interglacials von Elze. Palaont. Z. 29 (3/4).

[4] Lanzhou Institute of Geology, Academia Sinica, et al., 1979, A report of comprehensive expedition on Qinghai Lake. Science Press.

[5] Zheng Shou-yi et al., 1978, The Quaternary foraminifera of Dayuzhang irrigation area, Shandong Province, and a preliminary interpretation of its depositional environment. Studia Marina Sinica, 13.

[6] He Yan & Hu Lan-ying, 1978, The Cenozoic Foraminifera from the coastal region of Bohai. Science Press.

FEATURES OF DISTRIBUTION OF
THE QUATERNARY IN CHINA

CHEN MING

(Tianjin Institute of Geology and Mineral Resources, Chinese Academy of Geological Sciences)

ABSTRACT

The Quaternary strata in China are very developed, and include almost all the genetic types existing in the world. The marine sediments are mainly distributed in Southeast China, and the glacial deposits—in the West. The magnificent loess which is hardly seen in other places of the world is situated in North China. The laterite is spread in South China. The alluvial, diluvial and lake deposits distributed in lake and river areas are developed. The thickness of the Quaternary is changeable. The maximum thickness is more than 2,000 m. The loess and glacial deposits are thin gradually to the southeast, but, the laterite, on the contrary to the northwest.

The same genetic types of deposits in different geological periods are diverse in distribution, thickness and composition.

Besides tectonics, climate was the main factor which influenced the types, thickness and distribution of the Quaternary in China.

A REVIEW OF STUDIES ON THE QUATERNARY GLACIAL AND INTERGLACIAL SEQUENCES IN CHINA

ZHOU MU–LIN

(Tianjin Institute of Geology and Mineral Resources, Chinese Academy of Geological Sciences)

ABSTRACT

Studies of Quaternary glacial and interglacial sequences, at present, are confined only to some mountain regions, where the glacial landforms and sediments are well preserved, and its vicinities. They are subdivided into "Nangon Cold stage" (coinciding with Pretiglian stage in Europe), Poyang glaciation (equivalent to Menapian, Günz), Dagu glaciation (equivalent to Elsterian, Mindel), Lushan glaciation (equivalent to Saalian, Riss) and Dali glaciation (equivalent to Weichselian, Würm). Subsequently, several interglacials were subdivided between the glaciations, such as Nangou-Poyang interglacial (which is equivalent to Tiglian, Eburonian, Waalian interglacial), Poyang-Dagu interglacial (which is equivalent to Cromerian interglacial), Dagu-Lushan interglacial (equivalent to Holsteinian), and Lushan-Dali interglacial (which is equivalent to Eemian). Some previous subdivisions of the Quaternary, proposed by Davidson Black and his coworkers (1933), by Barbour G. B. (1931), by J. S. Lee (1937) and by Hullmut de Terra (1941), are, briefly described. On the basis of these subdivisions, the present Quaternary sequence is introduced by the author in this paper. However, several problems remain to be solved, for instance, the lower boundary of the Quaternary.

PRELIMINARY STUDY OF THE SUBDIVISION OF THE NIHEWAN FORMATION

WANG YUN–SHENG

(Tianjin Institute of Geology and Mineral Resources, Chinese Academy of Geological Sciences)

ABSTRACT

In this paper fifteen stratigraphic sections of the Nihewan formation in Yang-yuan-Yuxian Basin (Hebei Province) are analysed by means of structural analysis of sedimentary section; the sequence of the sedimentary facies of the Nihewan formation is established and the evolution of the palaeolake of Yangyuan-Yuxian is outlined. On the basis of analyses mentioned and the data of lithostratigraphy, geochemistry, palaeontology and palaeomagnetism, the Nihewan formation is subdivided into four members and a stratigraphic correlation of the whole region has been made. Based on the changes of the shrinkage and expansion of the lake, the climatic evolution of Early Pleistocene in North China is introduced.

POSTGLACIAL MARINE BEDS IN THE COASTAL AND DELTAIC AREAS IN EAST CHINA

LI CONG–XIAN, LI PING, AND WANG LI

(Department of Marine Geology, Tongji University)

Abstract—This paper mainly deals with the internal characteristics, and the formation of the postglacial marine beds, as well as the distribution regularity of the sand bodies, peats and barrier-lagoon systems within postglacial beds along coastal and deltaic areas in East China. The internal characteristics of postglacial beds along coastal zones over the world are the same as those in East China, and the other Quaternary marine beds are similar to postglacial ones in East China. The conclusion, therefore, obtained from the study of the postglacial marine beds may be useful to study the other marine beds.

In recent years, the research on the Quaternary in the coastal zones and deltaic areas of East China has been extensively developed, and several marine beds have been discovered, but the internal characteristics of marine beds have not been studied in detail. However, the internal characteristics of the marine beds are of importance in recongnition of transgression and regression, subdivision and correlation in stratigraphy, analysis of spatial distribution of facies, and prediction of the occurrence of oil, gas and coal seams in the beds of marine-continental transitional facies. Mainly according to the research in internal sedimentary characteristics of the postglacial marine beds in the coastal zones of provinces of Zhejiang, Jiangsu, Shandong, Hebei and in the Changjiang River and Luanhe River deltaic areas, this paper studies preliminarily the internal characteristics of the postglacial beds, and deals with some continental facies in the downstreams of the two mentioned rivers during the transgression and regression in order to compare with those of the marine beds.

INTERNAL CHARACTERISTICS OF THE POSTGLACIAL MARINE BEDS

The paleoland surface covered by the postglacial marine depositional beds in East China is undulatory, which consists of the bed rocks or weathering crusts along the coastal zones of Shandong Province and in Zhujiang (Pearl) River delta area; dark green clay, sandy clay in Changjiang River Delta area and Zhejiang Coastal Zone; tawny gravelly sand or tawny sandy clay in Changjiang River and Luanhe River valleys; and older marine clay in Nantong—Qidong Area of Jiangsu Province. There are two units, namely, the ancient river valley and the ancient interfluve on the paleoland surface. The slope of the ancient interfluve is seawards, and the percentage of the slope is about 0.018–0.031% in Changjiang River Delta area and nearly 0.86% in the coastal zone of Shandong Peninsula. Generally, the ancient river valleys are quite wide. The width of the late Pleistocene Valley of Changjiang River is about 30 km and the ancient valley extends more than 300 km in modern delta area*. The ancient valley of Luanhe River appears in the form of trumpet which is approximately 2–3 km in width at apical part and about 60 km at downstream near the coastal line**. The sediments on both of the ancient valley and the interfluve plain were deposited during the period of the sealevel rising. Therefore, the sedimentary beds seem to be very similar.

The postglacial marine beds in the coastal and deltaic areas are composed of the lower transgressive and the upper regressive sequences. The mariness of the transgressive sequence becomes higher, and that of the regressive sequence trends to be obscure in ascending order. As far as the facies is concerned, the postglacial beds include lower littoral, middle neritic and upper littoral facies. The middle neritic bed becomes progressively thin towards inland, and finally tapers off. As a result, the upper littoral bed overlies directly on the lower one, and then the littoral beds also gradually thin out.

In the ancient valley zones, the internal sedimentary characteristics of postglacial beds vary with river characters and tide intensity. In the macrotidal estuaries where the gradient is small and the slope is gentle, the postglacial depositional beds consist of the fining-ascending cycle in the lower part and the coarsening-ascending cycle in the upper part. The marine microfossils can be found in almost all the layers in the river-mouth area, but occur only within the upmost layers of transgressive sequence and the whole regressive sequence in apical area of delta, such as in Changjiang River Delta area. In the microtidal estuaries where the gradient is great and the slope is steeper, the postglacial depositional beds are composed of the fining-ascending cycle in the lower part and the coarsening-ascending one in the upper part, and the marine microfossils have been found in the upmost layers of the transgressive sequence and within the whole regressive sequence. This is similar to apical area of Changjiang River Delta. The postglacial beds consist of two superimposed fining-ascending cycles, in which marine fossils have not been found, for instance Luanhe and Zhujiang Rivers*** Deltas are of this case.

The sand bodies are formed and distributed in the coastal and deltaic areas. In the transgression, the coastal zones move landwards and seawards, so the sand bodies occur both in lower and upper littoral beds, between which there are neritic clay deposits. Thus, the distribution of sand bodies appears in

---

* Department of Marine Geology, Tongji University. 1978 The development and sand bodies characteristics of Changjiang River Delta.

** Department of Marine geology, Tongji University. 1981 The Research of fan-delta sedimentary systems of Luanhe River.

*** Department of Geography, Zhongshan University. 1977 On the development and evolution of the Zhujiang River Delta.

a two-storeyed structure in vertical sequence. Landwards, the upper and lower sand bodies are often directly superimposed in the place where the middle neritic clay bed thins out, and this makes the sand bodies thickened. This phenomenon can also be found in Haian, Taizhou and Hongqiao Districts of Changjiang River Delta. Marshes are developed generally in coastal zones. Therefore, the distribution of peat is the same as the sand bodies, and the peats also have a two-storeyed distribution in vertical sequence. Several peat layers occur in place where the middle neritic clay bed tapers off and the upper and lower littoral beds contact directly, for instance, there are three or four peat layers in the west of Taizhou and Haian, and even six peat layers in Zhenjiang area of Changjiang River Delta.

THE TIME DIFFERENCE OF POSTGLACIAL SEDIMENTARY BEDS

During the postglacial period, the sealevel rose and the paleoland surface was inundated by sea water. The arrival of the inundation seems generally later landwards and during the period the postglacial beds were formed also, the forming time of these beds were not the same. The sedimentation and duration of postglacial depositional beds in ancient valley and interfluve plain are different from each other. The postglacial sedimentary beds in the ancient interfluve plain started forming 12,000 years BP in Bohai Gulf, and about 10,000 years BP along the coastline of western Bohai Gulf, whereas the beds started forming 8,000 years BP in the Sidangkou Area from the modern coastline. Along the modern coastline in Shanghai Area that is located in the southern part of the Changjiang River Delta, the postglacial deposition started 10,000 years ago, while in Gangshen (Cheniers) Area, west of Shanghai, about 8,000 years ago.

After the transgression reached the maximum extent and the sedimentary rate surpassed that of sea level rising, the regression occurred. The regressive time of sea water was successively later from land to sea. It was 5,000–6,000 years BP when the regression of Bohai Gulf began to take place, four ancient coast lines were formed and the forming periods were 4,000–4,700, 3,000–3,500, 1,100–2,500 and 500–600 years seawards, respectively. Therefore, the depositional duration of postglacial marine beds gradually increases from land to sea, and the difference of depositional duration between areas along coastline and inland is about a few thousands years.

In the ancient valley the backwater and retrogressive aggradation took place due to the rising of sealevel during postglacial period. The influenced distance of backwater depended upon the fall of water table, and the extent of retrogressive aggradation was controlled by the distance of backwater. According to the model experiments, the distance of retrogressive aggradation is four or five times as long as that of the backwater in a wide estuary. The distance that tidal currents invade upstream is smaller than that of backwater in tidal estuary,

32

such as in modern Changjiang River, the tidal current only reaches upstream Jiangyin, which is about 150km from the river mouth, while the backwater caused by flood water-level change can reach Zhenjiang, about 300 km from the river mouth. In modern Huanghe River the tidal currents reach upstream for 2–3 km and the backwater influence reaches 30 km, however, the retrogressive aggradation that spread in the upstream can be found at a distance of more than 300km from the mouth. Therefore, a successive order of retrogressive aggradation-backwater-tidal current sediments can be found in the same lower river reach during the sealevel rising in the postglacial period. In Changjiang River Delta area the backwater arrived Zhenjiang about 14,650 years ago, and the marine bed in the same core in Zhenjiang was formed about 9,000 years BP so the retrogressive aggradation probably arrived Zhenjiang at least 6,000 years earlier than the tidal currents sediments, because the retrogressive aggradation must be earlier than the backwater sediments. It was about 15,000-18,000 years BP that sealevel began to rise in East China Sea area during the postglacial period. In this case, soon after the sealevel rose, the retrogressive aggradation might reach Zhenjiang Area, and the time difference of marine deposition in ancient valleys between up- and down-stream was considerably smaller than that in ancient interfluve plain. Besides, owing to the fact that distance of retrogressive aggradation exceeded that of the seawater invasion along the river, the river channel filling sequence can be observed, the deposition mechanism of which is just the same as that of the lower transgressive sequence in delta areas but without marine fossils, such as in the areas of Hugezhuang, Yangezhuang and Xiamaozhuang of Luanhe River fan-delta.

When the rate of sealevel rising was smaller than that of deposition, deltas developed rapidly, and the time of land-forming became successively later from land to sea. For example, six subdeltas of Changjiang River Delta system were formed during the recent 7,500 years. The forming ages are 6,000, 4,000, 2,000, 1,200 and 200 years ago, respectively. The depositional duration of sedimentary beds graduallg became longer from land to sea. But the time difference of the postglacial beds formed in the zones along the coastline and inland in the ancient valleys is smaller than that on the ancient interfluve plains.

POSTGLACIAL MARINE BEDS AND BARRIER-LAGOON SYSTEMS

The sedimentary characteristics of lagoon depend mainly on situation of closure The lagoons along the coastal zones in East China can be divided into three types, namely, bay-lagoon, semiclosed lagoon and closed lagoon. They represent different evolutionary stages. The different types of lagoon deposits are widely distributed in the postglacial sedimentary sequences in East China. However, some characteristics and evolution trend of lagoon are absolutely different in transgressive and regressive sequences. The transgressive barrier lagoon system is underlain by continental beds, and overlain by the marine

33

beds, and the sequence of development of lagoon is as follow: closed lagoon→ semiclosed logoon→baylagoon. The regressive barrier-lagoon system is underlain by marine beds, and overlain by the continental one, and the sequence of development is bay-lagoon→semiclosed lagoon→closed lagoon. The transgressive and regressive barrier-lagoon systems are usually superimposed each other in the same vertical sequence, as those in Xihu Lake, Hangzhou.

THE UNIVERSALITY OF INTERNAL CHARACTERISTICS IN POSTGLACIAL MARINE BEDS

The postglacial marine beds in Europe, Africa, America have the same internal sedimentary characteristics as those in the coastal zones of East China, because the changes of the postglacial sealevels are global. The Quaternary marine beds found in East China are also composed of the transgressive and regressive sequences which are similar to the postglacial ones. The conclusions, therefore, obtained from the study of the postglacial marine beds may be useful to the study of the other marine beds.

REFERENCES

[1] Li Cong-xian et al., 1980, Holocene transgressive-regressive sequence in Yangtze Delta area. Scientia Geologica Sinica, no. 4.
[2] Li Cong-xian et al., 1979, The characteristics and distribution of Holocene sand bodies in Yangtze River Delta area. Acta Oceanologia Sinica, vol. 1, no. 2.
[3] Zhao Song-ling et al., 1978, On the marine stratigraphy and coastlines of the western coast of Bohai Gulf. Oceanologia et Limnologia Sinica, vol. 9, no. 1.
[4] Xia Dong-xing, 1981, Whence came the high sea-level during the Holocene. Acta Oceanologica Sinica, vol. 3, no. 4.
[5] Zhao Xi-tao et al., 1979, Sea level changes of eastern China during the last 20,000 years. Acta Oceanologia Sinica, vol. 1, no. 2.
[6] Delta Research Group, Department of Marine Geology, Tongji University, 1978, Holocene formation and development of the Yangtze Delta. Kexue Tongbao, vol. 23, no. 5.
[7] Pang Jia-zhen et al., 1980, Evolution of the Yellow River Estuary Proceedings of the International Symposium on River Sedimentation, Beijing, China. Guanghua Press.
[8] Li Cong-xian et al., 1981, The time of Holocene transgression and sealevel changes in apical area of Yangtze Delta. Journal of Tongji University, 1981. No. 3.
[9] Wang Jing-tai et al., 1981, Evolution of the Holocene Changjiang Delta. Acta Geologica Sinica, vol. 55, no. 1.
[10] Li Cong-xian et al., 1982, Sedimentation and devolopment of the barrier-lagoon systems along coastal zone of East China. Marine Geological Research, Vol. 2, No. 1.
[11] Wang Pin-xian et al., 1979, Micropaleontological evidences for the history of the West Lake, Hangzhou, Zhejiang Province, China. Oceanologia et Limnologia Sinica, vol. 10. no. 4.
[12] Lagaaij R. et al., 1964, Typical features of fluviomarine offlap sequence. In deltaic and shallow marine deposits.
[13] Eppo Oomkens, 1970, Depositional sequence and sand distribution in the Postglacial Rhone Delta complex. Deltaic sedimentation: Modern and Ancient. ed. Morgan J. M.

[14] Eppo Oomkens, 1974, Lithofacies relations in the Late Quater-Miger Delta complex sedimentology, v. 21, no. 2.

[15] Wilkinson B. H. et al., 1977, Lavaca Bay transgressive deltaic sedimentation in Central Texas Estuary. AAPG. v. 61, no. 4.

[16] Wilkinson B. H. et al., 1978, Late Holocene history of the Central Texas coast from Galveston Island to Poss Cavallo. Geol. Soc. Am. Bull, v. 89, no. 10.

[17] Makaveev N. I., 1961, Experimental geomorphology published by Moscow Nat. Univ.

CLIMATOSTRATIGRAPHY OF THE SEDIMENTS IN THE PEKING MAN'S CAVE

LIU ZE-CHUN

(Nanjing Institute of Geography, Academia Sinica)

Abstract—The deposits in the Peking Man's Cave are more than 40 m thick. According to climatostratigraphy, the deposits of the Peking Man's Cave, the New Cave and the Upper Cave may be correlated to the L1–L12 of the loess section in Luochun, Shaanxi province, or $\delta^{18}O$ 16—2 stages of the deep-sea core V28 –238.

Peking Man's Cave, namely, locus 1 at Zhoukoudian, is 107 m across from east to west, 25 m from north to south and 128 m a.s.l. It is filled up with clastic sediments, ash layers and travertines and has been excavated as deep as 80 m. The bottom, however, has not been reached yet, indicating that the thickness of its deposits is far more than 40 m. This paper, based on the studies on the detailed geological sections, the animal fossils, pollen analysis and the isotopic chronology in recent years, aims at analyzing the palaeoclimatic changes inferred from the cave deposits and correlating the deposits in the cave with the loess layers in China and the $\delta^{18}O$ stages of deep-sea cores.

STRATIGRAPHY IN PEKING MAN'S CAVE

The deposits, consisting mainly of limestone breccia, fine gravel sand, fine sand, silt, clay, ashes and a small amount of travertine, can be distinguished into 17 layers (Fig. 1).

According to the palaeomagnetic study, the deposits above the 14th layer occurred in the Brunhes normal epoch and those below that formed in the Matuyama reversed epoch. Therefore, it is evident that most of these deposits took shape at 700,000 yrs BP. The 7th layer bears evidence of a reversed event at 380,000 yrs BP, corresponding to Lake Biwa E. in Japan. Several other methods were used to estimate the ages of fossil animal teeth, which found in the 1st and 2nd layers are referred to 250,000–230,000 yrs BP. Those in the 8th and 9th layers are of 420,000 yrs BP (U series dating) and those of aspidelite in the 10th ash layer are of 462,000 $\pm$ 45,000 yrs BP (fission track dating). It was determined in the past that the geological ages of the deposits in the cave could be referred to the Middle Pleistocene according to the correlation among the mammalian fauna fossils in the cave (neither fossils nor stone tools have been discovered below the 13th layer). The results are similar to those obtained from the latest isotopic dating.

36

PALAEOCLIMATE

The development of karst, apparently subjected to the influence of climate, is characterized by zonation. Changes in properties of deposits in karst caves also reflect, as a matter of course, changes in the paleoenvironment. For example, the deposits of the Peking Man's Cave constitute multiple layers of mixed limestone breccia—travertines indicating that they were influenced by palaeoclimatic changes.

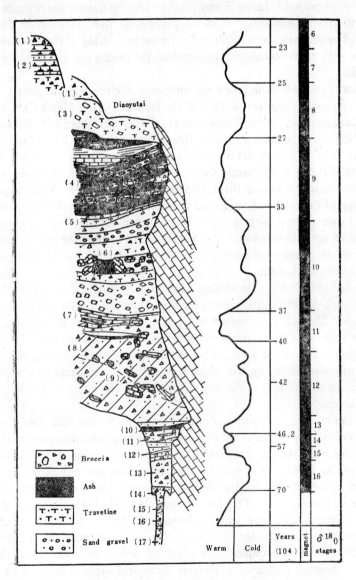

Fig. 1. Peking Man's Cave recording palaeoclimatic curve.

The limestone cave breccia at Zhoukoudian falls into two categories: bulky blocks and slabs resulted from the collapses of cave roofs and abris, and chips as a result of the denudation of cave walls. The breccia of the 6th layer in Peking Man's Cave and that at Locus 15 belong in category one. Pieces of breccia are different in size, more than half of the total were measured over 12 cm in diameter, many about 1–2 m and the maximum even over 10 m. Each piece contains more than 1–2 limestone bedding planes and it is subject to its joints taking a cubic or a cuboidal shape with quite straight edges which indicate the joint planes. Large pieces pile up, leaving among them much spaces filled with finer sediments. It is an observable that some huge limestone blocks fell or the floors as a result of fracturing or faulting of the limestone beds when the cave roofs collapsed. Sometimes, the blocks are still linked with the bed rocks.

The majority of the breccia layers are composed of chips with diameters of less than 12 cm, accounting for 70–75% of the total in Peking Man's Cave and the Upper Cave (containing late *Homo sapiens*—Upper Cave Man's fossils) and 80% in the New Cave (containing early *Homo sapiens*—New Cave Man's fossils) and more than 55% have diameters of less than 6 cm, while bulky blocks with diameters of 0.5–1 m are rarely found. Such breccia, with fine sand, silt and clay filling the spaces among them, was resulted on the whole, from the physical weathering of the cave walls in a cold climate. In peking Man's Cave, they are dispersed over mainly in the 1st, 3rd, 6th, 9th and 13th layers, attaining a total thickness of more than 18 m. The pollen found in those breccia layers bears evidence of the existence of conifer trees (*Pinus* and *Picea*), which demonstrates a cold climate at that time.

Between the breccia layers in Peking Man's Cave are layers of sand, ashes and travertine, where only a little breccia has been found, indicating relatively warmer climate. The pollen buried there are mainly composed of broadleaf trees and even those of *Cymplocos* growing in the south, while mammal fossils may be referred to those of the fauna of South China such as *Hystrix* and *Ailuropoda*. Montmorillonite makes up the main part of the clay minerals in some layers. However, the assemblage of the clay minerals is chiefly composed of illite in the breccia layers.

Although more than 100 species of animal fossils have been excavated, the study on palaeoclimatic changes is by no means thorough and researchers still hold different opinions on the analyses of these changes owing to the incomplete records of the fossils, the limited source of pollen, the unsatisfactory sparse collection of pollen samples, the multiple geneses of deposits and the different views on the modes of life of the animals and their relation to the environments. However, these layers, with strong evidences of remarkable changes, are able to demonstrate the equivalence between changes of deposits and palaeoclimatic oscillations. For example, the analysis of pollen diagrams shows that the 10th layer bearing pollen of a conifer—broadleaf—mixed forest with grasslands manifesting a climate which was similar to that of today. But in the 9th layer, a considerable amount of pollen of *Abies* and *Betula* as well as *Pinus, Quercus, Ulmus* and *Salix* exists. In the upper part of the layer a conifer—broadleaf—mixed forest with less *Abies, Pinus* and *Betula*, but more herbs is found and in the uppermost part there is a tendency of increase of *Pinus* and *Betula* and decrease of herbs, proving certain changes in the temperature. In the 8th layer, the pollen assemblage transforms into a broadleaf forest of the temperate zone, and *Symplocos* and *Corylus* have been discovered, thus, showing that the climate obviously turned warm or even warmer than Beijing today. Some changes in the composition of the cave deposits correspondingly took place, the 10th layer being made up mainly of ashes mixed up with a small amount of breccia while the 9th layer is composed entirely of breccia demonstrating a cold climate which caused intense physical weathering during its formation as compared with that of the 10th layer. Owing to a warmer climate, the 8th layer is composed of sediments intercalated with some travertine layers (about 10–18 cm thick) and pieces of breccia in this layer are small in size. The above—mentioned mammal fossils belonging to the South China fauna made their first appearance in this layer. From the evidences mentioned above we come to a conclusion that changes in the composition of the deposits in Peking Man's Cave indicate, to some extent, the palaeoclimatic changes in that area.

Furthermore, if the isotopic ages of the layers in Peking Man's Cave are correlated with those stages determined by the $\delta^{18}O$ of the Equatorial Pacific V28–238 core, it is clearly shown that nearly all the non-breccia layers (the thick ash layers and the travertine layers) are equivalent to the interglacial represented by the stages of the uneven numbers in $\delta^{18}O$ curve while the breccia layers are approximately equivalent to the glacial represented by the stages of the even numbers in the same curve. For instance, the 2nd layer (250,000–230,000 yrs BP) and the 4th layer (330,000–290,000 yrs BP) can be correlated with the 7th stage of the $\delta^{18}O$ (251,000–195,000 yrs BP) and the 9th stage of the $\delta^{18}O$ (347,000–297,000 yrs BP) respectively. The New Cave, lying to the south of Peking Man's Cave and several dozens of metres away from it, can be divided into two: the upper part being composed of stalagmites (75,000–68,000 yrs BP) and the lower one of travertine (98,000 yrs BP). They can be correlated with the 5th stage of the $\delta^{18}O$ (125,000–75,000 yrs BP). The 9th layer of breccia (425,000 yrs BP) is approximately equivalent to the 12th stage of the $\delta^{18}O$ (470,000–440,000 yrs BP) and the breccia layers of the Upper Cave of 18,340 $\pm$ 410 yrs BP (^{14}C dating) or 40,000 yrs BP (thermoluminescent dating) in age are corresponding to the 2nd stage of the $\delta^{18}O$ (32,000 – 13,000 yrs BP).

So it can be seen that the non-breccia and breccia deposit layers in Peking Man's Cave are able to infer the climatic changes of glacial cycles. Based on these positive results, a preliminary wavy curve expressing the palaeoclimates can be drawn. It shows that from 14th layer up to the top (700,000–230,000 yrs BP) there are four glacial cycles (Fig. 1).

Taken into account the appearance of fossils of some particular mammals and seeds of some particular trees, which exist only in South China today in a warm climate, the climate in Zhoukoudian area 400,000 years ago could be subtropical varying within the range of a warm interglacial and the temperature could be 5°C higher than the average 11.8°C in this area today which roughly corresponds to the present climate between the Changjiang (Yangtze) River and Huaihe River. In the period of low temperature which could be about 10°C lower than that of today, the pollen assemblage may contain a certain amount of pollen of conifer trees which grow along the margin of the northern forest belt located at the southern end of Dahinggan Mts. nowadays. Some frozen-cracked block fields, the rock sea of Late Pleistocene in the West Mountains of Beijing are found at an altitude of 1,700 m, but such weathering no longer exists at the highest peak there which is 2,000 m above sea level. In the Wutai Mt. some block fields, as a result of such physical weathering, can still be found on the slopes at the height of 3,000 m, which gives a clue to the fact that the average temperature during the glacial period could drop 7–8°C than that of today.

CORRELATION OF THE LOESS STRATIGRAPHY WITH THE $\delta^{18}O$ CLIMATIC CURVE

The pollen is proved to be that of a conifer-broadleaf-mixed forest of the temperate zone and the clay minerals also prove the climate during the sedimentation of the 17th–15th layers in the cave to be fairly warm and moist so that limestone breccia can hardly be found. Among the deposits in caves near Zhoukoudian, the sediments at Locus 12 which formed in Early Pleistocene, those at Locus 14 in Late Pliocene, and others which are equivalent to those two periods are all composed of layers of fine sand, silt and red clay, but no limestone breccia occurs. The clay minerals are mainly made up of kaolinite, and there are fish fossils *Barbus* and some mammals fossils such as *Orangutan*. All the evidences indicate that the Zhoukoudian Area enjoyed quite a warm and moist climate in which a red crust of weathering developed in Early Pleistocene. The limestone breccia deposits appear in the 14th layer upwards. Layers of limestone breccia sediment developed almost in all the sediments in other caves around the Zhoukoudian Area since Middle Pleistocene. So, based on the characteristics of these sediments, it can be assumed that the climatic variations in the Quaternary adhered to a certain orientation, in other words, from Middle Pleistocene (700,000 yrs BP) on, climatic variations became obvious and the continental climate became intensified, which corresponded to the reduction of the glaciation of oceanic type during the Quaternary glacial in mountains in West China after Middle Pleistocene, the enlargement of loess distribution, the increase of the thickness of the loess formation in the Loess Plateau of North China and the sluggishness in weathering. Loess in the middle reaches of the

Huanghe (Yellow) River may reach more than 100–200 m in thickness. On the basis of the loess section in Luochuan, Shaanxi Province, thirteen climatic cycles (L1–L13) and twenty-one subcycles can be distinguished, and the climatic variations indicated agree to the fact that the oscillation magnitude after the Brunhes/the Matuyama 700,000 yrs ago in the Early Pleistocene, was the greatest. The environment where the loess in China was deposited was moist forest-steppes but gradually turned to be dry steppes and desert steppes in Middle and Late Pleistocene.

The upper part of Luochuan Loess can be divided, in accordance with the different pedogenic processes, into four types as follows: the weakly pedogenic, the moderatly pedogenic, the comparatively strongly pedogenic and strongly pedogenic. The fossil soil between those fall into six types, namely, black loam, drab-like soil, calcareous drab soil, drab soil, leached drab soil and brown drab soil. They underwent, in various degrees, the dry and moist, cold and warm palaeoclimates during the deposition and evolution of loess in China.

And the deposits in Peking Man's Cave also give certain expression to those changes in palaeoclimate. The 13th layer consists of clay mixed up with a breccia bed being 2.9 m thick. The low stability coefficient of the heavy mineral assemblage and fossils of cryophilic mammals found there infer the inclination of a cold climate. The 12th layer, 1.5 m thick, contains gravels, sand mixed up with a thin layer of travertine where breccia comprises less than 10%, and the pollen is that of a conifer-broadleaf-mixed forest of the temperate zone and some fossils of *Dicerohinus zhoukoutienensis* which adapted to warm grasslands have been excavated. The 11th layer, 0.8 m thick, consists of coarse sand and limestone breccia, where the pollen is that of a conifer forest growing in mountainous areas and fossils of herbivorous animals occupy a large percentage, both inferring the inclination of a cold climate. The 10th layer consists of ashes and brown silt and is mixed up with two thin layers of travertine (6–7 cm thick), where the pollen is that of vegetation in the transition to a deciduous broadleaf forest of the temperate zone, inferring quite a warm climate. The layer is 462,000 yrs BP, or 520,000–610,000 yrs BP (thermoluminescent dating). Taken into account the fact that the climatic variations inferred from those layers (the 10th–13th) were not in a great magnitude, they can be correlated with L12, S12, L11 and S11 of Luochuan Loess. L12 is a layer of weakly pedogenic loess which formed in a dry and frigid climate, probably corresponding to that during the period of the development of breccia in the 13th layer in Peking Man's Cave; L11 is a layer of moderately pedogenic loess which was deposited in a less dry and frigid climate. But S12 and S11 are composed of drab soil of dry forests and grasslands, both demonstrating moderate climatic variations. Therefore, they can be correlated with the 12th and 10th layers respectively. In the $\delta^{18}O$ climatic curve, the curve of 16th–13th stages is also not in a great undulate magnitude, either. The ages are 627,000–502,000 yrs BP, so that they can be correlated to the 13th–10th layers.

The 9th layer composed of a thick layer of limestone breccia can be approxi-

41

mately correlated with the 12th stage of the $\delta\ ^{18}O$ according to its age as mentioned above. L10 is composed of strongly pedogenic loess which proves that the climate was not too dry and cold when the loess was deposited at that time. The Loess Plateau is characterized by grasslands or dry forests—grasslands. Therefore, a forest could cover the Zhoukoudian Area, which, lying at the eastern foot of the Taihang Mts. and facing the Great North China Plains, could be relatively warm and moist. And the pollen contained in the 9th layer, as mentioned before, supports the assumption and, moreover, a large number of seeds of cold—resistant trees mainly living in the north such as *Abies, Betula* and *Pinus* have been discovered. So the limestone breccias in the layer contains more silt and silty clay which account for more than 20—30%. Although the majority of the mammal fossils are those of the forest type, the existence of fossils of *Coelodonta antiguitatis yenshanensis* adapted to cryosteppes have been noted.

Travertine layers were deposited from the 9th layer upward, which indicates a transition to a warm and humid climate, and the pollen has been correspondingly found to be those of a broadleaf forest of the temperate zone, and what is more noteworthy is the sudden appearance of pollen of *Corylus* and *Symplocos*; the former grows within the reaches of Changjiang River and the latter in the subtropical zone which lay in the further southern part of the country. Mammals were *Myotis* sp., *Hystrix* sp. and *Bubalus tellhardi*, etc., all found in a benign climate. Such kinds of animals now only live in South China. The 7th layer consists of silt and sand as well as thin layers of travertine, indicating a quite temperate climate. According to the paleomagnetic study, a reversed event which took place at about 380,000 yrs BP has been recorded in the 7th layers and it has also been recorded in S9 of the Luochuan Loess section. It is evident that the 8th and 7th layers can be correlated with S9, corresponding to the 11th stage of the $\delta\ ^{18}O$ (367,000 yrs BP).

The 6th layer of well developed limestone breccia, composed of three layers of large breccia pieces and three layers of small pieces, attaining a thickness of more than 6m. This layer can be correlated with L9 in view of the nature of earth because L9 is composed of weakly pedogenic loess which formed in an extremely dry and frigid climate. And as the peak of the 10th stage of the $\delta\ ^{18}O$ (347,000 yrs BP) is fairly high, it most probably stands for the same age as those of the 6th layer and L9.

The 5th layer, which is the only thick one composed of travertine bed, attaining a thickness of 45cm, must be the sediments formed in a temperate and humid climate owing to the fact that fossils of *Ailuropoda* sp. and *Cynailurus* sp. have been discovered, both of them belonging to the fauna in South China, and the latter still living in Guangxi Autonomous Region today.

The 4th layer composed of ashes is the thickest of all (more than 6m thick), where limestone breccia can hardly be found. Montmorillonite makes up the main part of the clay minerals, the mammal fossils are chiefly those of mammals found in a warm climate and the pollen is that of a conifer—broadleaf mixed forest of the temperate zone. According to their ages, these two layers (330,000—290,000 yrs BP) can be correlated with the 9th stage of the $\delta^{18}O$ (340,000—297,000 yrs BP) which was a fairly long interglacial lasting about 50,000 yrs, or corresponding to S8—S6. There are two sublayers (4d and 4e) in the 4th layer, the first being 310,000 yrs BP. and the second 290,000 yrs BP according to the thermoluminescent dating. The sedimentation of the 4th layer, therefore, covered quite a long period which can be judged either from its thickness or from its age. During this period, the walls of Paking Man's Cave at Zhoukoudian underwent physical weathering yet intense dissolution, so it is evident that other caves probably went through the travertine sedimentation, too. For example, in Peking Man's Cave a thick layer of travertine sediment, which is 320,000 yrs BP according to U-series dating is found on the walls of a small lateral branch cave, near the cave roof and buried in the 2nd layer, and the old stalagmites in the Upper Cave were most probably also formed during that period.

The 3rd layer, which is more than 3 m in thickness, is another example of fully developed limestone breccia containing a small amount of pollen of *Pinus*, *Picea*, etc. The layer can be correlated with L6 and the 8th stage of the $\delta^{18}O$ in accordance with their sedimentary environment and stratigraphic sequence. L6 is another fairly thick layer of weekly pedogenic loess, proving the climate to be rather cold during the sedimentation. The formation of the 3rd layer of breccia was quite the same, which is characterized by fairly rounded pieces of eroded limestone breccia covered with furrows, accounting for 40% locally. Such limestone blocks were deposited in the 3rd layer and the formation was due to the rainwash weathering during the sedimentation of the 4th and the 5th layers.

The 2nd layer consists of travertine mixed up with silty clay and silt, attaining a thickness of 1.7 m. The fact that five layers of travertine were developed and the pollen is that of deciduous broadleaf forests mixed with grasslands demonstrates that a temperate climate occurred during the sedimentation. As stated above, these layers can be correlated with the 7th stage of the $\delta^{18}O$ according to their ages, the former being 250,000—230,000 yrs BP and the latter 250,000—195,000 yrs BP S5 of Luochuan Loess which is 220,000—180,000 yrs BP according to the thermoluminescent dating and particularly developed fossil soil which is 5 m thick, contains three layers of fossil soil, a kind of drab soil, so that it can be correlated with the 2nd layer as far as their sedimentary environment and ages are concerned.

The 1st layer is also a breccia bed, 1—2 m in thickness. At the site of Diaoyutai on the west wall of the cave, it overlies the 3rd breccia bed with an erosive surface in between, and the cement and weathering of the breccia vary in degree, so that it can be easily distinguished. This layer can be correlated with the 6th stage of the $\delta^{18}O$ (195,000—128,000 yrs BP) or corresponding to the period when L5 of Luochuan Loess underwent its sedimentation.

In the New Cave which is located at the southern slope of the Dragon Bone

Hill, there are several layers of travertine mixed up with those of mud, attaining a thickness of 3 m and old stalagmites have been found on the top layer. According to their ages, the travertine (98,000 yrs BP) and the stalagmites (75,000—68,000 yrs BP) containing pollens of a broadleaf forest of the temperate zone, can be correlated with the 5th stage of the $\delta^{18}O$ (128,000—75,000 yrs BP) and corresponding to S4—S2 of Luochuan Loess sections and the R—W interglacial of the classical Alpine glacial model.

The breccia layer in the Upper Cave attains a thickness of more than 5m, which mixed up with several ash layers covering the old stalagmites. The layers, being 40,000—18,000 yrs BP in age, can be correlated with the 2nd stage of the $\delta^{18}O$ (32,000—13,000 yrs BP), i.e., corresponding to the sedimentary period of Malan Loess. The age of S_o, the upper part of Malan Loess is $8,000\pm400$ yrs BP C_{14} dating and that of fossil soil of S1 which is below Malan Loess is $41,000\pm4,800$ yrs BP (thermoluminescent dating).

REFERENCES

[1] Pei, W. C., 1937, Les fouilles de Choukoudien en China, Bull. de la Soc. Préhistorique Française, 34 (9), p. 354—366.
[2] Pei, W. C., 1939, New fossil materlials and artifacts collected from Choukoudien during 1937—1939, Bull. Geol. Soc. China, 19, p. 147—188.
[3] Woo, J. K., Chai, L. P., 1954, New discoveries about *Sinanthropus pekinensis* in Choukoudien, Scientia Sinica, 3 (3), p. 335—351.
[4] Chao, Z. K., Li, Y. H., 1960, Report on the excavation of *Sinanthropus* site in 1959, Vert. Palas. 4 (1), p. 30—32.
[5] Huang, W. L., 1960, Restudy of CKT *Sinanthropus* deposits, Vert. Palas. 2 (1), p. 83—95.
[6] Chow Min-chen, 1955, *Sinanthropus* living environment inferred from vertebrate fossils, Kexue Tongbao, no. 1, p. 15—22.
[7] Kahlke, H. D., Chow, P. S., 1961, A summary of stratigraphic and paleontological observations in the lower layer of Choukoudien, Locality 1, and on the chronological position of the site, Vert. Palas. 3 (2), p. 212—240.
[8] Chou Pen-shun, 1979, The fossil *Rhinocerotides* of Locality 1, Choukoudien, Vert. Palas. 17 (3), p. 236—258.
[9] Kurten, B., Yassri, Y., 1961, On the date of Peking Man, Societus Scientiarm Fennica, 23, p. 1—10.
[10] Hsu, J., 1965, The climatic condition in North China during the time of *Sinanthropus*, Scientia Sinica, 15, p. 410—414.
[11] Chia, L. P., 1978, A note on the weather conditions in Choukoudien area of Peking Man's time, Acta Stratigraphica Sinica 2 (1), p. 53—56.
[12] Li Yan-xian, Ji Hong-xiang, 1981, Environmental changes during Peking Man's time as viewed from mammalian fossils, Kexue Tongbao, 26 (2), p. 170—172.
[13] Qian Fang et al., 1980, Magnetostratigraphic study on the cave deposits containing fossil Peking Man at Zhoukoudian, Kexue Tongbao, 25 (4), p. 359.
[14] Zhao Shu-sen et al., 1980, Uranium-series dating of Peking Man, Kexue Tongbao, 25 (5), p. 447.
[15] Gue Shi-lun et al., 1980, Age determination of Peking Man by fission track dating, Kexue Tongbao, 25 (6), p. 535.
[16] Pei Jing-xian et al., 1979, Thermoluminescence ages of quartz in ash materials from *Homo erectus pekinensis* site and its geological implication, Kexue Tongbao, 24 (18),

p. 849.

[17] Chou Shi-hua et al., 1980, Radiocarbon dating of fossil mammal bones from the upper cave, Kexue Tongbao, 25 (5), p. 448.

[18] Nicholas, J. D., Opdyke, N. D., 1972, Oxygen isotope and palaeomagnetic stratigraphy of Equatorial Pacific core V_{28-238}: oxygen isotope temperatures and ice volume on a 10^6 year scale, Quaternary Research, 3 (1), p. 39–55.

[19] Lu Yan-chou, An Zhi-sheng, 1979, Inquiry on natural environmental changes during the Brunhes Epoch, Kexue Tongbao, 24 (5), p. 221–224.

[20] Lu Yan-chou, 1981, Pleistocene climatic cycles and variation of $CaCO_3$ contents in a loess profile, Scientia Geologica Sinica, no. 2, p. 131–142.

[21] Chia, L. P., 1959, Report on the excavation of *Sinanthropus* site in 1958, Vert. Palas. 3 (1), p. 41–46.

PRELIMINARY PALEOMAGNETIC STUDY
OF LATE CENOZOIC BASALT GROUPS
FROM DATONG DISTRICT, NORTH CHINA

Liu Chun,
(Institute of Geology, Academia Sinica)

Kazuaki Maenaka
(Hanazono College, Kyoto 616, Japan)

and Sadao Sasajima
(Department of Geology and Mineralogy, Faculty of Science, Kyoto University, Kyoto, Japan)

ABSTRACT

In Datong district of North China, late Cenozoic basaltic rocks are widely distributed. These basalts are in contact with Lishi Loess and Nihewan beds which are the representative rock types of the Quaternary in North China. The age of Lishi Loess is a major problem in the Quaternary geology in China. In this paper, we describe a paleomagnetic study of these rocks, which was used to correlate with the radiometric ages, a total of 84 oriented samples were collected from the basalt, they underlying baked loess and unbaked loess. These samples were collected from: Heishan-Gelao Village (site B), Yujiajai Village (site Y) both in Datong County, and from Dongshuitou Village (site T), which is located in Yanggao County. The remanent magnetizations were measured by a Schonstedt Magnetometer (SSM—IA) at Kyoto University, Japan. The results of the alternating demagnetization test on pilot samples showed that the NRM of the samples from site B and site T were stable, however, the samples from Y were not stable. To isolate the stable direction of magnetization, all samples were demagnetized in a peak field of 200 Oe, (a value chosen from the pilot study). Based on the results obtained after magnetic cleaning, the following conclusions are reached:

(1) The directions obtained from the basalt, baked loess and the unbaked loess agree well with each other. Magnetic polarization so far reported on Lishi Loess seems to be established.

(2) The polarity of the measured samples, except one, show normal polarity before and after demagnetization. It is suggested that Lishi Loess was deposited during Brunhes Epoch.

(3) The magnetic declinations of the collected samples indicate eastward deviation and inclinations showing a shallower dip than the present value (59°) for Datong District. The amplitude of the secular variation seems to be rather

large.

(4)   The reversed magnetization observed in samples from Heishan Area (Black Mountain site B) records a short reversed episode during the late Brunhes Epoch. This could be correlated with the Blake event if the thermoluminescence date of the underlying baked loess (0.16 m.y.) is taken into account.

MAGNETOSTRATIGRAPHIC STUDY OF SEVERAL TYPICAL GEOLOGIC SECTIONS IN NORTH CHINA

LI HUA-MEI AND WANG JUN-DA

(Institute of Geochemistry, Academia Sinica)

Abstract—In the east of the Datong Basin, the beginning of loess accumulation is at about Middle Matuyama Epoch, the river-lake sediments of Yujiazhai section along the Shanggan River belong to the Middle Pleistocene. Datong volcanic rocks were mainly erupted in Brunhes epoch, at about 0.45 m.y. Lacustrine sediments in Nihewan Basin began at Kaena event with an age of 3 m.y. BP and came to the end at Brunhes Epoch, middle-late period of Middle Pliestocene, the stone implement-bearing bed within Nihewan formation is located under the basal boundary of Jaramillo event, its age is about 1 m.y. Polarity epochs and events since 3.0 m.y. were recorded in the 600 m long core from Bo 3 drill hole in Northern Bohai Coastal Plain. The stratigraphic division and correlation, and the basal boundary of the Quaternary system are discussed along with the data of isotopic chronology, paleontology and micropaleontology.

In the last two years, detailed paleomagnetic studies have been made on some sections and cores in Datong Basin, Shanxi Province, Nihewan Basin, Hebei Province and the Northern Bohai Gulf Coastal Plain. We attempt to further discuss the stratigraphic division and the basal boundary of the Quaternary in North China along with the data on isotopic chronology, paleontology and micropaleontology.

More than 1,200 oriented specimens were collected from the above-mentioned regions.

ANALYSES OF STRATIGRAPHIC MAGNETIC POLARITY OF THE REGIONS

Datong Basin

In the eastern area of the Datong Basin (39°56′—40°15′ N, 113°35′—113°50′ E) loess is mainly distributed in the northern part of the area. There are more than 10 volcanic cones in the basin. Fluviolacustrine sediments are overlain by basalts along the Sanggan River in the southern part of the area.

In this region, a well developed loess section is located near the Dongshuitou Village, Yanggao County. The loess section is about 34m thick and 9 layers of buried soil are developed in the section. There are fluvial facies sediments of 4 m thick at the base of the section. 82 oriented samples were collected from the Dongshuitou section. Samples from the upper unit, 15 m thick, interbedded with 4 layers of fossil soil were normally magnetized, those from the lower unit predominantly show reversed magnetization. The Brunhes-Matuyama boundary with the age of 0.73 m.y. lies at the base of the fourth layer of buried soil. In the section the Jaramillo event was also identified. Loess began deposition during the Middle Matuyama Epoch. There is a basalt layer of 1—4 m thick in Caiyuangou and Handong Loess sections. 42 oriented loess samples were collected from the two sections of 20 m thick. Measurements show that samples were predominantly normally magnetized. The $^{40}Ar / 39_{Ar}$ age of the basalt is 0.45 ± 0.02 m.y. Loess sampies collected at the depth of 1 m below the basalt layer have a TL age of 0.22 ± 0.02 m.y. It shows that the basalt and loess in the sections should belong to Brunhes normal polarity epoch. Near Huangjiawa Village of Datong County, basalt layers in the southwestern gully of Jinshan cone and the northwestern gully of Langwoshan cone are exposed. 16 lava and loess samples were normally magnetized. In consideration of that K–Ar age of basalt is about 0.45 m.y. in the sections and there is one layer of buried soil in loess overlying the basalt, the formation time should belong to Middle Pleistocene.

Yujiazhai section is a suite of fluvial and lacustrine sediments composed of sand, silt and clay, located at the northern bank of Sanggan River. The basalt, 4—5 m thick, is at the top of the section, its K–Ar age is 0.455 ± 0.036 m.y. Exposed thickness of the section is about 25 m and more than 30 oriented samples show normal magnetization. Middle Pleistocene antler fossils of *Megaloceros* were found in lacustrine sand layer of the Dongshan section near Xubao Town and the TL age of the baked soil underlying the lava is $340,000 \pm 21,600$ yrs BP[1]. Based on these data, these fluviolacustrine sediments should belong to Middle Pleistocene.

Nihewan Basin (40°00′—40°20′N, 114°15′—114°50′E)

The famous Nihewan formation is referred to a series of fluviolacustrine facies strata lying between the Pliocene *Hipparion* Red Clay and Late Pleistocene Malan Loess. Yangshuizhan and Nangou sections near the Hongya Village are located on the margin of the basin. They can basically represent the lower part of Nihewan formation. There is a very clear erosion surface between these strata and Pliocene *Hipparion* Red Clay with *Hipparion* and *Chilotheriun*[2]. There are brown-red sandy clay beds containing sand, gravel stripe and calcareous nodules between lake deposits and *Hipparion* Red Clay. They belong to diluvial facies and the age is considered as Late Pliocene. In the middle and upper parts, 96 m. thick, of the Nihewan formation the sediments are composed of variegated clay, sandy clay and silt layers, among which 7 thick layers of gravel and more than 10 thin layers of marlite interbedded in the section are also found. In the lower part of the section, shell fossils are recognized. There is an erosion surface between lacustrine strata and the underlying diluvium. 257 oriented samples were collected from the section. Paleomagnetic results show that samples from the upper section display mainly reversed magnetization, those from the lower section predominantly show normal magnetization. The polarity boundary corresponds to that between the Matuyama reversed polarity epoch and the Gauss normal polarity epoch. Its age is 2.48 m.y. which is roughly consistent with previous result[3]. The Olduvai and Kaena events were also recorded respectively. Although the magnetization of *Hipparion* Red Clay and diluvial brown-red sandy clay is not so stable, they are mainly characterized by normal polarity. Some reversed magnetic samples from the lower part may be a reflection of Mammoth event.

Paleomagnetic measurements on 100 samples taken from the lower Chengqiang section show that the boundary between the Matuyama and Gauss Epochs appears at the gravel layer near the top of a unit of gray clay layers and shell-bearing layer. The erosion surface between lacustrine strata and lower diluvial strata is also located at Kaena event.

Haojiatai section near the center of the basin is a good one of Nihewan formation, with the thickness of 112 m. Lacustrine strata are composed of abundant light colour calcareous marl, sandy clay and clayey sand layers. The top is a part of Late Pleistocene Malan Loess, 15 m thick. There is a distinct erosion surface between the upper and lower Nihewan formation. On the erosion surface there is a yellow silt layer, 14 m thick, in which fine sand and small gravel stripes were interbedded. *Equus sanmeniensis*, etc. was found in the lower part of the silt bed. 257 oriented samples were collected from Haojiatai section. The lower unit, 40 m thick, was essentially reversed in magnetization. The normally magnetized zones are in the middle unit, while some reversely magnetized zones in the same unit. In the upper unit, all samples from lacustrine strata of 24 m thick and Malan Loess of 15 m thick display normal magnetization except for a few samples. The middle and lower units belong to Matuyama reversed epoch, and the upper unit to Brunhes normal epoch. The boundary between Brunhes and Matuyama is located at the middle part of upper Nihewan formation. Above and under the interface between the Upper and Lower Nihewan formations along which an erosion surface was found, there are two normal polarity zones. On account of the existence of depositional hiatus, we suggest that the normal polarity zone above the erosion surface represents the Jaramillo event, and another zone under the surface represents the Olduvai event. A magnetic polarity event in light red buried soil, under Malan Loess, was supposed to be Blake event with an age of 0.1—0.11 m.y.

In Donggutuo section, the base of the Jaramillo event appears at 5 m over the basal boundary of the section. Therefore, the age of stone implement-bearing bed at a depth of 2 m over the basal boundary is about 1 m.y.

50

Northern Bohai Gulf Coastal Plain

A 600 m long core of Bo 3 drill hole (39°15′ N, 118°30′ E) in the northern Bohai Coastal Plain comprises mainly clayey sand, sandy clay, intercalated with
thin clay layers, thick loose silt layers, fine sand and coarse sand. Based on
foraminifera and ostracod analyses, Hu Lan-ying and others found that there
were three marine layers interbedded within a depth of 80 m in the core. The
first marine layer found at 73—78 m in depth, the second, at 32—46 m, and the
third, at 0—15 m. Among them the second layer is the most developed. A
lot of micro-fossils exsit in each marine layer. The three layers can be correlated with other three marine layers interbedded within 100 m depth in the western Bohai Gulf Coastal Area respectively[4].

254 samples orientated in up-down orientation were taken from the core. In
the light of the curves of inclination change of the samples after demagnetization, the core is divided into three subsections: the first subsection at 0—171 m
in depth samples taken from which are mainly normally magnetized except that
a few samples from 15 and 103 m in depth are reversed; the second, 171—
493 m inclinations of samples mainly are negative, but at 198—218 m and 361
— 392 m in depth samples are normal in magnetization; the third, at 493—
600 m, it is very obvious that inclinations of samples are positive. At 572 —
588 m there are several normal samples. Obviously, the three subsections
mentioned above correspond to Brunhes normal epoch, Matuyama reversed
epoch and Gauss normal epoch, respectively[18]. Two normal polarity horizons in the second subsection are related to Jaramillo and Olduvai events of
Matuyama Epoch. The reversed samples in the third subsection reflect the
Kaena event of the Gauss normal polarity epoch. Due to the third marine
layer representing the last transgression in postglacial period at 0—15 m in depth,
radiocarbon date (^{14}C = 8620 ± 250 yrs), and pollen information, a short
polarity event recorded at 15 m in depth of the core should be Gothenburg event
with an age of 10,000 — 12,000 yrs.

Basically, this suite of littoral and swamp sediments was continuously deposited. In the light of polarity ages and their corresponding thickness, from
the basal boundary of the Kaena event to the boundary between Brunhes and
Matuyama Epochs the deposition rates of each interval of the core were calculated to run from 19.5—23.4 cm per 1,000 yrs. And it is nearly close to the deposition rates of other areas in the North China Plain (20 cm / 1,000 yrs.)[5].

RESULTS AND DISCUSSION

The Paleomagnetic Results

The paleomagnetic results show that same polarity can be recorded by the rocks

51

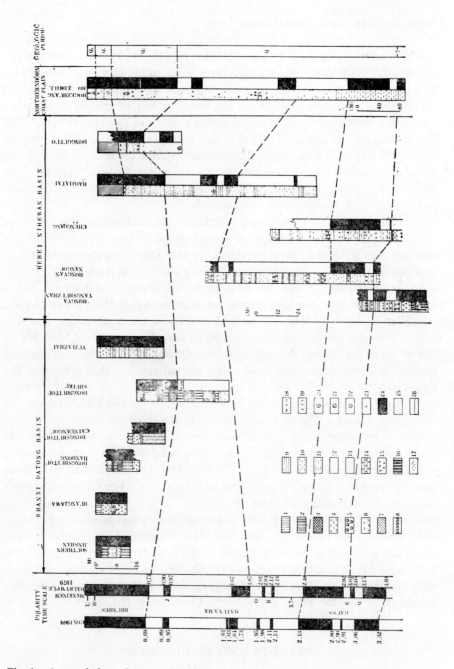

Fig. 1. A correlation of magnetostratigraphic column of three regions in North China
1–Malan Loess, 2–Lishi Loess or Wucheng Loess, 3–paleosoil, 4–basalt, 5–volcanic breccia,
6–cinder, 7–baked soil, 8–marlite, 9–clay, 10–sandy clay, 11–clayey sand, 12–silt, 13–sand,
14–gravel, 15–angular gravel, 16–*Hipparion* Red Clay, 17–intertongue, 18–calcareous nodule,
19–shell, 20–vertebrate fossil, 21–stone implement, 22–marine foraminifera, 23–plant remains,
24–normal polarity, 25–reversed polarity, 26–depositional interval.

formed in the same peried whatever their various lithologic categories are, such as volcanics, sedimentary and baked rocks, and also sediments of fluvial, lacustrine, marine and eolian facies. Therefore, it is possible that various geological sections can be correlated in a large area even or over the world by recording magnetic polarities. The results also show clearly the relationship between strata in those areas and changes of sedimentary type from eastern Loess Plateau through intermontane basins to the coastal plain (Fig. 1).

Some Views to the Nihewan Basin

1. The beginning of the deposition of lacustrine sediments in Nihewan Basin is at Kaena event of Middle Gauss Epoch, about 3.0 m.y. Since fossils of *Lynx variabilis, Zygolophodon* sp., *Paracamilus* sp., *Antilospire yuxianersis* and *Hipparion* which are relic species of Tertiary found recently in the lower Nihewan formation at east cliff section near Dongyaozitou Village coexist with Early Pleistocene *Coelodonta antiquitatis* and *Gazella sinensis*[6], and the fossil associations found in the layers near the bottom of lower Nihewan formation in Dongyaozitou, Chengqiang and Hongya sections located on the edge of the lake basin are similar the same case, it seems reasonable that the age of Quaternary bottom boundary should be referred to 3.0 m.y.

2. Nihewan formation extended to the middle period, or even late period of Middle Pleistocene. The mammal fossils of Early Pleistocene found in the past, such as *Equus sanmeniensis, Proboscihipparion sinense, Eqimachairodus* cf. *crenatidens,* etc. mostly exist at the bottom of Upper Nihewan formation. There is a facies change between Nihewan lacustrine layer and Lishi Loess of Middle Pleistocene in the Daheigou section. Paleomagnetic measurements on 125 samples from Hutouliang section near the center of the lake basin show that fluvial-lacustrine sediments more than 56 m in thickness were almost normally magnetized. Ostracod fossils such as *Limnocythere dubiosa* Daday, *L. sanctipatricii* and *L. binda* Huang etc[7] of Middle Pleistocene in the section can be correlated with those in the first marine layer at 73—78 m in depth in Bo 3 drill core. So Hutouliang Lake sediments must have been deposited during Brunhes Epoch, and can be related to the upper part of Upper Nihewan formation in Haojiatai section, which belong to Middle Pleistocene.

3. The middle-upper parts of Hongya section correspond with Lower Nihewan formation of the Haojiatai section. There is a break in the profile between Upper and Lower Nihewan formation. Hutouliang section is related to the upper part of Upper Nihewan formation in the Haojiatai section. Therefore, the contraction of Nihewan Lake Basin was from south to north (from the margin of the lake to the center) as well as from northeast to southwest.

Problems on Stratigraphic Boundaries

1. Holocene basal boundary

The marine layer at 0—15 m in depth in Bo 3 drill core which can be correlated with the first isotopic oxygen stage representing warm period from deep-sea core in the Pacific Ocean[8] shows the last transgression in postglacial period. Meanwhile, the Gothenburg polarity event was recorded at 15 m in depth in the core.   Hence, it is suitable to consider the event with an age of 10,000—12,000 yrs, as the Holocene bottom.

2. Late Pleistocene bottom

In the Haojiatai section, the Blake event was recorded at the first layer of buried soil underlying Malan Loess in Haojiatai section.   The second marine layer at 33—46 m in depth in Bo 3 core corresponds with the fifth isotopic oxygen stage in deep-sea cores.   The Blake event recorded in deep-sea cores is related to the early fifth stage.   So the Blake event with an age of 0.1—0.11 m.y. can be roughly considered as the bottom of Late Pleistocene.

3. Middle Pleistocene basal boundary

Mammal fossil-bearing strata below the Brunhes-Matuyama boundary with an age of 0.73 m.y. belong to Early Pleistocene.   And Middle Pleistocene Zhoukoudian mammal fossils belong to Brunhes Epoch[9], while the tenth layer being ash of cave accumulation of Zhoukoudian has a fission track age of 0.462 ± 0.045 m.y.[10] So, there is no dissentient at present that the Brunhes-Matuyama boundary recorded in the sections of the Datong Basin and Nihewan Basin and in drill core of northern Bohai Gulf Coast Plain, should be considered as the Middle Pleistocene bottom.

Problem on the Basal Boundary of the Quaternary

Where should the Quaternary bottom be placed ?   If the bottom part of Nihewan formation bearing mammal fossils of Neocene relic species associated with Early Pleistocene type species can be considered as the beginning of Early Pleistocene, the Quaternary bottom should be 3.0 m.y. in age, then, the result can be correlated with Villafranchian of France, the statum bearing 1470 ape-man fossils of Turkana Lake in Kenya[11] and also the Hadar layer bearing ape-man fossils in Ethiopia[12].   If based on pollen evidence found at 464 m in depth and ostracod data collected at 505 m in depth in Bo 3 drill core, the Quaternary basal boundary should be as old as 2.48 m.y. corresponding to the boundary between the Matuyama reversed and Gauss normal epochs. This viewpoint roughly agrees to the results obtained from Beijing plain Shun 5 drill core in which the layer bearing *Hyalinea balthica* is about 2.3 m.y. in age[13], also to that from the marine sedimentary layers near Rome in Italy in which the first appearance of *Arctica islandica* was 2 m.y. ago[14]. Global difference in time between marine and continental sediments has been acknowledged, so further study is needed to find out whether biological evolution following the sudden climatic change in ocean was slower than that in continent.

REFERENCES

[1] Yang Jing-chun, 1961, Acta Scientiarum Naturalium Universitatis Pekinensis, v. 7, no. 1, p. 87—100.
[2] Section of Nihewan Cenozoic, 1974, Vertebrate PalAsiatica, v. 12, no. 2, p. 99—108.
[3] Cheng Guo-liang et al., 1978, Scientia Geologica Sinica, v. 3, p. 247—252.
[4] Zhao Song-ling et al., 1978, Oceanologia et Limnologia Sinica, v. 9, no. 1, p. 15—25.
[5] Li Hua-mei et al., 1979, Scientific Papers on Geology for International Exchange, 5, p. 83—90.
[6] Tang Ying-jun et al., 1981, Vertebrate PalAsiatica, v. 19, p. 256—268.
[7] Huang Bao-ren, Kexue Tongbao, 1980, v. 25, p. 277—281.
[8] Shackleton N. J., 1973, Quaternary Research, v. 3, p. 39—55.
[9] Qian Fang et al., 1980, Kexue Tongbao, v. 4, p. 92.
[10] Guo Shi-lun et al., 1980, Kexue Tongbao, v. 25, p. 384.
[11] Fith F. J., 1974, Nature, v. 251, p. 213.
[12] Aronson J. L., 1977, Nature, v. 267, p. 323—327.
[13] An Zhi-sheng et al., 1979, Geochemica, no. 4, p. 343—346.
[14] Arias C., 1980, Quaternary Research, v. 13, no. 1, p. 65—74.
[15] Liu Tung-sheng et al., 1962, Acta Geologica Sinica, v. 42, p. 1—14.
[16] Cox A., 1969, Science, v. 163, p. 237—245.
[17] Mankinen E. A. et al., 1979, Journal of Geophysical Research, v. 84, p. 615—626.
[18] Li Hua-mei & Wang Jun-da, 1983, Geochemica, No. 2, p. 196—204.

AGE OF LOESS DETERMINED BY THERMOLUMINESCENCE (TL) DATING OF QUARTZ

Li Hu-hou

(Institute of Archaeology, Chinese Academy of Social Sciences)

and Sun Jian-zhong

(Institute of Geology, National Bureau of Seismology)

Abstract—The displaced electrons in quartz lattice can be transferred by sunlight. This is a basic characteristic of quartz for the use of TL technique to date the age of loess. We have tested this property of quartz and found it useful for the development of TL technical applications.

The main points in dating technique are mentioned, which include: 1. Shine fading; 2. Revision of the annual dose.

For explanation, an age determination of loess samples in Yuangou profile is presented as an example.

Baaed on the TL technique of quartz the main problems of loess dating are discussed:

BASIS FOR DATING THE AGE

One of the main basis to use the TL characteristics of a mineral is the electron displacement due to radiation effect in the crystal lattice. These transferred electrons will stay in a substable state to store energy. The stored energy will be released in the form of photoemission.

The use of TL technique dating the age of loess is based on another property of quartz. The stored energy electron will be transfered and releasing their energy not only by heating but also by the sunlight. If the quartz is exposed to the sunlight, the displaced electron will come back to the place in the crystal lattice where it is in a lower energy state. We call this phenomenon the "shine fading". Two kinds of electron transfer, TL and shine fading, are similar in character, i.e., the energy release takes place in the stored energy electrons themselves. Nevertheless, shine fading is different from TL, which has a certain frequency range and characteristic temperature in luminescence. The photon discharge in TL is a form of a coherent event which occurs in an instant. But, that in shine fading is another pattern in which the emission rate obeys the relation:

56

$$\frac{dN}{dt}=e^{-KI}, \quad I=\Phi t$$

the photon number of emission per time interval is the decay number of displaced electron in sample excited for a given time at constant luminous flux. There is an exponential decay with the luminous flux in the equation, k is coefficient of stored energy in the crystal; I is the intensity of the shine light; Φ is the luminous flux, t is the shine time.

Since stored energy electrons in quartz crystal have a characteristic of shine fading, all dating operations with quartz must be performed in a dark room. The sample can not be exposed in the sunlight either. Consequently, this characteristic is used in dating the age of loess. In order to obtain the evidence of the shine fading of quartz, we have performed the shine fading experiment with an artificial light source, an iodine-tungsten photographic lamp as the substitute of sunlight. It has a electric power of 1,300 watt and 3,000°K colour temperature. In the loess experiments of shine fading using this type of lamp, it is observed that in the quartz extracted from the loess, both the natural stored TL or the TL elicited by radiation would give rise to shine fading in different degrees which has a quantitative relation with the origin and size of the quartz grain tested. This relation can be expressed as coefficient k of the equation mentioned above. Therefore, for a specimen the regulation of shine fading from a fixed light source and in exposure conditions must be found. Then the optimum luminous flux in shine fading for different samples used could be determined.

These experiments are of paramount importance for loess dating with quartz having a 375°C thermoluminescent glow curve peak which decreases after light stimulus according to the rules of exponential. Combining the above-mentioned theory and the eolian genesis theory of loess[1]. It can be inferred that quartz had existed before the formation of loess. The process of shine fading by sunlight in quartz occurred either in the periods of transportation by wind or after the quartz arrived at the locus to be one of the constituents of loess. However, so long as the quartz was extracted from loess, it had already undergone shine fading by sunlight. Also that the stored energy electrons in quartz lattice all returned back to the valent band. Thereafter, the new stored energy electrons in the quartz lattice were produced by the environmental radiation in loess. Of course, after the quartz entered into the loess, the energy would progressively be stored during the later ages. Therefore, in principle, if the quantity of store energy in quartz is determined, the technique of loess dating can be established.

EXPERIMENTAL METHOD

Here, some practical steps of experimentation according to the above principles are given as follows:

Shine Fading Effect

The purpose of shine fading experiment is to find the responses of quartz from different origins that various specimens can be determined. Both the natural stored TL or artificial irradiated TL of samples must be determined. We hope that for all samples, more than 90% of stored energy by shine fading is an acceptable level, since it is impossible to give a fixed level of acceptability. One must assess each specimen with the actual operation to find what is useful in dating. Therefore, the most important of all is to determine the residual quantity of TL after shine fading, which will be obtained by using 8—12 samples to determine the responses of the specimen to shine fading. From this we will be able to select the optimum exposure condition. Another 8—12 samples are used to determine $N+\beta_i$ and $(N+\beta_i)_s$. Two groups of glow curves obtained give two straight lines, from which the beta equivalent dose can be calculated.

Revision of the Annual Dose

Since the loess is formed by wind, by which other silt dust was blown together. Consequently, the quartz is only one of the constituents. So, in loess, uranium, thorium and potassium, etc. are included. These natural radioelements could be available for the radiation dose.

Because the structure of the loess is not as compact as the rocks, due attention must be paid to the escaping radon in performing the experiment. There is a method of determining the α–particles of radon and thoron as well as their daughters with ZnS screen. In order to obtain the escape coefficient, the method is to bed down a loess sample in close contact with the ZnS screen and count the α–counts. Then, using the same loess sample, the α–count is made again with the sample sustained at a distance from the ZnS screen by an aluminium plate. All these experiments must be enclosed in a plastic box for two-weeks storage. Using these two α–count results by calculation, the radon escape coefficient can be obtained, which is used to correct the annual dose (Table 1).

CONCLUSION

The age of loess under the baked layer at Yuangou, Xiashenjing Commune, Yanggao County, Shanxi Province is determined by the method mentioned above. The result is 2.4×10^5 years old. From the strata correlation, it shows that loess belongs to Lishi Loess, which is undoubtedly an acceptable figure. The baked layer has an age of 2.1×10^5 years. This experiment points to the fact, that the age of loess could only be younger if there is any error, that might be due to the sampling in a location close to the baked layer. Yuan-

58

gou is one of the typical profiles, since it includes not only Malan Loess, Lishi Loess, black loamy soil and buried soil but also basalt and baked layer sandwiched in between. Below it, a layer of lake and river deposits is found. Such profile is very seldom met.

Table 1 The distribution of annual dose, when the contents of uranium and thorium are 1ppm each in loess

	$\alpha(\times 10^{-5}gy)$	$\beta(\times 10^{-5}gy)$	$\gamma(\times 10^{-5}gy)$	total$(\times 10^{-5}gy)$
thorium series	74.0	2.90	5.00	81.90
thoron and its daughters	43.0	1.85	3.17	48.02
fractions of thoron and its daughters	0.581	0.638	0.634	0.586
uranium series	278.3	14.64	12.68	305.82
radon and its daughters	152.1	8.71	11.99	172.8
fractions of radon and its daughters	0.547	0.595	0.946	0.565

Some determinations are being made with loess samples from various profiles to obtain the available data useful to the geologists.

In China, the loess is distributed over a vast area about 650,000 km², especially the loess plateau area in the mid-stream of Huanghe River being very typical and unique over the world.

For a long time, it is hoped to find a way to determine the age of loess but failed. The TL dating introduced in the present paper could be promising. The authors wish to thank Mei Yi for his kind cooperation during the experiments.

REFERENCE

[1] Liu Tung-sheng, Zhang Zong-hu, 1962, The loess of China, Acta Geologia Sinica, v. 42, no. 1.

PALEOMAGNET!C STRATIGRAPHY OF LOESS IN CHINA

WANG YONG–YAN AND YUE LE–PING

(Department of Geology, Northwest University)

ABSTRACT

The following ages are given only for discussion and reference.

1. The earliest loess in the middle reaches of Huanghe River was formed between 1,000,000 and 1,200,000 years BP.

2. Ages of loess strata in this district are listed in the following:

Geological age	Strata	Ages (1,000 yrs)
Q_4	Loess deposits and underlying black loam	0—5
Q_3	The 1st loess stratum	8—35
	The 1st reddish brown fossil soil	30—60
	The 2nd loess stratum	60—100
Q_2	The 2nd reddish brown fossil soil	100—110
	The 3rd loess stratum	110—131
	The 3rd reddish brown fossil soil	131—140
	The 4th loess stratum	140—165
	The 4th reddish brown fossil soil	165—174
	The 5th loess stratum	174—197
	The 5th reddish brown fossil soil	180—210
	The top of loess stratum beneath the 7th reddish brown fossil soil	400
	The top of the lowest loess silt	690
Q_1	Well cemented reddish loess strata in which Jaramillo normal event was recorded	690—1200

MICROTEXTURES OF LOESS AND THE
GENESIS OF LOESS IN CHINA

Wang Yong–yan, Teng Zhi–hong, and Yue Le–ping
(Department of Geology, Northwest University)

Abstract—On the surface of the quartz grains of loess mechanical microtextures are often observed such as dish-shaped pits, grooves with rounded and smooth bottom, rounded edge angles, development of pited faces, etc. These surface microtextures resemble those of desert quartz grains, which shows that wind action is supposed to have played an important role during transportation of loess original materials. As for the chemical microtextures such as silica precipitates, chemical itching and dissolution phenomena on the surface of the precipitates, they might have been subjected to weathering in situ, especially in the process of pedogenesis. Pedogenesis must not be neglected in the process of formation of loess.

Sometimes broken fissures in quartz grains with sharp edges and angles may be observed which might have been formed by the freezing weathering during glacial period.

Supporting-spaced and mosaic-spaced contact are the essential relation between the coarse minerals of loess. They often coexist with each other and the mineral grains are accumulated disorderly. Such special microtextures of loess may be regarded as the result of eolian accumulation.

On the basis of the above mentioned microtextures of loess, it may be considered that the genesis of loess in China is comprehensive but the main genesis is eolian.

By means of scanning electron microscope, we have observed and analysed the microtexture of loess samples of various geological periods and districts from 13 sections in Shaanxi, Gansu, Qinghai and Xingjiang. The content of this work includes studying the shape of quartz grains and its surface microtexture, contact relations between the coarse minerals as well as the cementing materials and its existing state. Such data are of great significance for the discussion of genesis of loess in China.

THE SHAPE OF QUARTZ GRAINS IN LOESS

The quartz grains of loess are rather complicated in shape, their approximate shapes which are observed may be divided as follows: slate-like, irregular, triangular, baton like, pillow-like, similarly round, semi-rounded, cubic, pointed, etc. The slate-like shape is predominant, next are the irregular, triangular and pointed grains, and still less are the baton-like, semi-rounded, similarly round, pillow-like and cubic grains, the general trend of grain shapes is that there are more edge-angle grains, and the grains of different shapes are mixed with each other disorderly. The edge angle of grains is generally obtuse due

61

to abrasion of blown sand and sharp edge-angle grains are very few. Some sharp edge-angle grains are characterized by fresh planes of fracture, which may be formed during the late period of transportation or after deposition. Such disorderly mixed condition of many kinds of grain shapes resembles the polymineral characteristics of loess, which may be explained by the result of transportation of grains from one place to another and the mixture of each other by the wind.

We have also observed the quartz grain shapes in deserts of Mu Us, Lingwu, and Gurbantunggut under scanning electron microscope. The shapes of desert quartz grain and its quantity are almost similar to those of loess, with only difference that the number of semiround and baton-like grains is a little increased, and sometimes the pointed grains are observed. The sphericity of edges of desert grains is higher than that of loess and the area of abrasion pockmark surface is larger. Such difference was produced by different energy of abrasion. The contact area between desert sand grains rolling along the earth's surface is larger than that of suspended loess grains, and for the rolling grains, there are more probabilities to contact each other and the abrasion engergy is higher. As for suspending grains drifting along the wind have less contact probabilities to contact and little abrasion energy. Therefore, although the quartz grain shapes between desert and loess are similar, yet, owing to the difference of abrasion energy, the area and degree of abrasion between desert and loess grains are somewhat different. Above all, the similarity between quartz grains of loess and desert shows that their genetic environments are alike.

PROPERTIES OF SURFACE MICROTEXTURE OF QUARTZ GRAINS IN LOESS

The microtextures on the surface of quartz grain in loess are formed in the process of transportation and pedogenesis of loess materials. On the surface of grains two types of microtexture—mechanical and chemical are often observed.

Mechanical Microtextures and Wind Action

The broken planes along the cleavage of quartz grain are the common microtexture with even surface and sometimes with step-like patterns. On the concave part of fresh broken plane the conchoidal fracture is usually observed. The fresh and neat cleavage planes or clear conchoidal fracture of grains may be broken in situ during later period and have not suffered long-period transporting and abrasion. Owing to the long-period abrasion and the precipitation of SiO_2, the cleavage planes and conchoidal fracture formed in earlier period become indistinct. On the grains of fossil soil, owing to the precipita-

62

tion of SiO_2 on the cleavage slices, sometimes smooth, round rises are produced. On the surfaces of quartz grains in loess there are some pits and grooves. Among the pits, dish-shaped ones are comparatively more besides the square pits. Generally, the pits are of rounded smooth bottom. The pits may be formed by colliding of larger grains. If a grain was dashed by larger pointed one, a funel dish-shaped pit may be formed and on the center of it a deeper point appears. Among the observed surfaces of grains there are some grooves with the same width, round head and round smooth bottom. Most grooves are straight but sometimes there are also arcuate and triangular grooves. As for the formation of grooves, it is considered that when the larger pointed grain rubbed against the smaller one, the surface of the latter may be curved, into a groove. With regard to the rounded and smooth bottom of grooves, it was formed by polishing of much smaller wind-blown sand. As for the grooves on the surface of quartz grains formed by water action, their bottoms are uneven and sometimes the directions of groove bent, evidently different from grooves on the loess grain. The pits and grooves on the quartz grain surface of loess show the evident influence of wind during the transportation of loess materials. There are also some broken grains in loess, some of which are completely broken and a few are partly broken. The broken fissure is generally straight and occasionally bent. The edges and angles of broken fissures are sharp, showing that they are new products. The broken fissure is mostly observed in loess of the western and northern parts of loess plateau and not discovered in the quartz grains of fossil soil. Such broken fissures may be formed in freezing weather during the deposition of loess materials. Sometimes the polished pockmark surface on the grains may be observed, and it is especially clear under high magnifying micrographs. The polished pockmark surface is the feature of wind abrasion.

The surface microtextures of silt quartz grains explain that the formation of materials might have passed through different processes, but the wind action is supposed to have played an important role during their transportation.

The type of mechanical microtextures on the quartz grains and its appearance ratio of the above-mentioned three deserts basically resemble those on the loess grains, only the polished pockmark surface is relatively more, the dish-shaped pits are increased and the clear conchoidal fracture disappears.

The similarity of these properties of surface microtexture on the loess quartz grains and on the desert grains expresses the role of wind action during the transportation of loess materials.

Chemical Microtexture and Pedogenesis

The chemical microtexture was formed during pedogenesis, basically includes silica precipitates, chemical etching and dissolution phenomena.

The silica precipitates are generally situated in the concave parts, pits and grooves

of the surface of loess quartz grains. The silica precipitates which reveal themselves in facial distribution mostly occur on the quartz grains of fossil soil,
in earlier loess and loess in the moist region to the east of Liupan Mts. The
silica precipitates on the convex part of grains are in rare occurrence and sometimes only odd pieces of them may be observed. The convex part of silica
precipitates sometimes appears smooth rounded surface. In silica precipitates,
sometimes recrystallized quartz grains and growth of crystals may be observed.
Silica precipitates distributed neatly are very rare, generally they have suffered
itching and dissolution. Microtexture of dissolution involves triangular pits,
round pits, pits with lid and inclined pits. If the precipitates passed through
different intensity of dissolution, a stalagmite-like protrusion may be formed,
if it passed through intense itching along both sides of silica precipitates ridge-
or tongue-like textures are formed. Sometimes some laminae turn up along
the edges of cleavage plane may be seen, the slices are often obscure due to
precipitation of silica and were formed by mechanical and chemical processes.
Owing to the dry climate, transportation of sand and no process of pedogensis,
on the surface of desert quartz grains the occurrence of silica precipitates is
very rare. The chemical microtexture is closely related to the degree of pedogenesis. The older geological time, the more matured pedogenesis and the
more developed the chemical microtexture.
Among 10 quartz grains of Late Pleistocene Loess in Lanzhou only 4 grains
provide silica precipitates, while in early Pleistocene Loess of the same section,
among 10 grains there are 8 grains which have silica precipitates on the grain
surface. The loess distributed in the moist region passed through deeper pedogenesis than that in dry climate, therefore, on the surface of loess grains in
the moist region, silica precipitates are more developed. In the loess section
of Pingliang which is situated in the southern loess plateau to the east of Liupan
Mts., among 50 quartz grains there are 26 grains having chemical microtexture. The above-mentioned relation between chemical mirotexture and pedogenesis explains the genesis of loess, apart from its special transportation,
sorting and accumulation, the original mateials of loess must pass through
the pedogenesis and only after the pedogenesis can the deposits of loess materials
be called loess.

CONTACT RELATIONSHIP BETWEEN THE COARSE
MINERAL GRAINS OF LOESS AND INTERGRANULAR PORES

Under scanning electron microscope the contact relationship between coarse
mineral grains of loess can be basically classified into four kinds—the supporting
contact, the mosaic contact, the bag contact and dispersed distribution.
The contact area between coarse minerals of supporting contact is small. The
minerals are usually point contact though sometimes are edge contact. The
skeleton grains support each other to form the intergranular macropores.

On the micrographs the mosaic contact is observed and the skeleton grains contact each other with large contact area and present mosaic mineral arrangement. Frequently, the longer grain interpenetrate into other grains, the larger contact area occur and the intergranular pores become smaller. The two kinds of contact do not exist singly, often they coexist in the same sample, even in the same sight. Sometimes the larger grains contact together and form a bag-like space filling up with smaller grains. Under such condition, coexistence of supporting and mosaic contacts may appear either between the small grains or between the small and large mineral grains. In some micrographs the coarse mineral grains are not in contact with each other but separated by smaller grains or cemented materials and appear dispersed distribution. The supporting and mosaic contacts are the most common relationships between coarse mineral grains in loess. Statistics of appearance ratios both in districts and geological ages shows the following condition: in the Early Pleistocene loess of arid district in the northern loess plateau and to the west of Liupan Mts., the supporting contact is 65% and mosaic contact 35%, in the Middle Pleistocene loess they are 76% and 24% and in the Late Pleistocene loess they are 65% and 35% respectively. The appearance ratios of supporting and mosaic contacts in Middle Pleistocene Loess to the east of Liupan Mts. are 70% and 30%, in the Late Pleistocene Loess they are 73% and 27% respectively. As for the Early Pleistocene Loess in this district, owing to the abundance of cemented materials, coarse grains are obscure in sight and no statistics has been made.

The above statistic data show that the contact relationships of coarse grains of loess in China resemble each other in districts and geological ages. Therefore, it could be considered that the mode of deposition of loess in China is basically the same in districts and geological ages.

Abundance of pores is an important characteristic of loess. The pores in loess observed under scanning electron microscope are mainly intergranular micropores of about 20 microns. Shapes of these micropores are various. The percentage of micropores in the Early Pleistocene Loess in regions such as Jiuzhoutai, Lanzhou is 29% of the total micropores larger than 20 microns in the whole section, in the Middle Pleistocene Loess the percentage is 33% and in the Late Pleistocene is 38%. Owing to the filling cemented materials, the pores larger than 20 microns of loess to the east of Liupan Mts. are greatly decreased in amount. Such variation of micropores shown not only in size and in quantity but also in districts and geological ages was caused under the influence of filling cemented materials instead of the difference of original pores. The total volume of intergranular micropores added to the volume of macropores larger than 20 microns often constitutes almost half of the physical volume of loess. Only when the materials of loess sink down from the air to the ground, such high porosity deposits can be formed. These micropores are preliminary pores which were formed during the accumulation of loess materials and the shapes of pores can be distinguished from those which

65

were formed by rapid evaporation.

The supporting and mosaic contacts between the coarse grains in loess, unoriented arrangement of mineral grains, abundance of pores in different shape and size,—all these properties are completely different from those of the sub-fluvial deposits. The microtexture of Early Pleistocene silt lacustrine deposits in Wuqi is another case. In this deposits, the mineral grains seen show evidently oriented arrangement, the mineral grains contact each other in facial contact and the intergranular pores are less and smaller. Such microtexture apparently differs from that in loess.

EXISTING STATE OF CEMENTED MATERIALS AND PEDOGENESIS PROCESS

Only silt deposits that have undergone pedogenesis may obtain loess features. The pedogenic process is controlled by temperature, moisture, soluble salts and clay minerals.

Clay minerals and soluble salts are the main cemented materials in loess and calcium carbonate and calcium sulphate are the main components of soluble salts. Calcium carbonate constitutes the largest contents in soluble salts. Generally, calcium carbonate in loess is of two types—the original and the secondary ones. The original coarse calcium carbonate minerals and original powder microcrystalline calcium carbonate were deposited together with other loess materials at the beginning. During the pedogenic process the original calcium carbonate leached, migrated and was enriched at certain horizons. The more the pedogenic process matured, the more the calcium carbonate enriched. In loess of arid region of northern loess plateau and to the west of Liupan Mts. calcium carbonate often appears to be some little slices and fragments situated in the pores between coarse minerals grains or cemented on the surfaces of them. In moist regions to the east of Liupan Mts. calcium carbonate is enriched at certain horizons forming calcium concretion beds and lithical loess. The main clay minerals in loess of northern loess plateau are illite hydromica, kaolinite and montmorillonite. Illite hydromica appears in scale-shaped fragments or belt-like plates, and kaolinite often appears plate-like. These clay minerals are often cemented on the surface of coarse grains, sometimes sandwiched between grains or extend into the macropores between the coarse grains. To the east of Liupan Mts. and in southern loess plateau, owing to leaching and concentrating of calcium carbonate, the original clay minerals of loess are mixed with calcium carbonate, and become indistinct, but in loess of this region halloysite is very common. Between the coarse grains of loess which is younger in geological age and has insufficiently passed through pedogenic process, aggregates formed by mixing of calcium carbonate and clay minerals often exist. In older loess or loess situated in moist region, owing to the continual supply of cemented materials, all coarse grains are ce-

66

mented.   The reason that the colour of loess in China appears light in the northwest and deep in the southeast is closely related to the degree of pedogenesis.

CONCLUSIONS

1.   The quartz grains of loess complex in shapes and grains of various types
are mixed up with each other.   On the quartz grain surface, dish-like pits,
grooves with smooth and rounded bottom; rounded edges and angles; polished
pockmark surface are often observed which resemble the mechanical microtextures of desert quartz grains, showing the wind is the main agent in transportation of loess materials of loess in China.
2.   On some surfaces of quartz grains in loess fresh cleavage planes and conchoidal fractures, with some sharp edges and angles of grains are seen.   Sometimes broken fissures on grains with sharp edges and angles are observed.
These microtextures could be formed during freezing weather after deposition
in situ and represent the cold period.   Therefore, it is considered that loess
formation was in the glacial period.
3.   The chemical microtextures were formed during the pedogenic process.
In loess to the east of Liupan Mts., especially in the fossil soil, the chemical
microtextures are well developed.   The chemical etching and dissolution
phenomena on the surface of silica precipitates are mainly in loess of this region.
The existing state and degree of concentration of cemented materials are also
related to the degree of pedogenesis.   Therefore, the pedogenesis which produced
the loess features from original loess materials must not be neglected in the
formation of loess.   In discussion of loess genesis usually we only pay due
attention to the process of transportation of loess materials yet ignored the
pedogenic process.   But, only after pedogenesis can the loess deposits be called
loess.
4.   After the transportation and deposition, most of the loess materials is
stable in situ under suitable landform and pedogenesis proceeds.   But sometimes due to suitable landform deposited materials may be retransported by
surface water and wind to another stable place and then pedogenesis begins,
but such retransported materials are uncommon that unlikely they can form
the principal body of loess.   However, the fact of retransportation of deposited
materials by water and wind should be considered.
5.   On the basis of the above-mentioned microtextures of loess grains and
process of loess formation, we may come to a conclusion that the genesis of
loess in China is comprehensive but its main genesis is eolian.

REFERENCES

[1] Lu Yan-chou et al, 1976, A preliminary discussion of the origin of loess materials in
China. Geochemistry, No. 1.

67

[2] Xie You-yu et al., 1980, Observation and comparison of some quartz grains. Hydrogeology and Engineering Geology, No. 6.

[3] Gao Gue-rui, 1980, Classification of loess microtextures and collapsibility. Scientia Sinica, No. 12.

[4] Bardon L. et al., 1971, Sample disturbance in the investigation of clay structure. Geotechnique. 12, No. 3, pp. 211–223.

A SATELLITE IMAGES STUDY ON THE DUST STORM AT BEIJING ON APRIL 17—21, 1980

LIU TUNG–SHENG, CHEN MING–YANG

(Institute of Geology, Academia Sinica)

AND LI XIU–FANG

(State Bureau of Meteorology)

Abstract—Using the meteorological satellite images, the authors describe the dust storm process occurred during April 17—21, 1980 in Beijing, China and discuss the eolian process of loess and the continuity of loess deposition in the geological past.

When the dust fell over Beijing on April 18, 1980, the surface wind speed was extremely low. The sky was in a yellow haze and the visibility was low. Pedestrians found their clothes dirted with yellow dust. Report from the Beijing airport said that the passengers on board had seen the dust haze in the high sky about 8,000 m. So it was quite a strange, rare phenomenon[1].

1,000 and more similar records of "dust rain" have been found in ancient Chinese literature. Based on some records, it is possible to retell the weather and fallen dust process in the past.

This event here discussed was already studied by Liu Tung-sheng and Fan Yong-xiang, who analysed the meteorological background during April 17–20 in North China. The study on the fallen dust samples collected in 10–17 hr., April 18 (Beijing time, windless), shows that the event was really a process of modern loess deposition[2].

In 1981, Liu Tung-sheng was presented the meteorological satellite images by the Arctica and Alpine Research Institute, of the United States and the State Bureau of Meteorology of China.

Both images are in two spectral bands (visible 0.4–1.1 μm or 0.5–0.75 μm and infrared 10.5–12.5 μm). Based on observation data of the meterological stations, further study on some problems of this event is given in the present paper.

I

Meteorological satellite image is one of the useful tools for observation of all kinds of weather phenomena, including sand storm and dust storm from the outer space. The characteristics of the sand and dust storms in the images were studied before. "It looks like cirrostratus, the tone is a little dark as a piece of film and its boundary disappears uniformly"[3]. The satellite images of sand and dust storms of this event are much like those mentioned above, only that in different stages of the dust storm, the images change slightly. In the visible and infrared spectra images, the tone are greyish white uniformly and the boundaries are very clear in the height of the dust storm on April 18. During April 19—20, as the dust storm moved to the south, owing to partly falling of the dust and the decrease of the dust density in air, the dust storm shown in images change to greyish white as a belt-like film, the boundaries disappear uniformly, and the tones of dust storm in those three days are lighter than those of the desert. The storm was seen by the passengers on board, people in Beijing and the staff members of the meteorological stations in the dust region. So, this dust storm showed in images was really a strong fallen dust process.

The occurrence, transportation and deposition of the dust storm were controlled by the occurrence and development of the cyclone. By the satellite images acquired from different time, the development of the weather and storm may be clearly shown as follows: At 12:30, April 16, the jet stream in the northern Siberia moved to the south. At 12:18, April 17, the embryo of cyclone was formed in the south of the Baikal Lake. The upper westerly around the Balk-hash Lake moved southeastward to Hami and during that time the dust storm occurred already. The dust storm appeared as a greyish white film in the images at 14:00. And at 19:00, April 18, the cyclone became an occluded cyclone, and spiral structures appeared. A piece of greyish white dust covered over the middle and lower reaches of Huanghe (Yellow) River, Bohai Sea and Northeast China. Its boundary became "W" form near the Liupan Mountain and its eastern part mixed with cloud and the boundary was not clear. At 8:52, April 19, the cyclone moved to the east and developed into its mature stage, the dry tongue stretched into the center of the cyclone at the back of the front cloud band. The dust storm zone was shown as a greyish white film at the tail of the cloud band. The northern boundary was along Xian, Zheng-zhou, Bohai Sea and the southern one was along Shenyang and central Korea, and a piece of dust over the middle and lower reaches of Changjiang (Yangtze) River moved towards the south slightly. On April 21, the dust became thinner so it can not be distinguished in the satellite images. The cyclone center moved from the south of Baikal Lake through Hailar to the Soviet Union, and the dust storm underwent processes of occurrence, development and disappearance. According to the images, the boundaries of the dust storm are from Hami to Erenhot in the north, Liaodong Peninsula to Korea in the east, the west of Qilian Mt. in the west and from Wuhan to Nanjing in the south respectively. Based on the temperature of dust transformed from the measurements of the density on the infrared images, the altitude of the dust in the air can be estimated. On April 18, the mean height of the blown dust was about 7,000—7,500 m in the strong wind region, and about 7,000 m over Beijing. On April 19, the

70

mean height of the dust was about 6,000 m over Beijing, and about 3,000 m over the middle and lower reaches of Changjiang River.

II

Based on the observed data from the meteorological stations of both China and her neighbouring countries the State Bureau of Meteorology compiled a series of weather maps of different time showing the weather, and the development of dust and sand storm during April 16—21.   On April 16, the cold air invaded from Siberia to the south of Baikal Lake.   At 2:00, April 17, over the southern part of the Lake, depression formed, the pressure center was 994 mb.   At the same time, the short wave trough over the Balkhash Lake moved to the east and made the depression deeper.   At 14:00, April 17, the strong wind region formed in Hami, Gansu Corridor and western Nei Monggol (Inner Mongolia) where the desert or gobi area lies.   The wind speed was from 12 m/sec to 20 m/sec, and the sand and dust were blown, so the storm was formed which caused a poor visibility, for instance, in general it was 2 to 8 km, but only 0.4 to 0.6 km in some areas of this region.   The cold front line was located in the south of the Balkhash Lake and the south of the Baikal Lake.   The Ri value of the westerly below 300 mb was 0.2 and smaller than the critical value 0.25, therefore, the vertical exchange of the turbulence occurred, strength-ening the surface wind and raising the dust up to the upper air.   At 14:00 April 18, the dust moved to the southeast, and its southern boundary reached Xian and Beijing.   At 10:00–17:00, April 18, the yellow haze covered over Beijing and from about 7,000 m a.s.l. the dust fell down to ground, and the surface wind speed was very low.   At 14:00 April 19, according to the dust records of the meteorological stations, the sand and dust storm could be divided into two regions, the northern one was over Jinan, Beijing, Hohhot and Chang-chun, while the southern one was over Xian, Zhengzhou, and the middle and lower reaches of the Changjiang River.   This phenomenon was probably caused by another cold front in Hebei and Shanxi Provinces.   The main cold front was located in Wuhan, Nanjing.   At 14:00, April 20, the dust moved toward south to Changde and Ningbo and to the east of Shenyang and Korea.   The cold front reached Wuhan, Nanjing, Harbin and Japan.   At 14:00, April 21, reports from the Japanese meteorological stations said that the dust appeared over southern Japan and Okinawa Island which was considered as the most eastern region the fallen dust reached and the cold front reached the Pacific to the east of Japan.

III

In view of the dust distribution, both the southern boundaries of the dust and cold front showed southeastward arcs and moved gradually to the southeast. The above observed data fully prove that the fallen dust in Beijing during that time came from a distant place. In a short few days, from Hami strong wind region to the middle and lower reaches of Changjiang River, i.e. from the source region of the dust and sand to the distant fallen dust region, the dust covered several million square kilometers, indicating a complete dust storm weather process, including dust occurrence, transportation and falling.

On April 17 in Hami strong wind region the mean speed of the surface wind was 12–20 m/sec and sand and dust were blown up. Bagnold R. A. has suggested that so long as the mean speed of current is 4 times greater than the falling speed of particles, the particles will be raised up, but, some geologists still hold that when the mean speed of current is 11 times greater than falling speed of the particles, the particles will remain suspended[4]. When Ri value of the upper westerly was smaller than 0.2, the turbulent flow exchanged vertically. As a result, the dust was brought to the upper air and moved to the falling area, which shows that the strong wind region is the source of the dust storm.

The distributive characteristics of the atmospheric dust in those days are very similar to those of loess. The middle reach of Huanghe River was the main fallen area of loess, where the thickness of loess is great and was also the comparatively dense area in this dust storm. The "Xiasu earth" in Nanjing is also an accumulation of dust storm, but thinner in thickness and smaller in extent than those of the middle reach of Huanghe River. The belt-like distribution of sandy loess, loess and clay loess and the boundaries among them present in arc forms toward southeast, which precisely correspond with those of cold front and southern boundaries of the dust mentioned above. The grain size and mineral analyses of fallen dust collected on April 18, (the surface wind speed was low in Beijing) indicate that the grain size and mineral composition of the fallen dust are much similar to those of the loess in Luochuan, Shaanxi.

The analyzed results of aerosol collected on April 18—19 in Beijing show that the concentration of aerosol in dust storm days is more higher than that in ordinary days, which indicates that aerosols in dust storm days come from a distant place[6]. Also this suggests that during the Quaternary time loess eolian process might be closely related with cyclone development of the weather in which the cold front was moved to the southeast.

Study on this dust storm process supplies many valuable information in this field. Gerald M. Friedman et al. have suggested "Because of the structure of the earth's atmosphere suspended can travel at two distinct levels, forming low altitude and high altitude supensions"[7], the atmospheric suspensions can be divided into low altitude suspensions which are those confined in the lower atmosphere from 2 to 5 km a.s.l. and high altitude suspensions from 10 to 15 km. Judging from this dust storm event, because the suspensions the cyclone carried raised to a height from 5 to 10 km, it is possible that the suspensions far away from the source area were relatively lower.

As the distributive area of this fallen dust was very large that its southeastern boundary reached the middle and lower reaches of Changjiang River and even Japan, and the composition of the dust was the same as loess, we consider that though a large amount of loess occurred in dry and cold climate, the loess may still occur under present climatic conditions, such as Holocene loess yielded in 5,525±200 BP was found in 1979.

A series of records in the Chinese historic literature also support this conclusion, such as the "Dust rain" occurred in 3,100 BP was recorded in "Zhu Shu Ji Nian" (Chronicles Recorded on Bamboo Slips).   All these tell us that the loess has continuously deposited for more than 5,000 years.   In many profiles of loess, a bed of paleosol with about 4—5 m in thickness is always found, which indicates that a little amount of loess could still deposit even in the warm and humid climatic periods when the paleosol was formed.

Since Tertiary, as the Qinghai-Xizang (Tibet) Plateau has uplifted, a large area in Asian Continent transformed into deserts or Gobi.   As long as the transformation has been uninterrupted, the desert and Gobi, will not disappear. Whenever the strong cyclone or the strong cold front is formed, the dust storm weather may occur. Therefore, since Quaternary there is the slightest possiblity that the climatic change could lead to an end of the deposition of the loess. As Bagnold R. A. suggests that after dust and loess falling, the dust loess cannot be moved and resist the eolian erosion.   It is possible that the dust sediments can remain their original sedimentary structures and textures.   The thick loess profiles in the Chinese loess plateau have a continuity through geological time. This character has been proved by the geochronological studies, such as by the paleomagnetic, thermoluminescent studies, etc. and plays an important role in the study of the evolution of Quaternary environment.   The authors wish to thank Fan Yong-xiong of the State Bureau of Meteorology for his valuable help.   They also want to extend their sincere thanks to Lao Qiu-yuan, Wang Chang-jiang, Lu Yang-chou, and Gao Fu-qing.

REFERENCES

[1] Zhang De-er, 1982, An analysis of "rain dust" in Chinese historic periods, Kexue Tongbao, no. 5 (in Chinese)
[2] Liu Tung-sheng et al., 1982, The dust fall in Beijing, China, on April 18, 1980, American Geological Society Special Paper, 189.
[3] Institute of Atmosphere Physics, Academia Sinica, 1972, The analysis and prediction of the weather, using the satellite cloud picture, Science Press, Beijing. (in Chinese)
[4] Babnold, R. A., 1941, The physics of blown sand and desert dunes, Methuen & Co. Ltd, London.
[5] Blatt, H. et al., 1972, Origin of sedimentary rocks, Englewood Cliffs, Preatice-Hall, Inc.
[6] Zhou Ming-yu et al., 1981, Some proporties of aerosols during the passage of a dust storm, Kexue Tongbao, 10. (in Chinese)
[7] Friedman, Gerald M. et al., 1978, Principles of sedimentology. Wiley, New York.

THE EVOLUTION OF CHEMICAL ELEMENTS IN LOESS OF CHINA AND PALEOCLIMATIC CONDITIONS DURIN LOESS DEPOSITION

WEN QI-ZHONG, YANG WEI-HUA, DIAO GUI-YI,
SUN FU-QING, YU SU-HUA, AND LIU YOU-MEI
(Institute of Geochemistry, Academia Sinica)

Abstract-Geochemical data of Luochuan Loess section have been analyzed by statistic method. The results indicate that the contents of elements of Al, Fe^{+++}, Mn, Ti, K decrease from bottom upward; whereas, the contents of Ca, Sr, Si and Fe^{++} show an opposite tendency. REE content in loess lies in the range of 160—210 ppm, and their distribution patterns in different loess samples show a similarity with each other. $CaCO_3$ and Sr in paleosols were strongly leached; whereas, Al_2O_3 and Fe_2O_3 show an obvious accumulation. Meanwhile, trace elements Mn, P, Zn, Cu, etc. and REE content also increase relatively in paleosols. The fluctuations of elements in 1st, 5th, 8th and 14th paleosols are especially evident, the boundaries of which are similar to those of sedimentary cycles or stratigraphic boundaries.

Periodicity in the evolution of chemical elements agrees well to those of insolation and isotopic oxygen changes in deep-sea sediments by analysis of autocorrelation coefficient and power spectrum $CaCO_3$ content, FeO/Fe_2O_3, $CaO+K_2O+Na_2O/Al_2O_3$ and Sr/Ba ratios well reflect climatic periods with the intervals of 80—100, 40—50, 20—30 thousand years, which demonstrates that the climatic variation was controlled by astronomical factors. Based on the established elemental periods loess sections are divided into sedimentary cycles and subcycles, providing important evidences for the division of climatic cycles since Quaternary.

Evolution of the chemical elements in strata can reflect the evolutionary regularity of nature conditions, also mark the evolution of the earth[1]. The regularity of the distribution, migration and accumulation of elements in loess may be used to explain the paleogeography, paleoclimatic conditions during loess deposition and the environmental variation after loess deposition as well.

This paper deals mainly with the samples taken from Luochuan section. On the basis of the past research work, the common elements (100 samples), trace elements (130 samples) and REE (23 samples) have been further analysed for loess and paleosols in this section. The common elements and trace elements are analysed by chemical analysis and atomic absorption spectroscopy technique; Sr and Ba by spectroscopy; and REE by neutro-activation method. All data have been analyzed statistically. And our aim is to inquire the evolutionary tendencies and periods of elements in loess section and their relationship with paleoclimatic conditions.

THE CIRCULATION AND RHYTHMS OF LOESS DEPOSITION

The section studied is located at Potou Village 5 km from Luochuan County, Shaanxi Province, with a thickness of 130 m. According to the lithologic characteristics, the distribution and development of paleosols, as well as vertebrate fossils, etc., four first-order sedimentary cycles can be divided from bottom to top, moreover, there are many rhythmic variations in each sedimentary cycle. The first-order sedimentary cycles are consistent with four loess stratigraphic units. Wucheng Loess, Lower Lishi Loess, Upper Lishi Loess and Malan Loess proposed by Liu Tung-sheng et al[2].

According to paleomagnetic imformation up to date the deposition of Wucheng Loess began in middle-late Matuyama Epoch[3]. The bottom of Lishi Loess in Luochuan section lies at the boundary between the Brunhes and Matuyama Epochs (0.71 m.y.). In addition, the ^{14}C age of black loamy soil at the top of Malan loess is 9830 ± 1300 years*. These data have provided the rudiment of the time scale for evolution of chemical elements in loess section.

THE CONTENT AND DISTRIBUTION OF ELEMENTS
IN LOESS SECTION

From analysis, it can be seen that the average values of Si, Al, K, Ti, Zn, Co, F and RE_2O_3 approach or correspond to Clark value in earth crust; however the contents of Ca, Sr, Ba and Pb are more than twice as high as Clark value, Cl is more than thrice as high as Clark; the other elements are all at lower values.

The average values of chemical compositions in Luochuan Loess show similarity with those of the middle Huanghe Valley, which indicates that the loess is a homogeneous material. However, obvious differences exist from those of other areas over the world.

At loess strata of different ages, the variation of element distribution is as follows: the contents of Ca, Sr, Fe^{2+} and Cl in the upper part of Malan Loess are higher than those both of lower Lishi and Wucheng Loess, whereas the contents of Fe^{3+}, K and Mn show an opposite tendency. These elements are usually similar to the normal distribution or logarithmic normal distribution in loess section

The distribution of elements is also not uniform in different kind of particle grains. The contents of SiO_2, TiO_2, FeO and Na_2O in clay fraction are lower than those in the whole rock; whereas the contents of Al_2O_3, Fe_2O_3 and K_2O are opposite which are higher. This is evidently due to the increase of illite content and other secondary minerals in fine grains. Moreover, in clay fraction, about one-third of Fe_2O_3 exists in a form of free state. In addition, the

---

* Qiao Yu-lou et al., 1979, Radiocarbon dating of loess sediments.

contents of trace elements such as Zn, Cu, Mn, P and REE in clay fraction
are also higher than those in the whole rock, which shows either replacement
of elements in crystal layers of minerals or adsorption of clay might exist.

EVOLUTION OF ELEMENTS IN LOESS WEATHERING PROCESS

Paleosols of different types[4] of different ages are found in Luochuan section.
Elements have undergone differentiation and reassignation during the change
from the loess into paleosol. In paleosols, the contents of Al, Fe^{3+}, K, Ti,
Mn, Zn, Cu, Ni and RE_2O_3 have increased to some extent as comparing with
loess parent rock; whereas Ca, Fe^{2+}, Sr, etc. are lower than those in loess.
Relevant ratio of oxides are lower than those in loess, too (except K_2O/Na_2O),
moreover, the values of oxide ratios and Sr/Ba have progressively decreased by
strong pedogenesis, whereas F/Cl value shows an opposite tendency.
By means of "t test analysis" (Table 1) it has been found that paleosol was
formed from loess by weathering process, and the elements of Ca, Al, Fe^{3+},
K, Si, Ti, Mn, Zn and Ni, etc. have more obvious variations than loess, but
Pb, Co, Ba do not show evident variations. In addition, by analysis of relative
values of oxides leach and accumulation[5] it has been proved that Fe_2O_3,
Al_2O_3, K_2O, etc. were accumulated relatively, the sequence of accumulation
is as follows: $Fe_2O_3 > K_2O > Al_2O_3 > SiO_2 > MgO$; but CaO, Na_2O, etc. were
leached relatively during loess weathering. At the same time a large amount
of Fe^{2+} were oxidized and reformed into Fe^{3+}. The intensity sequence of
leaching is as follows: $CaO > FeO > Na_2O$, in which, the accumulation of
Fe_2O_3 and the leaching of CaO show most obviously, and there exists a normal
relationship that between them[6].

Table 1 Tests for notable levels of elements in paleosol (and buried weathering bed)

Paleosol type	Element																
	Si	Al	Fe³⁺	Fe²⁺	Ca	Mg	K	Na	Ti	Mn	Zn	Cu	Co	Ni	Pb	Sr	Ba
Drab-brown earth	+++	+++	+++	++	+++	−	+++	++	+++	+++	+++	+	−	+++	−	+++	+++
Luvic drab soil	+++	+++	+++	+++	+++	+	++	−	+++	+++	+++	++	−	++	−	+++	−
Drab soil	+++	+++	+++	++ ·	+++	−	+++	−	+++	+++	+++	+++	+	+++	−	+++	−
Calcareous drab soil	++	+++	+++	+	+++	−	++	−	+++	+++	+++	+	−	+++	−	++	−
Buried weathering bed (1)	+++	+++	+++	+++	+++	+++	+++	−	+++	+++	+++	− · −	−	+++	−	−	⊥
Buried weathering bed (2)	+	+++	+++	+++	+++	+++	+++	−	++	+++	+++	+	−	+++	−	++	−
Buried weathering bed (3)	−	++	+++	+++	++	+++	+++	+	++	+++	+	−	−	+++	−	+++	−
Leaching (L.) or accumulation (A.) state		A.	A. oxidated		L.		A.	L.	A.	A.	A.	A.		A.		L.	

— no obvious differentiation ($|t| < t_{0.05}$);
+ slightly obvious differentiation ($t_{0.05} < |t| < t_{0.01}$);
++ moderately obvious differentiation ($t_{0.01} < |t| < t_{0.001}$);
+++ strong obvious differentiation ($|t| > t_{0.001}$).

During loess weathering, $CaCO_3$ and Sr were strongly leached, Fe_2O_3 and Al_2O_3 shows an obvious accumulation, the secondary clay minerals were increased, the textural properties were transformed, which made loess turn from grey-yellow or light yellow-brown bed rich in calcified layers to brown or red-brown paleosol with neutral layer rich in alluminium-silica. The frequent superpositions of loess-paleosol series in the section reflect the basic pattern of frequent, repeated climatic fluctuations from dry to humid during Quaternary. At the same time the values of oxide, F/Cl, leaching and accumulating degree, also reflect the differences of paleoclimatic fluctuation amplitude of each warm-humid pedogenesis during loess deposition and evolution.

THE TENDENCY AND ORIENTATION OF ELEMENT EVOLUTION IN LOESS SECTION

In order to inquire the tendency of content evolution of chemical elements in loess section, we have carried out the calculation by "five-term moving average analysis". From Fig. 1 it can be seen that the evolution of Si show its increase to a certain extent from bottom upward, whereas Al, Fe^{3+}, K, Mn and Ti are progressively decrease upward, and there are obvious increase in paleosols corresponding to the 1st, 5th, 8th, 10th layers with in a general tendency of decrease.

The evolutions of Ca and Sr are similar to each other, the contents of which are increasing from bottom to top. They show several regular decreases in a general tendency of increase, i.e. the clayey bed of paleosol has been strongly leached. And the content of Ba is in a minor fluctuation. The evolutional tendency of Fe^{2+} is similar to that of element Ca, but its fluctuation is not obvious as compared with Ca.

The content evolution of Zn is from low to high, then to low again, whereas Cu is just opposite. But both have shown several peaks in above-mentioned paleosols. It shows that the fluctuations of element contents in the 1st, 5th, 8th, 10th and the 14th paleosol are especially evident, in which the boundaries are consistent with those of sedimentary cycles or stratigraphic boundaries. The content distributions of Mg, Na, Pb, Ba, Co and Ni in loess section are comparative homogeneous so their fluctuations of evolutional curves present smaller or relatively smooth. REE amount in loess lies at the range of 160—210 ppm, there is a tendency of progressive decrease from bottom upward (from Wucheng to Lishi then to Malan Loess). But their distribution patterns in different loess beds show a similarity among them[5].

These evolutional tendencies of elements in loess section are related to the distribution and variation of mineral and grain components in section; to paleoclimatic environment and conditions of geochemical media. The paleoclimatic condition during the Wucheng Loess deposition is relatively more warm-humid than those of Lishi and Malan Loess deposition. The content evolu-

77

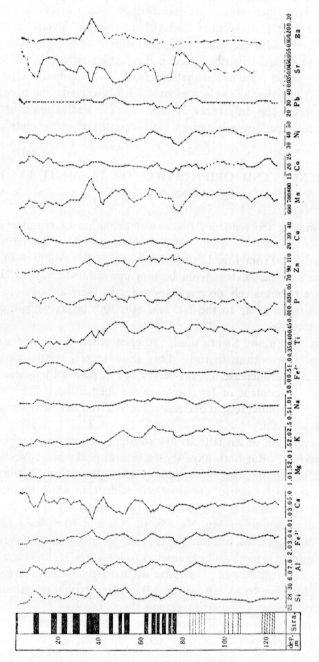

Fig. 1. Evolution of elements in loess strata of Luochuan.

78

tions of Ca, Sr, Fe²⁺, etc. in loess section increase progressively from bottom to top, the contents of $CaCO_3$ also show fluctuant increase with rhythms; whereas Fe^{3+}, Al, K, Mn, Ti, etc. decrease upward. It can be held that this evolutional tendencies of chemical elements in loess section are related to the climatic conditions, which have been shifting from wet to drier since Quaternary. And these tendencies show an agreement with evolutions of animal species and sporo-pollen in loess section. All mentioned above indicate that the evolution of the paleoclimatic conditions revealed by the loess deposits has a general tendency with rhythmic fluctuation since Pleistocene, i.e. from humid forest prairie to arid desert steppe[7].

PERIODS OF ELEMENT EVOLUTION AND PALEOCLIMATIC CHANGE

It has been found that some components and ratios of elements dominated by climate, for example, $CaCo_3$, Sr/Ba, FeO/Fe_2O_3 and $CaO + K_2O + Na_2O/Al_2O_3$, etc. exist periodic variations in contents of different magnitudes, and they show synchronism with cycles of loess-paleosol cycles. Accordingly, some new light have been shed that the variations of the above-mentioned element components and ratios in loess section can be used to learn the history of climatic changes.

The climatic fluctuations vary with the lapse of time, so we regard this variation as a function of time sequence. We have used "analysis of autocorrelation coefficient (Rp) and power spectrum (Sr)" as mathematic models[8,9]. Rp and Sr obtained are drawn in Figs. 2,3 and each period is listed in Table 2. From these figures and table, the main periods may be summarized as follows: 80—100, 40—50, 20—30 thousand years. The period-terms are relatively stable, and relevant to climatic changes dominated by some factors. In order to verify this hypothesis, we have calculated, by the same mathematic model mentioned above, the insolation changes at 65° N (converting it into latitudes of those years) for the last 600 thousand years which were calculated by M. Milankovitch[10] according to astronomical hypothesis of climate and oxygen isotopic (δO^{18}) values in deep-sea sediments gained from cores V12–122 (17° N, 74°24′ W)[11] and V28–239 (3°15′N, 159°E)[12]. It seems that whether insolation changes or variations of oxygen isotopes (δO^{18}) of deep-sea cores from different latitudes are all stable period-terms (Figs. 4, 5, Table 2), which are basically similar to the periodic variation of element components.

From these calculations and studies, several conclusions can be made: (1). The cycles of frequent superpositions of loess-paleosol series in Luochuan section were controlled by paleoclimate, which show synchronism with cycles of composition contents and ratios of elements dominated by climate, and they all have obviously three period-terms, which are 80—100, 40—50, 20—30 thouasnd years; (2). These period-terms are consistent with the periods calculated by M. Milankovitch on the basis of astronomical factors, and with

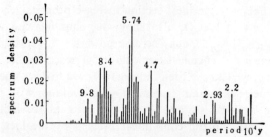

Fig. 2. Power spectrum of FeO/Fe$_2$O$_3$ ratio

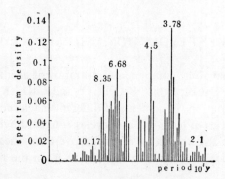

Fig. 3. Power spectrum of Sr/Ba ratio

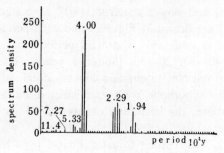

Fig. 4. Power spectrum of insolation change at 65°N

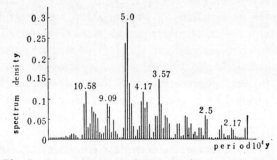

Fig. 5. Power spectrum of 0^{18} value in core V28—239

Table 2 Comparison of fluctuation periods of composition contents and ratio of elements with periods of insolation changes and oxygen isotope

Time sequence	Section (10⁴ yr.)	Period (10⁴ yr.)						
Alternative superposition of loess-paleosol	120	12.38**	8.43*	4.8–4.33	3.57*	2.98	2.65	2.38
	61	12.96–11.66**	8.33**	4.8*	3.33*	2.78		2.20
$CaCO_3$ content	120	11.81	7.85*	4.30*	3.00*			2.2–1.97
	47	11.32*	7.55**	3.94**	2.92*		2.45	2.26–1.81*
$CaO+K_2O+NaO/Al_2O_3$		12.86**	8.18**	4.50**	3.67		2.50*	
Sr / Ba		10.17	8.35*	4.50*	3.78**			2.10
FeO/Fe_2O_3		9.80	8.40	5.74**	2.93			2.20
Insolation change	60	11.40	7.27	5.33**	4.00**	2.29**		1.94**
Isotope oxygen (V28–239)	120	10.58*	9.09	5.00**	4.17*	3.57*	2.50	2.17
Isotope oxygen (V12–122)	46		7.50**	5.19	3.75*	3.07*		1.90

** The most significant,   * Significant

periods of oxygen isotopes ($\delta0^{18}$); (3). The content variations of elements governed by climatic factors can reflect the climatic periods well, which provides a new approach for studying Quaternary climatic changes in continuous continental deposits; (4). It has been further proved that climatic changes in Quaternary were controlled by astronomical factors. Establishing element periods and dividing loess section into sedimentary cycles and subcycles by them have provided significant data for the division of climatic cycles since Quaternary.

Acknowledgement — The authors wish to thank Prof. Liu Tung-sheng for his guiding. The analyses of common elements were made by the ninth laboratory of our institute. Sun Jing-xin and Wang Yu-qi, etc. (Institute of High Energy Physics, Academia Sinica) determined the REE value, the figures were drawn by members of map-making group in our institute. In addition, this study has been warmly supported and helped by Mr. Zheng Hong-han, Hong Li-wen, Cheng Qing-mu and Han Jia-mao, etc. We here express our sincere thanks to them all.

REFERENCES

[1] Hou De-feng, 1959, Geochemical concept of strata. Scientia Geologia Sinica, no. 3, p. 68—71 (in Chinese).
[2] Liu Tung-sheng, Zhang Zong-hu, 1962, Loess of China. Acta Geologica Sinica v. 42,

81

no. 1, p. 1—4, (in Chinese).

[3] Liu Tung-sheng et al., 1978, The geological environment of loess in China. Kexue Tong-bao, no. 1, p. 1—5, (in Chinese).

[4] Lu Yan-chou, An Zhi-sheng, 1979, Investigation of variation of natural environment in loess plateau since about 700 thousand years. Kexue Tongbao no. 5, p. 221—224, (in Chinese).

[5] Wen Qi-zhong et al., 1981, A preliminary investigation of REE in loess, Geochemica, no. 2, p. 153—155, (in Chinese).

[6] Wen Qi-zhong et al., 1981, The ratios of oxides and relative values of weathering leaching or accumulation in loess section and their geological significance. Geochemica, no. 4, p. 381—386, (in Chinese).

[7] Liu Tung-sheng, Wen Qi-zhong, Zheng Hong-han, 1980, The paleoclimatic records in loess of China and their reflection of the ancient climatic evolution, Proceedings of Geological Sciences for 26th Congress of I. U. G. S, Geology Press. Beijing, p. 78—79, (in Chinese).

[8] Ralston A., Wilf, H. S., 1960, mathematical methods for digital computers. Wiley, New York.

[9] Agterbery, F. P., 1974, Geomathematics, Elservier Scientific Publishing Company, Amsterdam, London, New York, P. 363—400.

[10] Yao Zhen-sheng, 1959, Climatological principle. Science Press, P. 361—379.

[11] Imbrie, J. etc., 1973, Paleoclimatic investigation of a late Pleistocene Caribbean deep-sea core; Comparison of isotopic and faunal methods. Quaternary Research, vol. 3, no. 1, p. 10—32.

[12] Shackleton, N. J., 1976, Oxygen-isotope and paleomagnetic stratigraphy of Pacific core V28—239 Late Pliocene to latest Pleistocene. Geological Society of America, Memoir 145, p. 449—462.

PALEOCLIMATE EVENTS RECORDED IN CLAY MINERALS IN LOESS OF CHINA

ZHENG HONG–HAN

(Institute of Geochemistry, Academia Sinica)

Abstract — 86 samples of loess and paleosol for study on clay mineral were taken from Heimugou Loess section at Luochuan and Nuanquangou Loess section at Longxi. The results indicate that the loess and paleosol of different ages show a close similarity in their clay mineral species with the illite as the dominant component, and they also contain more or less certain amounts of chlorite, kaolinite, montmorillonite, vermiculite, halloysite and a detectable amount of mixed-layer minerals.

Some indicators, the ratio of the illite (001) reflexion peak height and width at half height (H/W value) and the ratio of (001) and (002) reflexion intensities (I(001) / I (002)) are adopted to discuss the loess variation in the clay mineral crystallinity after pedogenesis.

The relative amount of clay mineral species and their crystallinity are adopted as indicators showing the climate fluctuation amplitude, and the time scale was controlled by data of biostrata, magnetostrata, radiocarbon and thermoluminescence dates. 13 climatic cycles with 26 stages have been distinguished in loess sections since loess deposition. An attempt to correlate climatic events recorded in loess sections of China with glacial / interglacial in Northern Europe and the isotope oxygen stages in deep sea sediments is presented in this paper.

INTRODUCTION

Loess was early developed at the beginning of Pleistocene and loess deposition continued during Holocene in China. Good sections are widespread in regions of Gansu, Shaanxi and Shanxi Provinces of North China. Loess section, composed of superimposed loess representing a cold-drought climatic condition and paleosol formed under a warm-moist climatic condition, is an ideal geological body for the study on evolution of Quaternary climate since the deposition of the section was basically continuous, with detailed record of the climate evolution from the beginning of the loess deposition.

It is undoubtedly an effective method to use the biogenic traces to distinguish the paleoclimatic conditions recorded in different beds during their developments. However, there exist many difficulties in studying the paleoclimatic records of loess strata and in the discussion of the paleoclimatic conditions for a given section due to insufficient fossils and very small content of sporo-pollen in general. The minerals, especially the clay minerals, are important materials for the study on paleoclimatic condition and on the fluctuation rule since deposition of loess, because different species of clay mineral have a close simi-

---

* Gu Xiong-fei, Han Jia-mao and Deng Bing-jun also took part in this investigation.

larity in their crystal structure, they can be easily changed under hypergene environment and their composition and crystallinity will reflect the characteristics of the environments they underwent.

This article is a preliminary result on the study of clay mineral and discussion of paleoclimatic conditions reflected by them both in Nuanquangou section at Longxi of Gansu and in Heimugou section at Luochuan of Shaanxi Province.

CLAY MINERAL SPECIES

Luochuan and Longxi sections are developed in different geomorphological types. the former is widespread on the loess "Yuan" area (a kind of region with an even loess topography) and the latter is developed in a basin region. Their stratigraphy shows a close similarity. The section can be subdivided into Late Pleistocene Malan Loess, Middle Pleistocene Lishi Loess and Early Pleistocene Wucheng Loess in descending order. Holocene Loess is found at the top of Longxi section. The loesses of different ages show disconformable contacts with each other (Fig. 1) Samples used for the examination of clay mineral were taken from each bed of the above described sections. The total number of the determined sample is 86.

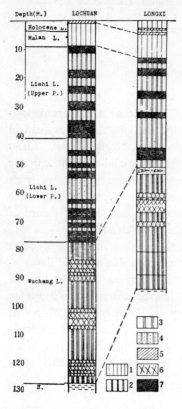

Fig. 1. Luochuan and Longxi Loess sections
1— Malan Loess and Holocene Loess;
2— Lishi Loess; 3—Wucheng Loess;
4— Silt bed, 5—Black soil,
6— Weathering bed, 7—Paleosol.

84

X-ray powder diffraction technique was adopted in the examination of clay minerals, some of them were examined by electronmicroscope. The particle size is less than 2 μm in the clay mineral analysis. The clay minerals either in loess or paleosol samples show a close similarity, their basal reflexion peaks lies at d=15.5, 14, 10 and 7.2 Å in diffraction patterns (Fig. 2). Based upon the changes of diffraction peaks after glycolated, NH_4Cl treated, heated under 550°C and HCl treated, it can be judged that the reflexion at 15.5Å is of montmorillonite, 14 Å peak is chlorite with some vermiculite, 10 Å peak is of illite, 7.2 Å peak is an overlapped reflexion by kaolinite (001) and chlorite (002) In addition, some samples have a small amount of halloysite and mixed-layer minerals.

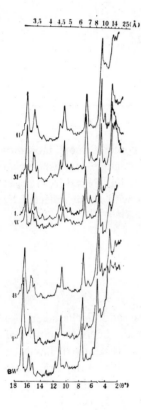

Fig. 2. X-ray diffraction patterns of Clay minerals in loess and paleosol.
H — Holocene Loess;   M — Malan Loess;
L — Lishi Loess;   W — Wucheng Loess;
B — Black soil;   P — Paleosol;
BW — Buried weathering bed.

Given in Table 1 is the main mineral species and their relative amount estimated by basal reflexion intensities. This result has shown that either illite or montmorillonite content in loess is commonly less than those in paleosol, indicating that the clay minerals in paleosol changed under warmer and moister climate conditions, and some of them may be produced from other clay minerals or from detrital minerals such as feldspar.

85

Table 1 Main clay mineral species and their relative amounts in Luochuan and Longxi sections

Strata	Sediments	Amount	Clay minerals and relative amount %			
			illite	chlorite	kaolinite	montmoril.
		Luochuan	section			
Malan 1.	Loess	3	68	12	11	9
Lishi 1.	Loess	4	67	11	10	12
(Upper)	Paleosol	6	72	8	8	13
Lishi 1.	Loess	9	63	11	11	16
(Lower)	Paleosol	8	74	7	7	13
Wucheng 1.	Loess	12	60	11	10	19
	Weathering B.	10	61	10	9	20
		Longxi	Section			
Holocene 1.	Loess	2	56	19	13	12
	Paleosol	2	62	15	14	10
Malan 1.	Loess	1	51	22	17	10
Lishi 1.	Loess	7	62	11	12	15
	Paleosol	10	64	10	11	15
Wucheng 1.	Loess	6	69	6	8	17
	Weathering B.	3	73	5	9	13

Luochuan section shows clearly that the older loess contains more montmorillonite and vermiculite with less lillite, whereas, the younger loess shows an opposite tendency. These indicate that the paleoclimate has shifted from moist to drought since Quaternary in general, which agrees with those reflected by data of biostratigraphy and sporo-pollen stratigraphy [10,11,12].

CRYSTALLINITY

Illite is the most important clay mineral in loess section, and its characteristics in different beds of loess section can be used to obtain the imformation of clay minerals in deposition and reformation processes. Recently, it has been commonly accepted that illite is actually a disorder mixed-layer mineral made up of mica layers with minor expansible layers[2], in which the quantity of expansible layer decides the peak site and shape of (001) reflexion. Illite "crystallinity" was discussed by C. E. Weaver in 1960, based upon the ratio of peak height at 10 Å to that at 10.5 Å[16] and the relationship between the crystallinity of illite and its formation conditions in a given geological environment was expounded by B. Kubler in 1966 in terms of the width at half height of the diffraction peak at about 10 Å[6]. In order to show the deflection degree of illite (001) reflexion from theoretical value of mica (001) (d=10 Å), the ratio of peak height and width at half height at about 10 A (H/W value) was adopted to describe illite characteristic in this paper. H/W value in loess is generally greater than 25 and that in paleosol less than 20. Comparison between younger and older loess indicates that the former has a greater H/W value while the latter has a minor one, which shows that the illite in older loess and in strata undergone pedogenesis has worse crystallinity. In other words, they contain more expansible layers.

Another marker for illite crystallinity is I(001) / I(002), which is an indicator of mica layer characteristic. Difference of intensities between the 1st and the 2nd order reflexions may be related either to iron content in its octaheadral layers[5], or to local weathering of illite leading consequently to the decrease in K content which weaken the (002) intensity[17]. It is obviously that the stronger the weathering process to clay minerals, the higher ratio of its I(001) / I(002). So it can be used as an important indicator showing the difference of diagenesis level and pedogenesis degree in separated samples in the study on climate events recorded in loess section.

From I(001) / I(002) curve of Luochuan section (Fig. 3), it can be seen that the value of loess is in a minor fluctuation range with a value of 2.5 except for a few samples, which reflcts the loess to be homogeneous materials. On the other hand, the ratio of I(001) / I(002) in paleosol has greater value of ca. 3.7 in general, which indicates that the pedogenesis effected deeply to interlayer ions of illite.

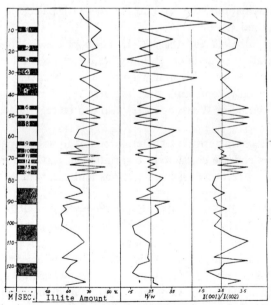

Fig. 3. Illite relative amount and its crystallinity indicators in Luochuan section of Shaanxi Province.

CLIMATE CYCLES AND SEA-LAND CORRELATION

Climate Cycles Recorded in Loess of China

1. **Climate fluctuation amplitude.** The data of clay mineral have shown that the variations of relative amount and crystallinity in loess sections are synchronous, the fluctuation amplitudes of different climate events have been shown by clay minerals from various angles, for example, in Lishi Loess under the 1st, 4th, 5th, 6th paleosols and upper silt beds, the crystallinity indicators show higher values, whereas, those in loess beds under 2nd and 3rd paleosols all show lower values.

In order to explore the climate amplitude by mineral indexes, the clay mineral data have been calculated into variation value for paleosols by following formula:

$$V = (L - S) / L \times 100\%$$

wherein V = Variation value of clay mineral indices, the greater the value, the stronger the pedogenesis; L = Determined value of clay mineral in parent material (underlying loess); S = Determined value of clay mineral indices in paleosol.

Thus, these data may better reflect the variation extent of clay minerals after pedogenesis. Following are the characteristics of paleosols observed:

1) The 3rd and the 10—11th paleosols, their illite relative amount and I(001) / I(002) have a slight lower variation value. If other paleosols are representatives of interglacial fluctuations, the 3rd and the 10—11th paleosols should be products in interstadial period.

2) The 7—8th paleosols and the 12—13—14th paleosols, their illite relative amount and variation value of crystallinity indicators lie in a higher range, and constitute two soil groups with similar amplitude respectively. It seems likely that each group only represents one interglacial. Similarly, each group of weathering bed buried in Wucheng Loess also can be regarded as one warmer stage respectively.

According to the characteristics of alternations of loess and paleosols in stratigraphic sections and the characteristic variation of clay minerals in loess section climatic cycles recorded in loess sections and their characteristics can be distinguished.

2. **Time control for climatic events.** Up to the present available chronological data are listed in Table 2, which have provided a rudimentary time scale for climatic events recorded in loess sections. It was urged at a Wenner-Gren symposium in 1975 on the Middle Pleistocene that the lower boundary should be defined at the Brunhes / Matuyama reversals, while the upper boundary should be defined at the beginning of the last interglacial marine transgression[3]. And data on Table 2 shows that these two boundaries are coincided with those between Malan/ Lishi and Lishi/ Wucheng in loess sections of China. How-

ever, the developing age of the 1st paleosol at the top of Lishi Loess, can be likely correlated to the 2nd phase of Eemian interglacial period in Northern Europe.

Table 2 Geochronologic data of loess in China

Strata	Fossils*	Magnetostrata**	^{14}C and TL age (Years, BP)
Holocene Loess	—	—	Upper black soil at Longxi Sec.=7,360±250; Lower black soil at Longxi Sec = 8,550±300; Black soil at Luochuan Sec = 8,000±400
Malan 1.	*Struthiolithus sp.* *Myospalax fontanieri*	Blake event (?): 11–1 core and Tingjiagou sec. at Luochuan; Zaitang sec. near Beijing	Upper part of Malan loess, TL=39,000±5,000
Lishi 1.	*Dali man* *S. lantienensis*; *Myospalax tingi*; *M. chaoyatseni*; *M. arvicolinis.*	Brunhes/Matuyama: Luochuan, Wuchen, Longxi; Jixian, Wuquanshan at Lanzhou.	—
Wucheng 1.	*Hipparion sinensis*; *Nyclereutes sinensis*; *Myospalax omegodon.*	Jaramillo event: Luochuan, Longxi, Wuquanshan at Lanzhou.	—

* Liu T. S. et al., 1962, 1964, 1965; Wang Y. Y. et al., 1979; Wu X. Zh. et al., 1979.
** An Z. S. et al., 1977. Cheng G. L. et al., 1978; Li H. M. et al., 1974, 1980; Wang J. D. et al., 1980; Wang Y. Y. et al., 1980.

Based upon the stratigraphy and the characteristics of clay minerals in loess sections, 13 climatic cycles with 26 stages have been distinguished, in which greater amplitudes in the even-numbered stages 6, 10, 12, 16 may reflect colder condition during corresponding glaciation, whereas, greater amplitudes in the odd-numbered stages 5, 9, 11, 15, 19 may indicate stronger interglacial climatic condition (Fig. 4).

Land-Sea Correlation of Quaternary Climate Events

Correlation of climatic cycles recorded in loess sections of China with glacial interglacial periods in North Europe and stages of isotope oxygen in deep sea sediments is presented in Table 3.   It is an attempt to apply the clay mineral indicators to stratigraphic subdivision and correlation.   The purpose is to

89

Table 3 Tentative correlation of climatic cycles recorded in loess sections of China with glacial / interglacial periods in northern Europe and stages of isotope oxygen in deep sea sediments

Periods	18O stages* (V28–238)	Ages (kyr., BP)	Gl/Interl.** in N. Europe	Climatic cycles in loess of China Cycles, Stages, Loess or paleosol		
Holocene	1		Post glacia- tion	I	1	Holocene Loess at Longxi section
		13	Wiechselian	II	2	Upper Malan 1.
	2 3				3	Black soil buried in Malan Loess at Jingning section
Late Pleistocene					4	Lower Malan 1.
	4 5a–e	128	Eemian		5	1st paleosol
	6		Warthe	III	6	Lishi 1.
	7a–c	240	Eemian		7a–c	2nd paleosol
	8		Saalian		8a–c	Lishi 1./3rd paleosol / Lishi 1.
	9	330	Eemian	IV	9	4th paleosol
	10		Saalian		10	Lishi 1.
	11	400	Holstein	V	11a–e	5th paleosol
Middle Pleisto- cene	12		Elster	VI	12	Lishi 1.
	13	480	Holstein		13	6th paleosol
	14		Elster	VII	14	Lishi 1.
	15	570	Cromerian		15a–c	7th pal./1./8th pal
	16		Elster	VIII	16	Upper silt
	17	630	Cromerian		17	9th paleosol
	18				18a–e	L./10th pal. 1. 11th pal. /1.
	19		Cromerian	IX	19a–e	12th pal. /1. / 13th pal. /1./ 14th paleosol
	20	710		X	20	Lower silt
	21				21	Upper superim- posed weath. bed
Early Pleisto- cene	22	810		XI	22	Wucheng 1.
	23				23	Middle super- imposed weath. bed
	24	900		XII	24	Wucheng 1.
					25	Lower superim- posed weath. bed
				XIII	26	Wucheng 1.

* Shackleton, N. J. et. al. 1973
** Bowen, D. Q., 1978; Kukla, G. J., 1977.

arouse more attention and interest to the study on clay minerals and the classification of climatic cycles in loess research.

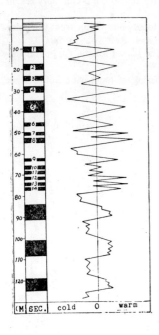

Fig. 4.   Climatic curve recorded, in clay minerals in
Loess section.

REFERNCES

[1]   An Zisheng, et al., 1977, Paleomegnetic research of the Luchuan section, Geochemica,
No. 2, pp. 234–249.
[2]   Bailey, S. W., 1980, Summary of recommendations of AIPEA nomenclature committee
on clay minerals. American Minerologist, vol. 65, No. 1—2.
[3]   Bowen, D. Q., 1978, Quaternary geology — A stratigraphic framework for multidisci-
plinary work, Pergamon Press.
[4]   Cheng Guoliang, et al., 1978, Study on paleomagnetism for several Quaternary sections of China (in press).
[5]   Grim, R. E. et al., 1951, The mica clay minerals.   In: Brindley, G. W., ed., X-ray identification and crystal structure of the clay minerals, Min. Soc. London.
[6]   Kubler, B., 1966, La cristalinité de lillite et les zones tout à fait supérieures de méta-
morphisme.   In: Colloque sur les Etages Tectoniques, A la Baconnière, pp. 105—122,
Neuchated, Suisse.
[7]   Kukla, G. J., 1977, Pleistocene land-sea correlations (I), Europe, Earth Soc. Rev., vol.
13, pp. 307—374.
[8]   Li Huamei et al., 1974, Preliminary result of research on paleomagnetism for Wucheng
Loess section.   Geochimica, no. 2.
[9]   Li Huamei et al., 1980, Magnetostratigraphy study for Quaternary Sediments of North
China.   Essays of geology for international exchange (5), Geology Press.
[10]   Liu Tungsheng et al., 1962, Loess of China.   Science Press. Beijiang.
[11]   Liu Tungsheng et al., 1964, Loess of Middle Huanghe Valley, Science Press, Beijing.

91

[12] Liu Tungsheng et al., 1965, Loess deposits of China, Science Press.

[13] Shackleton, N. J. et al., 1973, Oxygen isotope and paleomagnetic stratigraphy of Equatorial Pacific core V28—238: Oxygen isotope temperatures and ice volumes on a 10^5 and 10^6 year scale, Quaternary Res., vol. 3, pp. 39-55.

[14] Wang Yong-yan et al., 1979, Discovery of *Dali man* fossil and the preliminary results. Journal of Northwest University, no. 3.

[15] Wang Yong-yan et al., 1980, Exploring problems of loess subdivision from the paleomagnetic data for northern plateau of Weihe valley, Geol. Rev. vol. 26, No. 2.

[16] Weaver C. E., 1960, Possible uses of clay minerals in the search for oil, Clays and Clay Minerals, Bull. Amer. Assoc. Geol., 44, 1505—1518.

[17] White, J. L., 1962, X-ray diffraction studies on weathering of muscovite, Soil Science, vol. 93, pp. 16—21.

A PRELIMINARY STUDY ON THE CLAY MINERALOGY OF THE LOESS AT LUOCHUAN SECTION

HAN JIA-MAO

(Institute of Geology, Academia Sinica)

Abstract — Clay mineralogy of the loess at Luochuan section, Shaanxi, was investigated, which has been shown that clay minerals in loess exhibit great complexity in aspects of their compositions and crystallinities.

Illite, kaolinite, chlorite, vermiculite and montmorillonite are all found in the loess, and mixed layers as well. The main component is illite which is assigned to the mixed layer of mica-like layers with a few expanded layers. In general, there is a little more vermiculite and less chlorite in the paleosol than in the loess. It shows that pedogenic processes took place in a humid environment favourable to the formation of the vermiculite.

By means of H_w (the ratio of the height of the reflection at 10 Å and the width of the same reflection at 1/2 height above background), the crystallinity of illite is determined, which indicates that there was a great difference of H_w between loess and paleosol. It also shows that there existed different climatic conditions when loess deposited and paleosol formed. The value of H_w, on the other hand, increases gradually from Wucheng Loess (Lower Pleistocene) through Lishi Loess (Middle Pleistocene) to Malan Loess (Upper Pleistocene). It might be related to their depositional conditions and the degree of diagenesis.

INTRODUCTION

Study on the components of the loess in China made rapid progress in the past two decades but little has been done in the field of clay mineralogy of the loess. Only few relevant literature can be found so far. Variation in composition and structure of clay minerals recorded characteristics of the geological environment they formed. Study on clay mineralogy of loess will certainly enrich our knowledge about the origin of the loess materials, the loess formation and the Quaternary paleoclimatic condition as well.

The purpose of this paper is to determine the composition of the clay minerals in loess, to make quantitative clay mineralogical analyses from the bulk chemical analytical data of the samples and to discuss the paleoclimatic characteristics during the loess formation based on the variation of the structures of clay minerals, especially the "crystallinity" of illite.

Luochuan section was chosen to collect samples systematically since it is one of the type sections of loess in China; and a lot of data about bio- and magneto-stratigraphy, mineralogy, petrology and chemistry have been accumulated.

GEOLOGICAL BACKGROUND

Luochuan section is located in a branch of Heimu Gully near Potou Village five kilometers south of Luochuan County. It is in a Cenozoic depositional basin, loess with tremendous thickness of more than 130 m can be seen. Malan Loess over-lies Lishi Loess which, in turn, is above Wucheng Loess. At the bottom of the section exposes red clay. Lishi Loess is usually divided into two parts: Upper Lishi Loess and Lower Lishi Loess.

Malan Loess (8 m). Greyish yellow silt. no bedding, massive, friable with big pores, calcareous.

Upper Lishi Loess (26.1 m). Four paleosol layers alternated with four loess layers. The loess is light yellowish brown, silt, compact and massive with few pores, a few white calcareous mottles. Paleosol, dark brown, with much clay, angular blocky, matrix low calcareous, carbonate nodules at lower boundary.

Lower Lishi Loess (48.9 m). Ten paleosol layers alternated with ten loess layers. The loess is the same as the one of Upper Lishi but more compact. Paleosol is similar to the one of Upper Lishi.

Wucheng Loess (46.2 m). Reddish brown, silty, dense massive, hard, with multilayers of calcareous concretion and weatherings.

Red Clay (N_2), Dark brownish red, clayey, many manganese mottles and a small amount of carbonate nodules.

MATERIALS AND METHODS

Two kinds of samples were collected, twenty for determining main components of clay minerals and the other fifty-three for measuring the crystallinity of illite. Both representativeness and distribution in the profile are taken into consideration in selecting the samples.

The fraction smaller than 2 μm is separated by routine method. For X-ray diffraction analysis, three oriented glass slides were prepared from the 2 μm fraction. One slide was X-rayed without any treatment. Another was glycerinated and then X-rayed. The third was X-rayed after heating for an hour at 500°C. In addition, samples treated with HCl and NH_4Cl were separately prepared and X-rayed. The remains of clay suspension were partly used for SEM and others dried at 100°C for DTA, TG and chemical analysis.

94

RESULTS

X-ray Diffraction

A close similarity is shown in the X–ray diffraction patterns of all samples. Main reflections occur at 14, 10, 7.1, 5.0, 4.7, 3.5, and 3.3 Å, which means that there is a great complexity in the composition of clay minerals of the loess. 10 Å reflection exhibits no changes after glycerinated and heated, which shows that there exists illite. The existence of kaolinite can be confirmed by disappearance of reflections at 7.1 and 3.5 Å after heated. Based on the variation of 14 Å peak after glycerinated, heated and treated with HCl and NH_4Cl solution respectively montmorillonite, chlorite and vermiculite can also be detected. In addition, there may be some mixed layer minerals as well.

Loess in different ages shows a close similarity in their clay mineral compositions, in spite of little difference. Illite is dominant in all the loess. The difference is chiefly reflected by the relative content of montmorillonite. A detectable amount of montmorillonite was found in Wucheng Loess but hardly any in Malan Loess.

The diffraction patterns of paleosol differ from that of loess in relative intensities of the reflections, which indicates that the change of relative content of each clay mineral took place in the paleosol formation.

Another feature of the X–ray patterns is that the (001) reflection of illite looks asymmetric towards the low angle side, which will be discussed in detail later.

Thermal Analyses

Two kinds of differential thermal analysis curves can be distinguished. One has two endothermic peaks at about 125 and 570 °C respectively. The other exhibits another endothermic peak at about 700 °C and an exothermic peak at about 885 °C besides those two above. The former coincides with the curves of paleosol, the latter with those of loess. In the TG curves relevant changes occur i.e. the TG curves of loess show a larger weight loss in the moderate temperature region (from 500 to 700 °C).

The results of thermal analyses are comparable with those of X–ray diffraction.

Transmision Electron Microscopy

Transmision electron micrographs show that most of the particles are hypidiomorphic and xenomorphic in shape. The surfaces of a few particles suffered erosion and dissolution. The particles often appear in poor crystallinity.
Illite. Small poorly defined flakes, commonly grouped together in irregular

aggregates, with a diameter of 0.1 to 1 micron, poorly crystallized, with concave edges caused by erosion and dissolution.

Kaolinite.   Poorly formed six-sided flakes, flake surfaces less than 0.5 micron, poorly crystallized, with larger thickness, suffered partly from erosion and dissolution.

Montmorillonite.   Flake-shaped units without regular outlines, usually smaller than other particles, variable in thickness.

Chlorite and vermiculite.   Large thin flakes, uneven in thickness even in one particle.

Differing from the loess, in paleosol euhedral crystal of kaolinite with better crystallinity can sometimes be found and halloysite occasionally.

Chemical Analyses

The bulk chemical analyses of clay particles indicated that the contents of SiO_2, Al_2O_3, Fe_2O_3, K_2O in paleosol are more than those in loess and in contrast, there are less FeO, CaO and CO_2 in paleosol.   The contents of MgO and Na_2O remain constant.   The content of free iron oxide increases in paleosol too, which means that the pedogenetic process of paleosol destroyed some iron silicates which became free iron oxides remained in the soil profile and at the same time carbonates were leached to lower boundary of the soil profile, as a result, carbonate nodules, were formed.

DISCUSSION

Quantitative Estimation of Clay Minerals from Bulk Chemical Analyses.

Because of the X-ray technique, chemical analysis has been ignored increasingly.   But by any means chemical analysis is significant because any compositional and structural changes in clay minerals are mostly caused by the variation of element composition and proportion.   So more and more chemists and mineralogists pay full attention to this topic.

The chemical composition of any rock type can be related to its known mineral constituents and composition by a same set of equations of the type:

$$a_i\, x_1 + b_i\, x_2 + \cdots\cdots \cdots\cdots + n_i\, x_n = 100\, K_i \qquad (1)$$

where a_i, b_i, $\cdots\cdots$ n_i, and K_i are the percentages of the element i in mineral phases X_1, X_2, $\cdots\cdots$, X_n and the bulk rock respectively. x_1, x_2, $\cdots\cdots$, x_n are the percentages of phases X_1, X_2, $\cdots\cdots$, X_n present in the rock.   The equations can be solved if there are more elements determined than the phases and so far as all phases are concerned, if the content of the elements determined can be fixed within the experimental precision of chemical analysis.   Pearson (1978) published a Fortran IV Programme for solving the equations[7].   In his method, n elements were chosen from m elements determined (n is usually less than m), and made n-variate equations. Xu Jian-guo (personal communication) uses all m elements determined and proposes a Basic Programme to solve the equations. He uses optimization method to gain optimum solution which is nonnegative and with the minimum of sum of squares of deviations. Xu's method is adapted in this study. We account for Ti, S, P, and CO_2 in estimating anatase, pyrite, apatite and calcite respectively. The trouble we have to face is the variation of the compositions of clay minerals, with exception of kaolinite, over wide ranges. So the more authentic the known composition (a_i, b_i, ... n_i) we choose is, the more satisfactory results we will obtain. The data of known composition used are selected from the literature[5,6,9].

The results calculated (Table 1) show that the major component of loess is illite and then kaolinite, montmorillonite, chlorite and vermiculite successively. From Malan Loess through Lishi Loess to Wucheng Loess, the amount of montmorillonite increases. Chlorite shows its maximum in Lishi Loess. On the contrary, vermiculite is the minimum. The data above are quite similar to the results of XRD. In comparison with loess, paleosol contains much more illite, a little more vermiculite and less chlorite. There may be a little increase in kaolinite too, which indicates clearly that pedogenic processes took place in a humid and warm environment.

Table 1 Quantitative clay mineralogical analyses from the bulk chemistry of samples

Number of sample	Name	Clay minerals K	I	M	Ch	V	Nonclay minerals C	G	Others
81-G-01	Malan Loess	16.80	48.70	3.74	5.74	4.69	16.10	2.85	1.37
02	Malan Loess	14.80	45.45	4.34	5.42	5.02	20.90	2.74	1.33
03	Paleosol	18.20	57.41	7.92	2.60	7.93	0.00	4.55	1.36
04	Lishi Loess	18.49	42.09	10.77	5.26	4.82	14.42	3.05	1.20
05	Paleosol	20.80	53.60	8.41	5.35	4.82	1.30	4.42	1.31
06	Lishi Loess	12.97	55.74	5.31	4.50	5.30	11.62	3.13	1.43
07	Paleosol	17.35	63.21	1.14	6.41	5.56	1.00	4.23	1.11
08	Lishi Loess	23.38	42.87	12.52	9.22	1.33	6.07	3.25	1.36
09	Paleosol	19.23	61.31	3.18	9.22	1.99	0.00	3.92	1.12
10	Lishi Loess	16.38	51.24	7.29	7.06	3.83	10.30	2.66	1.26
11	Lishi Loess	20.73	45.03	7.51	7.78	3.29	11.21	3.18	1.27
12	Paleosol	21.32	50.87	11.68	6.97	2.54	1.82	3.62	1.18
13	Paleosol	21.36	53.26	10.27	8.03	2.30	0.00	3.65	1.11
14	Lishi Loess	17.74	55.17	13.00	8.10	1.85	0.00	2.92	1.23
15	Wucheng Loess	19.77	48.95	8.14	6.60	4.06	8.39	3.02	1.06
16	Wucheng Loess	20.79	49.23	12.37	7.62	3.36	3.09	2.50	1.03
17	Wucheng Loess	20.28	50.02	11.14	5.34	4.76	4.34	3.04	1.09
18	Wucheng Loess	14.68	52.72	7.29	4.33	6.81	10.71	2.23	1.24
19	Wucheng Loess	23.12	40.70	13.99	2.71	6.89	8.10	3.39	1.10
20	Red clay	21.51	51.50	11.75	5.89	3.76	1.41	3.21	0.98
	Malan Loess	15.80	47.08	4.04	5.58	4.86	18.50	2.80	1.35
	Lishi Loess	18.28	48.69	9.40	6.99	3.39	8.94	3.03	1.29
	Wucheng Loess	19.73	48.32	10.59	5.32	5.18	6.93	2.84	1.10
Mean value	Paleosol	19.21	56.61	7.10	6.43	4.19	0.69	4.07	1.20
	Red clay	21.51	51.50	11.75	5.89	3.76	1.41	3.21	0.98

K—Kaolinite, I—Illite, M—Montmorillonite, Ch—Chlorite, V—Vermiculite, C-Calcite, G—Goethite, Others include Apatite, Anatase, Pyrite and MnO.

On the Illite in Loess

The term illite was proposed by Grim et al. in 1937 for the mica-type mineral occurring in argillaceous sediments[3]. Owing to its wide spread in rocks and sediments throughout the geological periods and some uncertainty in the original definition as a result, the use of the term illite was thrown into confusion. Recently, most clay mineralogists tend to consider illite as disordered mixed layer minerals. Just as indicated in the latest classification table for clay minerals "The status of illite... must be left open at present, because it is not clear whether or at what level they would enter the Table: many materials so designated may be interstratified[1]". In fact an asymmetry of the (001) reflection of illite has come into notice for a long time but some mixed layer structures with few expanded layers are often ignored. The term illite used here is considered precisely to be mixed layer of mica and montmorillonite with less montmorillonitic layers.

Taking the asymmetry of illite's (001) reflection into account, many mineralogists defined crystallinity indices of illite i.e. measurement of the degree of (001) peak's asymmetry, to discuss some geological problem and obtain a lot of good results[2,4,8]. The author thinks that the variation of the shape of the (001) peak at about 10 Å is principally dependent on the number of montmorillonitic layers in illite structure. And the shape itself can be controlled by the height and width of the peak. In this paper the author uses H_w, i.e. the ratio of the height of the reflection at 10 Å and the width of the same reflection at 1/2 height above background, as the measurement of the crystallinity of illite.

The H_w values of all samples show that there is a great difference between loess and paleosol. H_w values for loess are greater than 25, and for paleosol less than 20. This is particularly evident for Upper Lishi Loess. It is clear that there existed different conditions during loess deposition and paleosol formation. The loess deposited in arid and cold condition and the paleosol formed in humid and warm condition. It is difficult to determine absolute values of humidity and temperature. But according to the degree of pedogenesis the author holds that difference in humidity is more important than that in temperature.

Origin of Clay Materials in Loess

The materials of loess over the world are derived from both glacial and desert regions. In China the materials are mainly from extensive deserts in the Northwest because there were no continental ice sheets in Quaternary. Where did the clay materials come from ? Whether it formed in situ or not is still a problem argued. Based on the features of clay mineral composition and morphology of particle in SEM the author tends to deem that clay materials in loess are mainly clastic and came from source regions together with quartz, carbonates,

98

etc., even though some changes in composition and structure have taken place after deposition.

CONCLUSIONS

Clay mineral composition of loess is quite complicated. Illite, kaolinite, montmorillonite, chlorite and vermiculite are all found in loess, and mixed layers as well, and illite is dominant. Wucheng Loess, which is older in age, contains a little more montmorillonite, and more montmorillonitic layers in illite structure. It may be related to the diagenesis and environmental conditions in which it formed.

Paleosol exhibits the tendency of increase in illite and decrease in montmorillonite, and a little more vermiculite and less chlorite, which indicates that pedogenesis took place in humid environment favourable to the formation of vermiculite.

The crystallinity of illite H_w was defined as the ratio of the height of the reflection at 10 Å and the width of the same reflection at 1/2 height above background. It is quite useful in comparison with loess in different ages and between loess and paleosol. It can also be used as an indicator of fluctutions in paleoclimate.

Clay minerals in loess are mainly clastic and they derived from source regions even though some changes in composition and structure took place after deposition.

The paleosol was formed in more humid and warm condition if comparing with the loess depositional condition. The difference in humidity is more important than that in temperature.

ACKNOWLEDGEMENT—The author wishes to thank Prof. Liu Tung-sheng, my director, for his kind guidance throughout this study and for his critical comments, many of which were incorporated into the final manuscript, and Prof. Gu Xiong-fei for his encouragement and for critically reviewing the manuscript. The author has benefited a lot from comments and by discussions with He Zhu-wen, Zheng Hong-han, Wen Qi-zhong, and is also indebted to the colleagues of Institute of Geochemistry, Academia Sinica, for their assistance in experimental work.

REFERENCES

[1] Bailey, S. W., 1980, Summary of recommendations of AIPEA Nomenclature Committee, Clay Miner. 15, 85—93.

[2] Chamley, H., 1967, Possibilité d'utilisation de la cristallinité d'un minéral argileux (illite) comme témoin climatique dans les sédiments recents, C.R. Acad. Sci., Paris, 265, 184—187.

[3] Grim, R. E., Bray, R. H., Bradley, W. F., 1937, The mica in argillaceous sediments. A.M., 22, 813—829.

[4] Kubler, B., 1964, Les argile, indicateurs de métamorphismes, Revue Inst. Fr. Pétrole., 19, 1093—1112.
[5] Liu Tung-sheng et al., 1964, Loess of middle Huanghe (Yellow) River Valley, Science Press. Beijing.
[6] Liu Tung-sheng et al., 1965, Loess deposits of China, Science Press, Beijing.
[7] Pearson, M. J., 1978, Quantitative clay mineralogical analyses from the bulk chemistry of sedimentary rocks, Clays Clay Miner., 26, 423—433.
[8] Weaver, C. E., 1960, Possible uses of clay minerals in search for oil, Clays Clay Miner., 8, 214—227.
[9] Weaver, C. E. and Pollard, L. D., 1973, The chemistry of clay minerals, Elsevier Scientific Publishing Company, Amsterdam.

METEOROLOGICAL CHARACTERISTICS
OF DUST FALL IN CHINA SINCE
THE HISTORIC TIMES

ZHANG DE–ER

(Academy of Meteorological Sciences, Central Meteorological Bureau Beijing)

Abstract—Based on 1,156 historical records about the dust rain in China and some of the meteorological reports, the temporal and spatial distribution of dust fall and its origin, climatic background and synoptic process since the historic times are discussed, which shows that the genesis of loess in China is closely related to synoptic situation.

All kinds of dust fall occurred in China since the historic times. They were recorded in a lot of historical literature and some may be also read from weather reports. These data may furnish evidences for the discussion of genesis of loess in China.

The events of dust fall were recorded as "dust rain", "dust haze", "yellow fog", etc., which were usually consisted of dust and yellow sand, sometimes mixed with water and mud rain. In general, the thickness of a fallen dust was about from a few millimeters up to several centimeters. The earliest event recorded occurred in 1150 B.C.[1], but the information of the last 1,000 years is much abundant than that of the earlier centuries. In this paper, the author cited 1,156 historical records about the dust fall (exclusive of the local dust storm) and some present meteorological information to discuss the temporal and spatial distribution of dust fall, its climatic background and synoptic process as well as the origin of dust, and to explain that wind is the main transport force in genesis of loess in China.

THE TEMPORAL AND SPATIAL DISTRIBUTION OF THE
DUST FALL

Spatial Distribution

The places where dust fall occurred in the historic times are shown in Fig. 1. Their distribution covers an area from Xinjiang to the coast of East China Sea and from Nei Monggol to south of the Changjiang River. But also some dust rain happened in Fujian, Guangdong and Guangxi Provinces. It is notable, the areas where dust rain occurred nearly coincide with the distribution of loess in China[2]. Some places where dust rain occurred during the historic times in Japan[3] are also shown in Fig. 1. It seems that they were the expansion of dust rain area of China.

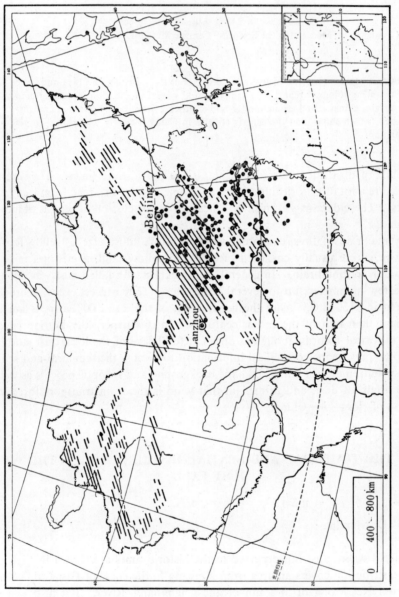

Fig. 1. Distribution of dust rain and loess in China.

102

Temporal Distribution

The year that dust rain fell was called the "dust rain year", and a "chronological table of typical dust rain year[4]" is obtained. From the table the number of dust rain year in different decades, or the frequency, has been calculated, which shows the alternations of periods with frequent and rare occurrence of dust fall. Fig. 2 shows the frequency curve since 300 A.D. The periods of frequent occurrence are: 1060—1090 A.D., 1160—1270 A.D., 1470—1560 A.D., 1610—1700 A.D., and 1820—1890 A.D., respectively.

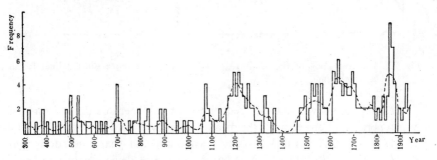

Fig. 2. The frequency curve of dust fall since 300 A. D.

CLIMATIC BACKGROUND OF DUST RAIN

Seasonal Characteristics

The dust rain could occur in different months of the year, but concentrated mostly in winter half year. Based on 508 cases quoted from historical literature, the number of dust rain in each month and its frequency are calculated. The results show that the high frequency of dust rain concentrates in February to May, especially in April, which accounts for 26 per cent of the annual value.

The Relationship Between the Dust Rain and the Temperature

Winter temperature index series of some regions in China, for the last 500 years have been presented by the author[5]. In order to show the temperature condition of China in the last 500 years, the statistics of mean series of winter temperature index has been made from the same index for five regions (1470—1969). A comparison between the mean series of temperature index and the the frequency of dust rain year shows that they are in opposite phase (Fig. 3).

103

The calculation shows that the correlation coefficient between the temperature index and the frequency of dust-rain year is –0.59, at a certainty level of 0.001.

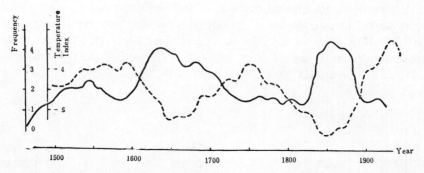

Fig. 3. The frequency of dust–rain year and temperature index (dotted line) for the last 500 years.

Further analysis shows that the frequencies in cold periods differ from those in warm period. For example, the mean frequency in cold periods (1621—1700 A.D., 1811—1900 A.D.) is 3.7 year / decade and 2.1 year / decade in warm periods (1511—1620 A.D., 1720—1780 A.D.). The deviation of winter temperature between the warm and cold periods is about 0.5°C in the Changjiang River Basin.

As compared with the curve of temperature fluctuation for the last 5,000 years which presented by Zhu[6], it shows that the high frequent spell of dust rain still concentrates in the periods of low temperature curve, but some frequent spells such as 690—710 A.D. correspond to the warm period of Zhu's curve, which needs to be further verified. The phenomena mentioned above provide evidences for a detail study on the temperature changes in the historic times.

The Relationship Between the Dust Rain and the Humidity

The comparison of frequency of dust rain with curve of humidity index in 35—40°N of China for the last 2,000 years[7] shows that most of the high frequency spells of dust-rain correspond to dry spells, which coincides with Liu Tung-sheng's expanation that "loess is silt deposits under the dry climate condition"[8].

THE SYNOPTIC ANALYSIS OF DUST FALL

The dust fall may be divided into two types. The first type, or dust fall in gale, usually occurred with strong cold front, and shifted from north to south accompanying with strong wind (northwesterly wind). It was recorded as

104

"dust fall scattered as the yellow wind from the northwest arrived". The sketch maps of moving paths of several events have been reconstructed based on the historical descriptions. In these maps, the shifting speed and path of the dust fall are similar to that of the typical cyclone cold front in the same region. Such kind of events still occurs at the present time. For example, from April 18 to April 20 in Beijing the author witnessed "yellow wind all over the sky, and formed thin layer of dust on the ground". This event was only a light dust fall if compared with those in the historic times, but its development was similar to those in historical cases. Some studies on this event[2] indicate that the dust haze and dust fall in the process swept over northern and central China. The dust haze was resulted mainly from a vigorous development of a Mongolia cyclone. The strong development of the Mongolia cyclone and the downward transport of the momentum of the upper westerlies appeared on April 17 formed the sandstorm in central and western Nei Monggol and Gansu Corridor. Through the mixture of upward motion and turbulence, the sand and dust were rolled up into the westerly jet stream of the upper air. Thereafter, dust and sand carried in the upper air jet stream rapidly shifted and fell down southeastward over loess plateau.

The second type of dust fall, or dust fall in calm, was usually described as "sand flying downward in calm", "dust rain as fog". This type of dust fall could occur anywhere. But it is noted that all the dust rains in lower latitude regions belong to this type, such as in Putian (1627 A.D.), Wuzhou (1768 A.D.), Xiangshan (1721 A.D.). The typical event happened in March 1958, which Liu[8] described as "In March 20—24, 1958, in Beijing, it was calm and cloudless, but the sky was tinged with yellow. Then a large amount of dust setted slowly on the trees, houses and streets." The weather records show that the occurrence of the dust haze was initiated at 14:55 (Beijing time), March 19, in Beijing, and from March 19 to 24, the daily average surface wind-speeds were lower than 1.8 m/sec. especialy on March 21 the whole day was calm. According to the synoptic maps, it is found that the dust originated mainly from the desert of southern Xinjiang under the influence of cyclone and trough and were transported eastward by upper air stream. When dust fall started, Beijing was precisely situated in the rear of the upper trough which was an area of sinking flow. According to the initial time of the dust storm and dust-haze recorded in 11 stations along 40°N in China, the shifting speed of the sand and dust was about 1.2° (longitude) / day, which approached the mean westerly component on 500 mb in the that time. In general, the regions in China where dust rain occurred were situated in the rear of the trough of East Asia, where the sinking current prevailed, which was favourable for the dust to settle down.

DISCUSSION ABOUT THE ORIGIN OF DUST

The two cases analyzed above show that the source of the dust was in the northwestern part of the Gansu Corridor and western Nei Monggol, where the arid desert and Gobi, as well as the Taklimakan desert of South Xinjiang are distributed.

In addition, from the climatic background it may be seen that when the dust fall of high frequency occurred from March to April, the depressions or troughs always appeared in the area mentioned above, where upward flow prevailed favourable for the dust and sand to be rolled up into the upper westerlies.

Acknowledgement—The author wishes to thank Prof. Liu Tung-sheng for his kind guidance and encouragement, and also Profs, Wang Xian-zao and Zhang Jia-cheng for their valuable suggestions.

REFERENCES

[1] Wang Jia-yin, 1963, Historical records of Chinese geology, p. 262, Science Press, Beijing. (in Chinese)
[2] Liu Tung-sheng et al., 1982, The dust fall in Beijing, China on April 18, 1980 American Geological Society Special Paper, 189.
[3] Central Meteorological Observatory et al., 1976, Historical records of Japanese meteorology. (in Japanese)
[4] Zhang De-er, 1983, Analysis of the dust rain in the historic times of China. Kexue Tungbao no. 3, p. 361—366.
[5] Zhang De-er, 1980, Winter temperature change during the last 500 years in southern China. Kexue Tongbao no. 6, p. 497—500.
[6] Zhu Ke-chen, 1973, A preliminary study on the climatic change fluctuation during the last 5,000 years in China. Scientia Sinica, no. 2, p. 226—255.
[7] Zhen Si-zhong, 1973, Humidity condition in southeastern China for the last 2,000 years Proceedings of climatic change and ultralong range weather forecasting. p. 29—32, Science Press. Beijing (in Chinese)
[8] Liu Tung-sheng et al., 1964, Loess deposits of China p. 230, p. 71, Science Press, Beijing (in Chinese)

THE CHARACTERISTICS OF SEDIMENTS AND LANDFORM EVOLUTION OF THE LATE CENOZOIC DOWNWORPED BASIN IN YUSHE AND WUXIANG DISTRICTS, SHANSI

CAO JIA–XIN AND WU RUI–JIN

(Department of Geography, Beijing University)

Abstract—This paper deals with the lacustrine sedimentary facies of the Pliocene and the Lower Pleistocene strata in Yushe-Wuxiang Basin, and discusses the features of a shallow downwarped lake basin. The development of Yushe-Wuxiang basin is divided into five stages.

GEOLOGICAL AND GEOMORPHOLOGICAL BACKGROUND

The Late Cenozoic downwarped basin of Yushe-Wuxiang is in the Zhuozhang River drainage area, eastern Shanxi Plateau. Within the basin is a sequence of fiuviolacustrine deposits, ranging from Pliocene to Lower Pleistocene, which are characterized by their colours and sedimentary sequences, and differ a great deal from those of graben-faulted basin in sedimentary features.

The Yushe-Wuxiang Basin is a dendritic saucer-like synclinal basin, trending NNE-SSW. The basement and the margins of the basin are composed of Triassic sandstones and shales. Late Cenozoic fluviolacustrine sediments are spread in the basin. This ancient lacustrine basin has been dissected into undulating hills due to subsequent uplift of the area. As a result, the Late Cenozoic deposits are widely exposed.

The Zhuozhang River flows southward across the basin and then drains southward through the Taihang Mountains to the North China Plain.

Structurally, the area under investigation belongs to the northeastern part of the gentle Qinshui River syncline. This syncline is regarded as the structural basement of the Late Cenozoic downwarped basin.

During Triassic the area was the center of subsidence and received about 1,500 m sediments of terrestrial sandstones and shales. During Late Triassic the area began to be affected by the Yanshan movement and the Taihang and Taiyue Mountains were intensively folded. The Yushe-Wuxiang Basin between these two mountains also rose slowly, but the synclinal structure persisted. Considerable erosion and deplanation took place from Late Triassic to Late Cenozoic. It is during the Himalayan movement that the tectonic movement in the area became active again. The two mountains were uplifted, and Yushe-Wuxiang Basin was comparatively downwarped. The Late Cenozoic downwarped synclinal basin was thus formed and was covered with water. This ended the erosion and deplanation processes, and, the fluviolacustrine sediments of Plio-Early Pleistocene were subsequently deposited in the lake.

THE LATE CENOZOIC STRATIGRAPHY

The distribution of the Late Cenozoic fluviolacustrine deposits along the valleys of present Zhuozhang River and its tributaries is practically controlled by the down-warped basin. The age of these deposits ranges from Early Pliocene to Early Pleistocene. The stratigraphic sequence shows the deposits become younger from east to west. The deposits form a monocline dipping to NW and are overlapped westward. In addition, the lacustrine strata also dip centripetally toward the center of the lake.

The Late Cenozoic fluviolacustrine deposits grouped into the Yushe group which includes Renjianao formation of Lower Pliocene, Zhangcun formation of Middle-Upper Pliocene and Louzeyu formation of the lower part of Lower Pleistocene. The detailed description of the stratigraphy has to be omitted because of the limited space of the present paper.

SEDIMENTARY FEATURE OF THE DOWNWARPED
LACUSTRINE BASIN

Sedimentary Features

1. Thickness

As the basin was slightly downwarped. The lake water was shallow. Only several hundreds metres of deposits accumulated in the basin in such a long geological period. The thickness of Yushe group ranges from 200 to 400 m, which shows a great difference from the sedimentation of graben-faulted basin of Fenhe and Weihe Rivers, where the thickness of the Late Cenozoic fluviolacustrine deposits may reach several thousands metres.

2. Rhythm of sedimentation

Some of the rhythms of sedimentation of Yushe group are imcomplete. Each

108

rhythm begins with the sediments of sand beds gradually to interbeddings of sand and clay, then to grayish green silty-clay beds, and ends up with grayish green clay beds or grayish white marls. Several cycles have been recognized, basically beginning with lakeshore facies and ending with central lake facies. Facies changes of the deposits recorded the expansion and contraction of the ancient lake, though the extent and duration of each facies change are not exactly the same. Moreover, there are remarkable lateral facies variations. Different sedimentary facies zones are distributed in forms of irregular circles around the centre of the lake. This is typically shown by the facies zones of Zhangcun Basin.

3. Color

The colors of the fluviolacustrine deposits of Yushe group are highly variegated. The variation in color depends firstly on the original colors of the Triassic bedrocks, such as violet sand and gravels, yellow sand and violet sandy clay. The sedimentary environments of rivers and shallow-water make the Late Cenozoic deposits maintain easily the original colors of bedrock materials under a good oxidation condition. Secondly, it depends on the lacustrine environment which made the colors of deposits utterly different in various parts of the lake basin. For example, in the center of the lake where the water was relatively deep and quiet, sediments were often under the reducing environment and the chemical sedimentation prevailed. Thus, ferrous compounds, carbonate, sulfide and organic matter were deposited on the bottom of the lake, such as grayish green or grayish white clays and marls. In the bays and inlets along the shoreline where the water was shallow and stagnant, plants and animals flourished, and grayish black and organic-rich silt were accumulated.
To sum up, They were regularly deposited in a certain sedimentary sequence and at relevant deposit sites. It is obvious that the sedimentation of the shallow-water lacustrine basin is charcterized in lithofacies by the variegation of colors.

4. Sedimentary facies

It is worthy to note that the downwarped lacustrine basin exhibits various types of sedimentary facies. In the following sections, a detailed description of these facies will be given.

(1) Shore facies

Because the water of the ancient lake in Yushe and Wuxiang Districts was comparatively shallow, the lake shore was subjected to rather weak wave erosion, consequently the flattened pebbles of shore facies have rarely been found. Along the gentle-slope coast especially in the lake bays, yellow sand and sandy clay are often seen lying directly over the bedrock. But along the steeper coast, avalanche debris are mixed up with mud and sand. The occurrences of shore deposits vary with the shore slopes and generally dip toward the centre of the lake.

(2) Deltaic facies

The deltaic sand bodies of Yushe group has a fan-shaped distribution and a great thickness of more than 20—30 metres. Sand has been finely sorted and sand beds exhibit many types of stratifications, such as large cross-bedding, and oblique bedding. The sand beds become thinner toward the centre of the lake with decreasing grain size and the deposits gradually become fine silt and sandy clay.

Abundant mammalian fossils and tortoise fossils are found in the deltaic sand bodies.

The deltaic deposits of Yushe group are mainly found in lake inlets where the ancient river entered the lake.

(3) Shallow-water facies

Shallow-water deposits are composed of the interbeddings of thick bedded yellow sand and violet sandy clays with intercalated bands of grayish green clay and exhibit locally horizontal small wavy beddings in fine sand beds. An individual bed may reach as thick as several metres.

The shallow-water facies seems to have deposited in an environment, where water level fluctuated frequently. The grain size and color of the deposits changed regularly and periodically.

Abundant animal and plant of fossils are contained in the deposits indicating an environment full of sunshine, rich in nourishment, and favorable to the inhabitation and propagation of organisms.

The shallow-water deposits of Yushe group have a wide distribution, which is an important feature for the shallow-water downwarped lacustrine basin.

(4) Central facies

Deposits of the central facies are characterized by grayish green, grayish blue, grayish yellow, grayish white and white silty clays, marls and calcareous precipitates, sometimes intercalated with thin bands of fine silt.

110

The dull colour of central facies deposits is a distinctive feature of the deposits in standing water and reducing environment.

Stable and large horizontal beddings and laminated clay are well developed in the sediments. The lamina is as thin as paper, alternating between black and white, and each lamina pair being only 0.11 mm in average thickness. The occurrence of the lamination is either horizontal or folded. The folding is due to the gravitational sliding and plicating of the lake silty clay. The appearance of the fold varies greatly. The deposits yield rich fossils of fresh-water fishes, plant leaves and microorganisms.

(5) Swamp facies

Swamp deposits of Yushe group were accumulated in the lake bays, where the water was undisturbed, and essentially made up of grayish black and black silty clay with yellow stains and organic-rich matter, some even developed into oil shale or peat, which has a sapropelic odour and is oily when damped. The deposits, with organic content as high as 11—12 percent, contain numerous plant remains including stems, leaves and trunks and a large number of fossils of hydrophilous insects and their larvae, loaches, small fishes, and giant mammalian bones.

The sedimentary deposits of various facies described above are basically distributed in irregular concentric belts around the centre of the lake, but most of the swamp facies are localized

Fossils

The paleobiota is commonly regarded as one of the most valid indicators for paleoenvironments, not only for paleoclimates but also for ancient landform and paleohydrochemical conditions.

Ever since the beginning of the twentieth century, the area under investigation has been well known both at home and abroad for its abundant "Longgu" —mammalian fossils. Principal mammalian fossils contained in Yushe group are mainly of elephants, rhinoceros, horses, deers, gazelles, pigs and carnivores. The reason for so many animal remains in these deposits is the undoubtedly favorable natural environments for inhibitation and propagation of animals.

The shallow water along the long and tortuous lake shore provided plentiful drinking water, and the lush vegetation provided plentiful foods. The shallow-water lake shore was a suitable place for animals to live. Their remains were embedded in the sediments on the bottom of the lake and fossilized afterwards. We have already noted that there are abundant fish fossils in silty clay of Yushe group, insects and loach fossils in swamp deposits, and turtle of tortoise

111

fossils in shore sand beds. We have also noted that a great number of plant fossils found there are seldom seen elsewhere. These facts indicate that shallow-water lakes and swamps were favorable for the existence of organisms.

It may be concluded that the Late Cenozoic downwarped lacustrine basin in Yushe and Wuxiang Districts has a distinctive shape, striking sedimentary features and unusual fossils, and synclinal downwarped lacustrine basin is of a unique type in the evolution of lake basins in North China.

EVOLUTION OF THE DOWNWARPED YUSHE-WUXIANG BASIN

Developing Period of River Valleys prior to the Formation of Lake Basin

The Himalayan movement put a termination to the peneplanation in Yushe and Wuxiang Districts after a long period of denudation and deplanation. However, the synclinal structure was preserved and developed further through successive downwarping, and a synclinal valley was formed. River developed in this synclinal valley and beds of alluvial gravels of Upper Miocene to Lower Pliocene deposited, namely, the Renjianao formation. Owing to the decline of the Himalayan movement, the lateral erosion of river was intensified, the river valley was widened and a series of tributary valleys were formed, which laid the topographical and structural foundation for the formation of subsequent Yushe-Wuxiang Lake.

Developing Period of Lakes and Swamps

In Early Pliocene, along with the continuation of downwarping in this area, an uplift began to take place to the southeast of Wuxiang, which resulted in sluggish drainage in the synclinal valley, the formation of the lake and the disappearance of river, and thus marked the beginning of the period of lake development. The incipient lake covered only a small area and the water was very shallow, mainly occupying the present Zhuozhang River valley. And in enclosed depressions, bays and inlets swamps developed.

Full Matured Period of the Lake

During Middle to Late Pliocene, affected by the continuing downwarping of the basin and the uplifting to the southeast of Wuxiang, the ancient lake of Yushe-Wuxiang shifted northwestward and expanded. The lake water became deeper and the area covered by the lake became larger. In addition to structural fac-

112

tor, climatic condition was also an important factor for the expansion of the lake.

Declined Period of the Lake

During Early Pleistocene downwarping of Yushe-Wuxiang Basin came to an end. As a result, the lake became shallower and smaller, and was gradually filled up. New streams emerged on the former lake bottom and formed wide valleys. Flowing down onto the dried lake – bottom plain, the rivers slowed down and a large amount of sand was deposited rapidly forming thick-bedded large sand bodies, and these deposits are referred to the Louzeyu Formation in the lower part of the Lower Pleistocene.

Based on the available information, two factors are explained for the extinction of the lake. One is that up to Early Quaternary, downwarping was less active or even came to an end, and uplifting began in the southeastern part of Shansi. Gradually, the lake was filled up and the differential movement between the basin and the uplift ended. The downwarping structure of the basin became less distinct and the lake extincted finally. The second factor is that the Quaternary climate became arid and there was not enough water to maintain the lake.

Period of the Occurrence of Zhuozhang River and the Prevailed Period of Loam Deposition

This area was uplifted once again during middle-late period of Early Pleistocene. The erosional dissection of the fluviolacustrine deposits of Yushe group was intensified and sediments of alluvial gravel beds were formed along the present Zhuozhang River valley, lying unconformably over the Yushe group and forming the IV terrace of the present Zhuozhang River which began to occur and drained across the Taihang Mts. into the North China Plain.

The sediments covering the alluvial gravels on the IV terrace are found to be Red Loam. In fact, Red Loam was simultaneously deposited on hilltops and gentle slopes in the area lying, unconformably over the underlying deposits of Yushe group and Triassic strata. The Red Loam is homogeneous lithologically, commonly nonstratified, with paleosol beddings, and is widely distributed on hilltops.

The sedimentation of Red Loam marked the beginning of prevailed period of loam development. During the following period, namely, from Middle Pleistocene to Upper Pleistocene, the deposition of reddish loam (Lishi Loess) and loess was prevailing.

Subsequent to the sedimentation of Red Loam, another uplifting of the whole area resulted in the incision of valleys, development of numerous gullies and increasing relief of hills. The reddish loam of Middle Pleistocene is found to be spread on undulating hilltops as well as valley-slopes, and is much more widely distributed than the Red Loam. The Malan Loess of Upper Pleistocene was deposited in the same way as the reddish loam, but a sharp erosion surface is found between them. Thenceforth, the features of natural conditions of southeastern Shansi has borne a strong resemblance to those of the present environment there.

The authors would like to thank Mr. Zhou Li-ping and Prof. Zhang Jiang-zhe for translating the Chinese manuscript into English.

REFERENCES

[1] Cao Jia-xin, 1980, A study of Cenozoic strata and sedimentary environment in Taigu-Yushe Wuxiang region, Shanxi. Quaternaria Sinica, v. 5, no. 1
[2] Teilhard P. de Chardin, Young C. C., 1933, The Late Cenozoic formation of S. E. Shansi, Bull. Geol. Soc. of China, v. 12.
[3] Licent E., Trassaert, M., 1935, Pliocene lacustrine series in central Shanxi, Bull. Geol. Soc. of China, v. 14.

A SEDIMENTARY MODEL OF THE CHANG-
JIANG (YANGTZE) DELTA

Xu Shi-yuan,
(Department of Geography, East China Normal University)

Li Ping
(Department of Marine Geology, Tong–Ji University)

and Wang Jing-tai
(Lanzhou Institute of Glaciology and Cryopedology, Academia Sinica)

Abstract—Based upon the hydrodynamics, sediment characteristics, paleomicrofauna assemblage and other data obtained from the numerous cores, the Changjiang Delta can be classified into four facies, i.e., delta plain, delta front, prodelta and deltaic associated facies, which, according to the local environmental difference, may be further divided into at least 13 subfacies. The vertical facies sequence of the Changjiang Delta in decending order corresponds well with successive facies order shown in plan from land to sea.

The Changjiang Delta complex comprises six subdeltas. They regularly range from NW to SE in en echelon pattern in plan and regressively overlap one another in vertical section. Their development is evidently affected by the tidal current direction, the Coriolis force and the southward moving longshore current. The presence of the deltaic associated facies–the offshore radiating tidal sand shoal near Jianggang and the sand bar in the Qiantang Estuary indicates that the Changjiang Delta is unique in sedimentary model with no counterpart found elsewhere in the world.

The Changjiang River is 6,300 km in total length. Its yearly runoff and mean sediment discharge are estimated at $924 \times 10^9 m^3$ and $486 \times 10^6 T$ respectively. The mean tidal range of its estuary is 2.6 m and the development of its deltaic sedimentary system is evidently affected by the tidal action. This paper deals chiefly with the criteria of sedimentary facies of the Changjiang Delta, its vertical sequence and facies model.

Situated in a region of tectonic subsidence, the Changjiang Delta experienced the processes of Holocene eustatic transgression and the later depositional regression caused by the progradation and construction of the delta. At the Holocene maximum transgression, the entrenched valley of the Changjiang River was drowned as a triangular bay widening seaward and with Zhenjiang – Yangzhou as its apex, then, the Holocene Changjiang Delta developed, emerged and gradually advanced southeastward stage by stage. Now the whole delta system comprises six subdeltas of different stages and covers an area of about 51,800 km².

MAIN CHARACTERISTICS OF THE DELTAIC SEDIMENTARY FACIES

In determining the sedimentary process and the pattern of distribution of the sedimentary facies, the runoff plays a dominant role in its interaction with the tidal current. According to such criteria as hydrodynamic and sediment characteristics as well as that of paleomicrofauna assemblage, the sedimentary system of each subdelta can be divided into 4 facies, i.e. the delta plain, the delta front, the prodelta and the deltaic associated facies, which, according to the local environmental difference, may be further subdivided into at least 13 subfacies (Fig. 1).

The main characteristics of each facies and subfacies are listed in Table 1.

Table 1　The main characteristics of the Changjiang Delta facies criteria

Facies	Subfacies	Mean grain size (Φ)	Standard deviation (δ_1)	Sedimentary texture	Fauna assemblage
Delta Plain	Lake-swamp	<9		Horizontal bedding, lenticular bedding	Plant rootlets and fragments relatively rich
	Valley flat	5—6		Horizontal bedding	
	Natural levee	4—6		Ripple cross-bedding	Predominantly terrestrial, but locally a small amount of shells of marine microfauna
	River bed	2—3	0.4—1.1	Wavy bedding, cross-bedding	
Delta Front	Mouth sand bar	2—4	0.5—0.7	Cross-beding, wavy bedding gas-swollen texture	Both terrestrial and marine, marine ostracoda with minor terrestial species, including echinoderm fragments and relatively abundant plant debris
	Distributary bed	4—6	0.9—2.7	Wavy bedding, cross-bedding scour-filling texture	
	Flank side beach	5—7	2.0	Horizontal bedding, wavy bedding, involute bedding	
	Interdistributary shallow water	4—8	1.5—2.4	Horizontal bedding cross-bedding	
	Delta front slope	4—7	1.2—3.0	Wavy bedding cross-bedding	
	Prodelta	6—8	1.8—2.9	Horizontal bedding	Gel-shelled foraminifera, *Globigerina*, etc.
Deltaic Associated	Radiating sand shoal	3—5	0.8—2.0	Cross-bedding, horizontal bedding	Foraminifera, marine ostracoda increased
	Mouth sand bar	3—6	1.0	Cross-bedding. horizontal bedding	Similar to that of delta front
	Coastal plain	7—8	2—8	Low angle crossbedding	Foraminifera, marine ostracoda relatively few

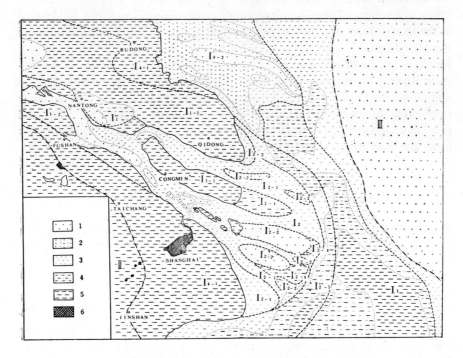

Fig. 1. Map of the Changjiang Delta sedimentary facies
       1. Fine sand; 2. sandy silt; 3. silt; 4. clayey silt; 5. silty clay;
       6. bed rock.
     I—Delta sedimentary system
        I_1 Delta plain facies
           I_{1-1} Marine–alluvial plain
           I_{1-2} Alluvial plain
        I_2 Delta front facies
           I_{2-1} Distributary bed subfacies
           I_{2-2} Mouth sand bar subfacies
           I_{2-3} Mouth flank side beach subfacies
           I_{2-4} Interdistributary shallow water subfacies
           I_{2-5} Delta front slope subfacies
        I_3 Prodelta facies
        I_4 Associated facies
           I_{4-1} Littoral plain subfacies
           I_{4-2} Radiated sand bar subfacies
     II—Continental sedimentary system
     III—Marine sedimentary system

From land to sea as the environment changes, the sedimentary facies changes gradually instead of abruptly from delta plain to delta front and then to prodelta. This is reflected in the following facts: the sediments become finer grained and worse sorted, the organic content higher and the shade darker, while the sedimentary structure changes from cross-bedding to horizontal bedding and the

117

marine fauna becomes more abundant in species with the plant debris diminishing in amount. These facts indicate that the Changjiang Delta is identical
in the lateral facies distribution to other deltas of such sea-debouching river as
the Mississippi.

Materials of the deltaic associated facies were mainly supplied by the Changjiang River. Their transportation and deposition and the resultant landforms
and fabrics are chiefly controlled by sea currents. However, the tidal action is
chiefly responsible for the formation of the Jianggang offshore rediating sand
shoal and the Qiantang Estuary sand bar, a unique sedimentary model for their
huge size and peculiar morphology.

VERTICAL SEQUENCE OF DELTAIC DEPOSITS

The detailed analysis of the sedimentary structure, texture, fauna assemblage
and the contact relationship according to the data from cores shows that the
Holocene Changjiang Estuary experienced a whole cycle of transgression-regres-
sion and developed a suite of littoral-shallow marine deposits. Its lower half
part represents the channel-filling transgressive sequence and the upper part
the deltaic progradation sequence. The vertical sequence of delta shows distinct 3–layered structure, duplicating the sedimentary facies pattern from
land to sea in plan, i.e., from delta plain to delta front and then to prodelta.

Though gradual, the vertical variation of the Changjiang Delta is by no
means ambiguous. From surface downward, the grain size of the sediment is
fine-coarse-fine, the sorting is poor-moderate-poor, the shade, dark-light-dark,
the sedimentary structure, horizontal bedding — cross-bedding—horizontal
bedding, the terrestrial fauna diminishing and marine fauna increasing. These
features are compatible with what the Mississippi's[3], Niger's[4] and other sea-
debouching river delta show in their vertical sequence. Thus, the 3–layered
structure seems to be the general character of the delta structure and may serve
as the main criterion for the recognition of both deltaic systems and the distribution of their facies.

The associated coastal plain of the Changjiang Delta underwent the same history as the delta proper, but is different from the latter in vertical sequence.
Its lower part is a barrier — lagoon system, the transgressive littoral sand and
the upper part is either a muddy system (its mud content 30%) produced by the
stage-by-stage seaward advance of the muddy coastal plain south of the Changjiang River, or a sandy system, the regressive radial sand bars, capped by the
tidal flat deposits north of the river.

EVOLUTION OF THE DELTAIC SEDIMENTARY MODEL

The above analysis of the sedimentary facies, combined with the abundant his torical, archaeological and drill-hole data, indicates that in general, each subdelta system underwent initial, mature and old evolutional stages. In the initial stage, the mouth shoal and bar were formed and the main channel bifurcated into a northern and a southern distributary. This is because when the river entered its estuary area and got into touch with the sea water, its gradient decreased, its flow broadened and mixed with the sea water. This caused a reduction in flow velocity and a rapid deposition of the river-borne sediments which contributed to the formation of the mouth bar and the bifurcation of the channel. Meanwhile, a part of sediments continued moving eastward and deposited to form the delta front slope subfacies, prodelta facies as well as the associated subfacies, such as sand bars, lagoons and swamps, etc., in the coastal area bordering the estuary.

Long in length and huge in size, the Changjiang River mouth bars show certain features which are characteristic of tidal sand bodies. At the early stage of their development, the water was relatively deep. The deposits were accumulated in high water and scoured in low water. They were poorly sorted and fine grained. With the growth of the mouth shoal and bar, however, the water depth decreased and the wave action increased. The result is that the deposits became better sorted and coarser grained, and that the content of fine sand is higher (up to 80—90%).

The emergence of the mouth sand bar marked the beginning of the mature stage. In this stage, the upper reach of the estuary narrowed, the southern distributary further grew at the expense of the gradual decaying of the northern one and the deltaic system developed into its full swing. The mouth bar then began to receive finer grained flood materials and gradually emerged as sand island. The whole deltaic system moved seaward, producing in succession the delta plain, delta front and prodelta facies. This agrees well with what the facies shows in the vertical sequence from surface downward.

At the later mature stage, new bars were developed in the mouths of both the northern and the southern distributaries. The further growth of the southern one made it as the major discharging channel of the Changjiang River and its mouth bar grew rapidly to become the main part of the next stage's subdeltaic system.

The old stage of the delta development is characterized by the silting up of the northern distributary, the attachment of its mouth bar to the northern bank, the extension of the river channel, the shifting of the river mouth southward and the initiation of a new subdeltaic system. The Changjiang Estuary is evidently affected by the tidal current direction, the Coriolis force and the southward-moving longshore current. The northern distributary played a much less important role than the southern one in discharging water and was mainly the flood channel; the sediments carried into the channel could not be entirely evacuated because of the fact that the flood speed is generally larger than that of ebb tide. This led to the silting up of the northern distributary and the ultimate attachment of its mouth bar to the northern bank of the river, giving rise to

119

an extensive marine-alluvial coastal plain, the eastward moving of the shore-line and the narrowing and the final shifting southeastward of the estuary. This marked the ending of an old subdelta development. Now, the sand bar at the mouth of the southern distributary had grown to such an extent that it became a main body of a new subdeltaic system.

With the declining of the old subdelta, it passed on from the aggradational period to that of degradation, i.e., it began to be reworked and destructed

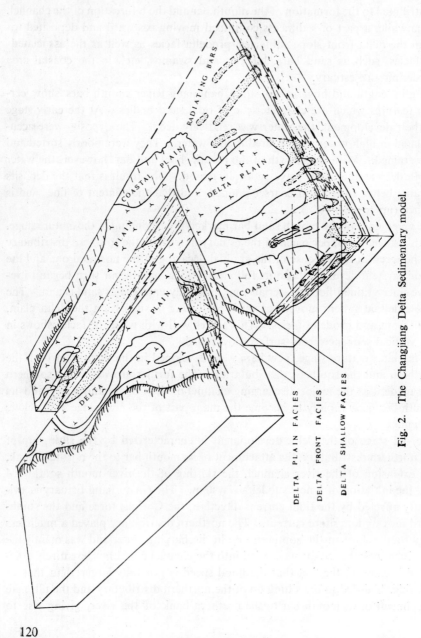

Fig. 2. The Changjiang Delta Sedimentary model.

120

by the marine agents. At the mouth of the abandoned northern distributary, a bay was formed and brought about the favourable condition for the convergence of tidal action and hence for the formation of a series of radiating ridges. With the mouth bar as their developmental center, all the six Holocene sub-deltas underwent the above mentioned evolutional process (Fig. 2). The northern distributary mouth bar was smaller in size and relatively short-lived in age. And as a result of the gradual silting up of the distributary itself, the mouth bar ultimately attached to the northern bank of the river. In contrast, the southern one was the major discharging channel, its mouth bar grew larger increasingly, forcing the distributary to re-bifurcate into a northern and a southern branches. In this way, the Changjiang Delta as a whole moved southeastward stage by stage. The mouth bars of the subdelta of various stages range from NW to SE in en echelon pattern in plan and regressively overlapped one another orderly not only in space but also in time. The cyclicity of the deltaic development is quite clear. So, the Changjiang River is different not only from the Niger[4] and the Mekong[5] which have only one main delta lobe, but also from the Mississippi[3] and the Huanghe (Yellow) River[2] whose subdeltas crisscrossed one another at random. This proves that the Changjiang Delta is a peculiar type in the world's deltaic sedimentary models.

REFERENCES

[1] Delta Research Group, Department of Marine Geology, Tong-Ji University, 1978, The formation and development of the Holocene Yangtze Delta. Kexue Tongbao vol. 23, no. 5.

[2] Quaternary Geology Research Group, Guiyang Institute of Geochemistry Academia Sinica, 1972, The ancient river channels and their fills in Yellow River estuarine region since 1855. Geochemica, no. 2.

[3] Gould, H. R., 1970, The Mississippi delta complex "Deltaic sedimentation: modern and ancient" pp 3—30.

[4] Allen, J. R. L., 1964, Sedimentation in the modern delta of the Niger river of West Africa "Deltaic and shallow marine deposits" pp. 26—34.

[5] Kolf, C. R., Dorubusch, W. K., 1975. The Mississippi and Mekong deltas—A comparision "Delta" pp. 193—206.

[6] Frazier, D. E., 1967, Recent deltaic deposits of the Mississippi river: Their development and chronology transaction of Gulf Coast Asso. of Geol. Soc. vol. 18.

FEATURES AND GENESES OF THE DIAMICTONS IN THE LUSHAN REGION

ZHANG LIN–YUAN AND MOU YUN–ZHI

(Department of Geology and Geography, Lanzhou University)

Abstract—On the basis of a series of characteristics of the diamictons of the Lushan, we believe that no glacial genesis exists; however, they can not be explained by any single non-glacial genesis. The diamictons of different ages, in different geomorphological positions and with different features were formed by different mechanisms. The Dagu diamicton (corresponding to Mindel glaciation) which in the past was said to be typical till, is really a piedmont fan deposit. Its main features can be explained by the mechanism of mud-rock flow accumulation.

In the 1930s Li Si-guang (J. S. Lee) affirmed that part of the Quaternary deposits and geomorphological phenomena in Lushan were due to the genesis of glaciation. Since then, the glacial problem of the Quaternary in the eastern part of China, especially in the mountains at low latitudes and low altitudes, have been debated among domestic and foreign researchers. The points in debate are those of glaciation and non-glaciation[1-9].

A GENERAL DESCRIPTION OF THE GEOGRAPHY AND GEOLOGY

The Lushan Mountain are located in the northern part of Jiangxi Province, China, within the subtropical zone. Climatic features of monsoon type are obvious.

According to the lithology of the formations, the mountain body can be divided into two parts.

The northern part is composed of the Sinian beds, mainly hard quartz sandstone, and partly soft muddy sandstone and sandy shale. The southern part belongs to the Proterozoic, mainly slate, schist, gneiss, etc, which is weak in its resistance to weathering. In the drainage area where quartz sandstone crops out huge boulders are spread both on the slopes and at the mountain foot. But, the source of the basins is limited merely in an area exposed of the Proterozoic, where no huge boulder is seen. This is a notable feature of depositional differentiation in this region.

122

FEATURES OF THE DIAMICTON

The Macroscopic Feature

The characteristics of the diamicton are mixture materials of large and small grains, in poor direction of arrangement, and low psephicity. The mean diameters of the largest boulder and the gravels in the diamictons of the same period show a trend of decrease from the top to the foot.

The original geomorphological shapes of the diamictons in the piedmont region were destroyed seriously in the early period, which is unable to reconstruct. Those of the later period, obviously like a fan after reconstruction, for example, those along Gaolong and in the Yangjiaoling region.

The Features of the Fine-Particle Fraction

1. Grain size characteristics

From the particle component of the fine fraction ($3-12\phi$) in the diamictons, the regular relation between the content percent of particle component and the period and transporting distance of the deposits is not clear. This is obviously different from the fact that the coarse fragments gradually become finer from the mountain outward to the piedmont.

This phenomenon can only be explained reasonably by the transportation of mud-rock flow involves selective sorting of the fine and coarse grains. The characteristics of the mud-rock flow are that selective sorting occurs in the population of large fragments, but does not in the fine-particle fraction. (10) The peak with the value of $4-5\phi$ is the result of mechanical abrasion, which appears in all the till. But the samples show valleys instead of peaks when the peak value is $4-5\phi$. All this shows that prior to the accumulation they did not undergo any glaciation abrasion.

The above-mentioned feature of grain size is similar to that of the mudrock flow deposits in non-glaciated regions.

2. Features of mineral association

A feature of the diamicton is that the mineral contents of orthoclase and hornblende are minor or even absent, which is similar to that of plinthitic red soil on the mountain. The degree of warm, humid weathering is rather high, which shows that the diamictons in the piedmont region came from the weathered mantle of red soil on the mountain.

123

The unstable minerals continuously went through warm, humid weathering after the weathered red soil had been transported to the piedmont. A considerable amount of them were transformed into clay minerals.

Chemical Features of the Clay Mineral and Clay Size Particles

1. The results of X–ray diffraction analysis

The X–ray diffraction curve of the diamicton is fundamentally similar to that of the plinthitic red soil. The older the deposits are, the higher degree of warm, humid weathering reflected by their composition of clay mineral is. In general, they are mainly composed of kaolinite, but illite is found in the later period. Though the degree of warm, humid weathering is different, both of them, old and young, have obviously to the been affected by it. The composition of the Middle Pleistocene clay mineral on the mountain is similar to that at the foot. For instance, the diamictons at the foot are composed of "illite-kaolinite" while the plinthite on the mountain is composed of "illite-hydrokaolinite", which also shows that the diamictons at the foot came from the weathered mantle of red soil on the mountain.

2. Microscopic features of undisturbed soil samples

The content of clay mineral is quite high, which is mixed with ferriferous materials. In the clastic mixture, the content of quartz is very high, and mainly angular to subangular in shape. The earlier the age of the deposit is, the more obvious the hydromication phenomenon of muscovite presents. The distribution of clastic particles is in disorder and uneven. In the diamictons of different periods, the feldspar content is very low, but the quartz content is very high. The muscovite has partly or totally become hydromica, but still keeps up its original granular shape. All these show that the diamictons came from the materials which had borne through warm, humid weathering. They kept on subjecting weathering in a humid-thermal climate after accumulation.

3. Chemical features of clay particles

So far as the ratios of SiO_2/Al_2O_3 and SiO_2/R_2O_3 are concerned, the ratios of the Poyang Period (N_2 or Q_1) deposits are a little higher. The ratios of the Lushan Period (Q_3) deposits are the highest (average 2.45 and 1.98). The ratios of the Dagu Period (Q_2) deposit are lower (average 2.04 and 1.54). The

124

ratios of SiO_2/Al_2O_3 and SiO_2/R_2O_3 of clay particles in the soil of China are: Yellow soil, 2.3—2.7 and 2.0—2.2; red soil, 1.80—2.20 and 1.70—1.90; laterite, 1.7 and 1.6[12,13]. The deposits of the Dagu Period are 1.87—2.19 and 1 26—1.80 respectively, fundamentally similar to the values for the red soil. The red diamicton and the weathered crust of plinthitic red soil on the mountain are not in coordination with the present climatic conditions of the mountain. The variation sequence, from "Middle Pleistocene red deposit and vermicular red soil—Late Pleistocene yellow deposit of the Lushan Period—modern mountain brown soil on the Lushan Mountain", reflects the descent of vertical natural zones in successive stages on the Lushan Mountain. It is in conformity with the general trend of climatic changes since the Middle Pleistocene Inter-glacial Period in which the humid-thermal degree is the highest. Since that period, the climate in the monsoon region has become dry and cold in China[8]. To be sure, the periodic uplift of the Lushan Mountain by stages is also a factor which has brought about the descent of the abovementioned natural zone.

THE HUMID-THERMAL PERIOD AND AN ANALYSIS OF THE GENESES OF THE DIAMICTON

The phenomenon that the diamicton is allitizedly weathered is very obvious in this region. However, determination of the periods of humid–thermal process, is a key problem to understand the sedimentary environments and geneses of the deposits. Those who hold the theory of glaciation in the Lushan region have already pointed out the fact of humid-thermal weathering after the deposition. They have determined three ice ages in the Lushan region according to the different degrees of humid-thermal weathering[1]. From the above facts, we have proved: The composition of the diamicton of the Dagu Period at the foot came from the weathered crust of plinthitic red soil, that is to say, it had gone through strong humid-thermal weathering before deposition, which means that the humid-thermal weathering in the early period is also a fact in this region. When we made a chemical analysis of the profile of red soil with as many as eight iron pans (with a thickness of 31 m) near Yejialong in Xinzi County, we discovered that the humid-thermal weathering decreases by degrees from bottom to top (Table 1).

It is evident that this suite of deposits did not go through only one stage of warm, humid weathering after deposition. They went through warm, humid weathering by stages in the process of aggradation. The molecular ratios of clay particles and the content of Al_2O_3 are all in conformity with the standards of laterite and laterized red soil. It appears that the period of deep weathing was the hottest period in Lushan in the Quaternary.

According to the geneses, the diamictons of different ages and different colours can be divided into three types.

Table 1 Chemical analysis of the clay particle of the red earth profile near Yeiialong, Xingzi Country

Sample	Lithological character-istics	Weight of chemical composition (%)								Molecular ratio of clay particles	
		SiO_2	Al_2O_3	Fe_2O_3	CaO	MgO	TiO_2	MnO	P_2O_5	SiO_2/Al_2O_3	SiO_2/R_2O_3
Yj1-14	Plinthitic red clay between 7th and 8th iron pans	39.10	28.68	10.65	0.00	0.81	0.53	0.05	0.45	2.32	1.87
Yj1-13	7th iron pan, brittle iron colloform film, no iron-concretion	40.80	29.40	13.65	0.00	0.99	0.53	0.08	0.67	2.36	1.82
Yj1-8	Plinthitic red clay between 4th and 5th iron pans	39.10	29.41	15.22	0.00	0.91	0.51	0.03	0.40	2.26	1.70
Yj1-2	Plinthitic red clay between 1st and 2nd iron pans	37.85	31.61	15.07	0.07	0.40	0.59	0.03	0.36	2.04	1.56
Yj1-1	1st iron pan with iron-concretions of a diameter of about 1cm	36.29	36.91	16.61	0.30	0.31	0.59	0.00	0.40	1.67	1.30

1. **Creeping type of the modern slope diposits laid down by creep** This type of deposit is in accordance with the modern slope. The psephicity of gravels is the poorest. Their flat surface and the plinthitic declination are in conformity with the slope.

2. **Discordant creep deposits** The typical ones are the deposits of the Dagu Period around Baishizui and Gutang. It is difficult to explain the features of their lithological characteristics and their fabrics of gravel orientation according to the conditions of modern relief and surface run-off. They were formed by the nearby deposits which crept slowly along the slope cut by streams in the late period.

3. **Mud-rock flow type deposits** The typical ones are the deposits around Caifengling inside the Wangjiaopo Valley, at Gaolung outside the Valley and at Yangjiaoling in the northwestern piedmont of the Lushan Mountain. Owing to the fact that their macroscopic features are similar to those of till, it is easy for us to mistake them for glacial genesis. Taking the profile of Yangjiaoling, which is mistaken for the typical till profile of the Dagu Period, for example, we have analysed the geneses of the diamicton of this type.

The gravels have none of the features of till stones, and their fabric features

126

are obviously different from those of till. They are in conformity with those of the modern viscous mud-rock flow. The original accumulation relief here was cut by the erosion of flowing water in the late period and became hilly. Nevertheless, to link-up iso-height sites on the divergent ridges shows the existence of concentric palaeo-contour. The blind landform which remains on the alluvial fan also shows that the original accumulation relief was a mud-rock flow fan[14].

The conditions, namely, steep relief in the surrounding denudation area, abundant red clastics and concentrative precipitation in the monsoon region in summer, were all very favourable to the occurrence of mud-rock flow of a storm-type.

To sum up, we come to the conclusion that the geneses of the diamictons in Lushan region are neither due to glaciation, nor due to a single genesis of non-glaciation. The diamictons in different geomorphological positions and with different features were formed by different mechanisms. The Dagu diamicton, which was said to be typical till in the past, is actually a piedmont fan deposit. Its main features can be explained by the mechanism of mud-rock flow accumulation.

Here, we wish to express our gratitude to Zou Liang-cheng (Department of Chemistry, Lanzhou University) for his chemical analysis, to Xing Ze-min (Lanzhou Institute of Glaciology and Cryopedology, Academia Sinica) for his X-ray diffraction analysis, to Tang Yong-yi for her microscopic determination, and to Ma Zheng-hai for his analysis of light and heavy minerals. Those who participated the field and indoor work with us are Zhu Jun-jie, Chen Huai-lu, Yao Tan-dong and Zhou Shang-zhe.

REFERNCES

[1] Lee, J. S., 1933, Quaternary glaciation in the Yangtze Valley, Bull. Geol. Soc. China, vol. 13, no. 1, p. 15—62.

[2] Lee, J. S., 1937, Quaternary glaciation in the Lushan Area, Central China; Monograph of the Geological Survey, Academia Sinica, vol. 2 (Published by the Survey in 1947).

[3] Barbour, G. B., 1934, Analysis of Lushan glaciation problem, Bull. Geol. Soc. China, vol. 13, p. 647.

[4] Sun Tien-ching, Yang Huai-jen, 1961, The Great Ice Age glaciation in China, Acta Geol. Sinica, vol. 41, no. 3—4, p. 233—244.

[5] Huang Pei-hua, 1963, Problem of glaciation traces on the south of Changjiang (Yangtse) River, Kexue Tongbao. (in Chinese)

[6] Zhou Ting-ru, 1979, Progress in the research work of palaeogeography in China during last thirty years. Acta Geographica Sinica, vol. 34, no. 4, p. 291—303.

[7] Shi Ya-feng, 1981, Is there really Quaternary glaciation on Lushan, Journal of Dialectics of Nature, vol. 3, no. 2, p. 41—45. (in Chinese)

[8] Chang Lin-yuan, 1981, The influence of the uplift of Qinghai-Xizang Plateau on the Quaternary environmental evolution in China, Journal of Lanzhou University, no. 3, p. 142—155.

[9] Xie You-yu, Wu Shu-an, 1981, Preliminary discussion on Quaternary sedimentary environment of Jiujiang-Lushan Area. Collected Papers of Institute of Geography,

Academia Sinica, vo. 13, p. 106—132.

[10] Bull, W. B., 1963, Alluvial fan deposits in western Fresno Country, California, Jour. Geol., vol. 71.

[11] Dreimanis, A., Vagners, U., 1969, The dependence of the composition of till upon the rule of bimodal distribution, Etudes Sur le Quaternaire Dans le Monde, vol. 2.

[12] Hwang, S. T. et al., 1957, Characteristics of the soils of the Lushan Area, Central China, Acta Pedologica Sinica, vol. 5, no. 2, p. 117—135.

[13] Li Qing-kui, Chang Xiao-nian, 1957, The chemical characteristics of red soil of China. Acta Pedologica Sinica, 5 (1), p. 78—96.

[14] Chang Lin-yuan, Mu Yun-zhi, 1981, Genesis of the epigenetic structure at Yangjiaoling ridge of Lushan, Kexue Tongbao, no. 16, p. 1006-8. (in Chinese)

128

PRELIMINARY STUDIES ON THE "LITHOLAC" SOAKING ON THE GOBI ROCK SURFACE IN GOBI DESERT

ZHU XIAN–MO, WANG JING–QIN, AND CHEN JUN–QING
(Northwestern Institute of Soil and Water Conservation, Academic Sinica)

ABSTRACT

We have made some investigations on the "Litholac" soaking on the Gobi rock surface in Gobi desert, Xinjiang by electron microscopy, X–ray analysis, microbiological analysis and chemical qualitative analysis. In addition to light weathered rock exposure with rare "Litholac" substances and partly stained iron rust at the bottom of the gravel, the rust rock surfaces have been wholly covered with swarthy plastic-like films (black and lightened), the thickness of which can reach 0.5 mm or so. The results obtained are as follows in brief: The "Litholac" substances formed by the rock-bearing organisms (including bacteria, alga and so on) are very alike between "Litholac" soaking in Gobi desert and the typical "Litholac" during the process of primary soil formation, and there are similar microfloras, in which there are not only a large number of iron bacteria, but also some kinds of nitrogen-fixing alga. It is thus clear that "Litholac" substances of Gobi desert is also an obvious sign of organisms growing on the rocky surfaces and in the initial stage of the process of soil formation. In fact, it is just only a specific form and colour under the arid and desert conditions.

In "Litholac" substances of Gobi desert, apart from well crystalline iron oxide, there have occurred some clay minerals, such as kaolinite, montmorillonite, illite in biological origin. As generally considered that the loess, desert and Gobi desert in China are homologous. Then the clay minerals in loess, particularly iron oxide, kaolinite, montmorillonite, etc. which cannot be formed at the same time under such environments of the arid desert in the traditional process of soil formation, could have arisen from the process of primary soil formation.

THE COMPARISON OF THE PSEUDO-TILL OF LUSHAN MT. WITH THE TILL OF TIANSHAN MTS. OF CHINA

XIE YOU–YU

(Institute of Geography, Academia Sinica)

AND CUI ZHI–JIU

(Department of Geography, Beijing University)

Abstract—By comparison of sedimentary features (granularity, mineral composition, chemical composition, quartz sand surface textures, shapes and fabrics of gravels and sedimentary structural features) of the pseudo-till of Lushan Mt. with the till of Tianshan Mts., we found that the main genesis of "boulder clay" (pseudo-till) of Lushan Area is mud flow instead of glaciation.

INTRODUCTION

The origin of diamicton which is widely distributed in Lushan Mt. and certain regions of eastern China is still an important topic which has been paid great attention to by the scientists of geomorphology and Quaternary geology of China.

In order to recognize correctly the genesis of diamicton, we must establish indicators to identify various kinds of diamicton step by step. We have obtained some results on the study of the sedimentary features and the identification of present till, mud flow and proluvium, in recent years. In this paper a preliminary relevant introduction is given.

CONSTITUTION OF GRANULARITY

From the angle of dynamics, we have made some progress in studying sedimentary process of sediments of different genesis. Between till and sediments transported by water flow, features of granularity show obvious difference. Prof. Li Si-guang (J. S. Lee) named the Quaternary sediments widely spread in Lushan Mt. region as "boulder clay". We observed diamicton mixed up of large blocks and clay from Shang Qingshan and Xinghua sections and else where[1]. Li Si-guang identified this kind of diamicton as till. But we have never found such glacial "boulder clay" in the Urumqi River source region at Tian Shan Mts. and other mountain areas in western China. There are some

130

difference between the granularities of these two regions. Based on the results of granularity analysis, we found that in fine grained composition, sandy clay and clay make up 78%, and the so called "boulder clay" is composed of clay or sandy clay bearing large blocks and blocks. For example, the mean diameter of "boulder clay" in Yangjiaoling is 0.049.mm. Huge silicarenite blocks are mixed up in fine grain component. All these belong to suspension sum of two-passages type and turbidity can be observed in C–M patterns of sediments.

The granularity components of tills in Tianshan Mts.; Southeast Xizang, and Karakorum Mts. are composed mainly of coarse grains, among which grains larger than 20 mm. are predominant. Among the matrices, the sand content are around 40 to 80%. The general features of the till granularity are reflected by the cumulative curve which is gentle and diffuse, and the frequency curve which has double-peaks or obscure multi-peaks in the shape of saddle. Normally the first peak of coarse granularity predominates. The sorting coefficient is within 2.0 to 3.5. So the matrix of fine grain among the till is mainly sand and silt, instead of clay. In this viewpoint of classification it is quite a great divergence to define the "boulder clay" as till. The glaciation mainly involves mechanical abrasion, and the peak values of sedimentary frequency appeared always within $4-5\phi$[2]. But the component of granularity of "boulder clay" in Lushan region turned out contrary to above-mentioned, which is within the range of $4-5\phi$, it is not a "peak" but a "valley". The distribution of such grain grade coincides with the granularity frequency curve of mud-flow in Wudu region, Gansu. Moreover, the sorting coefficient of "boulder clay" in Lushan region (S_0) is high, as shown in section of Yangjiaoling, Gutang, Baipeizhui the maximum reaches 29, and the minimum 12, similar to the results of mud-flow. But the results differ greatly from sorting coefficient of till (generally, 2.00 to 4.00 or 9.49)*. W. B. Bull (1963) pointed out that mud flow was a fluid which produced sorting only to coarse grained materials and not to fine grained ones[3]. The "boulder clay" of Lushan has precisely such feature. For example, the material content over 5 mm in grain diameter would increase with distance, while the median grain diameter would decrease wavily. But there is no decreasing trend of the median grain diameter in grains smaller than 0.1 mm.

MINERAL COMPOSITION

Since the real till was caused by cold weather during the deposition, and the chemical weathering was weak, the heavy minerals maintain all the assemblage features of bed-rock region. But no such regularity in Lushan Mt. region is found where a lot of sub-weathering aerugo minerals exist. Among the netted

---

* By Lanzhou Institute of Glaciology and Cryopedology, Academia Sinica.

red soil. the content of such minerals reaches 50—60%. The content of stable mineral is even higher reaching over 70%. The content of unstable mineral is lower. Feldspar of light mineral has been weathered completely, only a small amount of quartz grain is left. The till in Urumqi River source glacier at Tianshan Mts. only contains 5% stable minerals, and 40—50%* unstable minerals. This difference can be used as one of the indicators to identify till and pseudo-till.

Clay mineral composition is useful index reflecting climatic conditions during sedimentation. Through many identifications, we found that the clay minerals of Quaternary sediments in Lushan Mt. area are nearly all composite minerals, composed of illite-montmorillonite, kaolinite, chlorite and edellite. Among the till in Urumqi River source area at Tianshan Mts., the clay minerals are composed of illite and chlorite with a little kaolinite or montmorillonite. Clay mineral in Lushan is composed mainly of kaolinite prior to Late Pleistocene. It shows that this clay mineral was formed in humid and hot climatic condition, which is quite different from typical till containing only a large quantity of illite and montmorillonite. These examples show that the climatic condition under which the Quaternary sediments were formed in Lushan Mt. completely differed from the climatic condition under which the till is formed in recent glacier.

CHEMICAL COMPOSITION

We have analyzed the clay in the grade smaller than 0.02 mm. by bulk chemical analysis.

The remarkable feature of chemical composition of the Quaternary sediments in Lushan Mt. area is the very high contents of SiO_2 and Al_2O_3 which almost make up 80%, SiO_2 content varies from 40.05—78.92%, a few loess contents vary from 12.72—16.57%. Al_2O_3 content varies from 12.7 to 45.27%, and only a few make up 8.91%. Fe_2O_3 content is also very high, and varies from 2.22 to 14.77%, mostly are about 10%. The FeO content is about 0.22 to 1.38%. In addition, CaO content is very low which is about 0.07% to 0.65%. Ratio of SiO_2 and Al_2O_3 is about 2, which reflects the features of red soil crust of weathering. It is typical products of humid and hot climate. The ratios of SiO_2/Al_2O_3 of till in Urumqi River source Mts. and outwash sediments at Tianshan are very high about 4—6, showing the feature of desert soil type in cold environment. This is quite different from weathered red soil crust in Lushan the ratio of which is 2.

We have also found that calcium carbonate content in Lushan is very low; normally lower than 1%, some of them have no $CaCO_3$ content at all, because precipitation exceeded evaporation capacity, sediments were washed and some soluble weathered products were also eluviated under humid and hot climatic conditions in which $CaCO_3$ was easy to be washed because of unstability.

132

Above-mentioned facts shows that sediments were formed in humid and hot climatic condition.

The chemical compsitions of sediments in the above-mentioned two regions reflect two different environments. For a long time the scientists who insist Lushan Area was undergone glaciation have always used "the humid and hot process in the later stage to explain the existence of netted red soil. They think that purplish red and red soils in Lashan Mt. area were caused by later humid and hot environment after the deposition of the till. If this was true, the same section of deposits would show later humid and hot process stronger in the upper parts and gradully weaker toward the lower part. Meanwhile the extent influenced in the section should be limited (several or dozens of meters). But in fact, red soils are found 60—70 or even over 100 m in depth in cores at the piedmont of Lushan Mt. and along the Changjiang River Valley, which cannot be explained as later humid and hot process. Moreover, in the so called "fluvioglacial deposits of Dagu Period", which is totally 50 meters thick in Xingzi County and Yiejialing section. From 3 to 31 m, in the section the ratio of $SiO_2 : Al_2O_3$ and ratio of $SiO_2; Al_2O_3; Fe_2O_3$ have been reduced from 2.36 and 1.97 to 1.67 to 1.30 respectively according to Zhou Shang-zhe's analysis showing that changes of chemical element are not caused by later humid and hot climate but by climatic changes from warm to cool during the deposition.

THE SHAPE OF QUARTZ SAND SURFACE AND MICROTEXTURE OF SEDIMENTS

The quartz sand and fluvioglacial quartz sand in so called terminal moraine of Shangqingshan long in Lushan Mt. area and in the section of Yangjiaoling have been restudied by scanning electron microscope and strong chemical dissolution phenomena, on the quartz surface, such as dissoluted caves, furrows honeycomb and scale-off, etc are observed. And the SiO_2 developed into short thick deposits in lump shape after strong dissolution[3]. Above-mentioned phonomena of strong dissolution have not been observed in Tianshan Mts. till. On the contrary, quartz surface in the till shows irregular smooth surface, but some quartz surface with shell-like fractures, deep grooves, mechanically formed indentations upturned flakes and even a few with thin and long quartz crystals, All these are products caused by cold and alkalized environment[4]. Naturally, above-mentioned difference can still be explained as humid and hot climate of later stage, but we have never found any difference of dissolution from the upper part to the lower part of the Yangjiaoling sections in Lushan Mt. by scanning electron microscope. In this point of view, we can conclude that these deposits were not resulted in humid and hot climate of later stage.

Using scanning electron microscope to study microtextures of sediment is one of the methods to identify the genetic types. The microtextures features

133

of matrices of recent mud flow sediments (such as in Wudu, Gansu, Xishan Mt. Beijing etc.) show schistosity texture on plane, also reflecting the schistosity was developed in fine grains composed mainly of clay during the deposition. A lot of deep and irregular cavities formed by rapid deposition during transportation are also found in the sections. All these features can be seen in the Yangjiaoling section of Lushan Mt. which convincingly proves a genesis of mud flow.

THE SHAPE AND FABRIC OF GRAVEL

Some statistics on psephicity of gravel from 2 to 20 cm. or over 20 cm., in tills from West China including Urumqi River source at Tianshan Mts. have been made. About 70% are subangular, 25% angular, 2 to 3% subround, 1—2% round, but, no extremely round one is found. The case in Lushan Mt. differs from the above-said, most of the gravels collected in sections of Caifenling Yangjiaoling, Gaolong, Gutang are subround to round, and subangular makes up only 2 to 3%, while few are angular.

By comparison of fabric, we also found that the gravels in Lushan Mt. were not formed by glaciation. The A axis of till block concentrated in certain area and in one direction can be seen clearly in the fabric pattern. The orientation of A axis and the running direction of the glacier are basically similar by the statistics of till gravel ocourrences in Urumqi River source at Tianshan Mts. and Qilian Mt. and Luojie Mt. etc.

The dip angle of A axis normally is smaller which is about 0° to 20°. The dip angle of the largest flat (face AB) of till is always smaller than 20°. The dip angle of face AB has no preferable orientation.

The features of fabric in Yangjiaoling section a type section in Lushan Mt. is contrary to those of above-mentioned and face AB dips toward the upstream direction in good regulation and most dip angles are around 35°—45°, few reach 52°, the largest would reach 78°. The strike of A axis is northeastern–south–western which shows features of vertically or obliquely intersecting with the orientation of material source.

Above-mentioned fabric features shows that there are northing in common between gravel layers in Yangjiaoling and glacial deposits. Since the fabric features differ considerably from those of pluvial-alluvial deposits, in which the dip angles of face AB in gravels are generally around 20° to 30°, and A axis is vertical to flow direction, so we consider this kind of sediments as typical mud flow deposits. The largest dip angle of face AB shows, the feature of rapid depositing is more distinctively reflected. Most of the dip angles of face AB in gravels of the mud flow deposits in Wudu County, are over 30°. We have also observed small sand-gravels lenses distributed among or by the side of the large blocks in Yangjiaoling section in Lushan Mt. These are honeycomb-shaped texture of small flow sediments among the large blocks or in the sideback waters during the ordinary periods. This is one of the fabric features of mud flow deposits.

Prof. Li Si-guang pointed out a phonomenon that the larger blocks lie farther from the mountains and the smaller blocks which was caused by glaciation. Nevertheless the current studies have found no such features in the distribution of till gravels. On the contrary, the frontal flows of mud flow the "dragon's head" often bring large blocks to far away places and accumulate, while the smaller blocks stop nearby the mountains when the "dragon's head" flows slow down. Such features are found in all the recent mud flow areas. Prof. Li also pointed out that "flat and long gravels among the boulder clay straighten down into the mud or slant into the mud....", and also considered these as the result of till melt sliding down, such as in Xingqiao section[7]. Based on the current conception, there are some flow tills and melt-out tills, etc. in tills[8]. They are developed in marine glacial areas. The gravels in the above-mentioned tills can form larger angle when melting and sliding down along the ice slope, but their dip angles of face AB are always less than 30° which can be seen in Wangfeng moraines of Urumqi River source. The blocks slant into the "mud" with good regularity instead of in random isolated distribution. In Chaifengling section we found that most of the flat blocks trend to dip toward the upstream in the direction of Xiao Tian Chi Lake. Their dip angles are around 40°—50° reflecting the features of mud flow deposits as well.

THE COMPARISON OF SEDIMENTARY STRUCTURAL FEATURES

The till have following structural features:

1. Shear thrusting such as the Wangfeng tills Tianshan Mts. which has two shear planes with lower angles (12°—15°).

2. Inclusion. There are two kinds of block structures in Wangfeng tills at Tianshan Mts., one is large-scale bed-rock inclusion structure; another is large-scale block inclusion structure;

3. Inclined bedding or dome bedding. Above-mentioned structures can often be seen in tills, but not in Lushan Mt. sections where only small lenses of mud flow deposits are found. Though some people think Gutang section as thrusting structure, the authors has once seen a mylonite belt which is one meter wide, leads directly to a place where oriented gravels are found, in the bed rock in the lower part of Gutang section. So the authors consider this as a neotectonic phenomenon developed upon the revival of an old tectonic belt.

In addition, in a section of dark purplish red boulder clay of Poyang Glacial which was thrust by the glacier of Dagu Period mentioned by Prof. Li Shi-guang[7], we found that the section was located in front of slump, and thrusting and reversed sloping were all caused by slumps.

By comparison of sedimentary features, we found that the main genesis of boulder clay in Lushan Mt. area is mud flow instead of glaciation. Tropical, sub-tropical climate prevailed and corresponding geomorphology was formed at the piedmonts of the mountain while temperate climate and geomorphology occurred in the upper region of the mountain. Large gravels from principal rock found in Lushan Mt. could be formed, no matter what climate prevailed. Weathered clay, high and precipitous slope and rich precipitation could provide the basic conditions to form mud-flow and rock-flow age. Most of the fans and tongue-like deposits in front of the mountains are products of mud-flow, the boulder clay on slope and at the lower portion of the mountain was formed by mud flow, rock-flowage slope wash, pluvial and solifluctional deposits.

REFERENCES

[1] Xie Y. Y. et al., 1981, "Initial approach to Quaternary sedimentary environment in Jiujiang-Lushan Areas", Collected Papers of Geography, v. 13.

[2] Mills, H. H. 1977, "Textural characteristics of drift from some representative Cordilleran Glaciers", Geol. Soci. Amer. Bull., v. 80.

[3] Xie Y. Y., Chui Z. J. 1980, "Observation of some quartz sand by scanning electron microscope and the comparison", Hydrogeology and Engineering Geology, v. 6.

[4] Xie, Y. Y. Chui, Z. J. 1981, "Surface features of some till quartz sand of China observed by scanning electron microscope", Glacier and Permafrost, v. 3, no. 2.

[5] Wu An-ping. 1980, The characteristics of sedimentary fabric of modern till of No. 5 Gangnalou Glacier., Shulenan Mt. Journal of Lanzhou University v. 3.

[6] The Surveying Team of Quaternary Glacier, 1977, "Fabric analysis of Quaternary gravel layer in Luojieshan Xichang, Shichuan Province", Collected Papers of Quaternary Glacial Geology.

[7] Lee J. S. 1947, "Lushan In Ice Age", Monograph of former Geologica Survey, Academia Sinica" B no. 2.

[8] Studying class of glacial sedimentology (Chui Zhi-jou et al.), 1981, Some suggestions from discussion on moraine terminology, conception and translated terms, v. 3, no. 1.

THE FORMATION OF LOESS IN CHINA

WU ZI–RONG AND GAO FU–QING
(Institute of Geology, Academia Sinica)

ABSTRACT

According to the loess distribution, stratigraphy and rock composition, the formation of the loess in China is discussed in this paper.

The loess is one of the most important Quaternary deposits in China. It is widely distributed in the large areas north of Changjiang River (Yangtze Valley), and it is well developed in the areas along the middle reaches of the Huanghe River, forming the famous loess plateau lying in the arid and semiarid regions. Its southern boundary (parallel with 34—35°N) is marked by a series of mountain ranges, such as Kunlun Mts., Qilian Mts., and the mountains in central Shandong and Liaoning Provinces. To the south of this boundary the loess is less distributed, and there is change in its properties. In North China the loess spreads from west to east as a long belt. To the north of the loess region, there are some large desert regions, and to the north of the deserts there are rocky desert regions called "Gobi". The loess, deserts and Gobi form three successive zones, spreading out in the latitudinal direction.

According to the study of loess of China, we are aware of the fact that since Pleistocene at least there have been alternations of four humid climatic and three dry climatic periods, which have been substantiated by four loess formations, three erosion surfaces, the chemical and mineralogical compositions and biological characteristics of the loess as well. Minor climatic fluctuations in a dry period can also be detected from the presence of buried soils in each unit of loess.

The thick loess formation in China can be divided into the "Malan Loess" of the Late Pleistocene, "Lishi Loess" of the Middle Pleistonce (upper and lower), and "Wucheng Loess" of the Early Pleistonce.

With regard to the loess, especially the "Malan Loess", in the area along the middle reaches of the Huanghe River, the following viewpoint could be acceptable that the matierals of the loess were transported by the winds. The available evidences for the eolian origin are: (i) The occurrence of loess is not related to the underlying bedrock; (ii) the loess, deserts and Gobi are distributed as belt zones; (iii) the textures of loess are similar, and the compositions are homogeneous in the whole loess area; (iv) the loess is abundant in terrestrial zooliths and phytoliths; (v) the thickness of the loess becomes thinner and the median diameter of their grains gradually decrease from north to south, showing probably that deserts and Gobi are the source of loess matierals;

137

(vi) multiple overlappings of the buried soils are present in the loess layers. The formation of such a deposit with the above-mentioned characteristics can hardly be explained by any other agents than the winds. It is believed that the wind from the north and northwest of the loess plateau is the earlier in transporting the dust from its source, the desert regions and Gobi, or farther north to the deposition site of the present loess belt.

QUATERNARY LAMELLIBRANCH FAUNA PROVINCES IN EAST CHINA AND THEIR CHARACTERISTICS

Huang Bao–yu and Lan Xiu

(Nanjing Institute of Geology and Palaeontology, Academia Sinica)

Abstract—The Quaternary strata are well developed in East China where they yield abundant fresh-water and marine lamellibranch fossils. Based upon the lamellibranch assemblages the writers have divided the Quaternary lamellibranch fauna into three continental lamellibranch provinces and three marine lamellibranch provinces.

The Quaternary strata are well developed in East China where they yield abundant fresh-water and marine lamellibranch fossils. But, owing to the different environment conditions, the Quaternary fresh-water and marine lamellibranchia are quite different in composition as clearly seen distributionally. Based on different lamellibranch assemblages, three continental lamellibranch provinces and three marine lamellibranch provinces may be recognized.

Due to the influence of Himalayan Movement during the late Early Tertiary, continents uplifted one after another. The wide area of East China was also affected by the neotectonic movement, forming large and small depression basins and coastal fault-depressions. The forming of fault-depressions and warpings continued to the Late Quaternary . Since the migration of animals is limited by the changes of land and seas on earth's surface and the differences in water areas and barriers of mountains, the lamellibranch provinces become relatively evident. Besides, affected by the glaciation from Late Tertiary to Early Quaternary, the lamellibranch provinces of Quaternary are quite different from those of Tertiary.

CONTINENTAL LAMELLIBRANCH PROVINCES

1. *Lamprotula-Cuneopsis-Arconaia* province from Changjiang (Yangtze) Delta includes those deltaic areas from Jiaodung Peninsula towards south to Qiantangjiang River in Zhejiang and from Dingyuan, Suixi in Anhui towards east to Chuansha and Nanhui of Shanghai, etc. The lamellibranch fauna in this province is mainly represented by the following genera and subgenera: *Lamprotula* (*Lamprotula*), *L.* (*Cuneolamprotula*), *L.* (*Scriptolamprotula*), *L.* (*Parunio*), *L.* (*Sinolamprotula*), *L.* (*Odhnerella*), *Cuneopsis, Arconaia,* etc.,

139

associated with *Unio, Acuticosta, Schistodemus, Chamberlarinia* and a small number of *Corbicula*.

In this province, the distribution centre of the Pleistocene Unionids is in Sihong of Jiangsu. It is characterized by a large number of genera and species. Especially, a good number of Quaternary species have been found for the first time such as *Lamprotula* (*Lamprotula*) *fibrosa* (Heude), *L.* (*Seriptolamprotula*) *triclava* (Heude), *L.* (*Parunio*) *spuria* (Heude), *Cuneopsis shanghaiensis* Huang et Lan, *Chamberlarinia hainesian* Lea, all of which have not been found in other places. Although some species as *Lamprotula* (*Cuneolamprotula*) *bazini* (Heude) are also found in Nihewan in Hebei, belonging to the Nihewan Period of the Early Pleistocene and *L.* (*Scriptolamprotula*) *chiai* Chow in Linfen of Shanxi belonging to the Sanmen Period of the Early Pleistocene as well as in Xiangfen belonging to the Late Pleistocene, the faunal features in the Changjiang River Delta Province differ evidently from those of Unionids of Nihewan Period and Sanmen Period. In this province, neither *Lamprotula* (*Parunio*) *antiqua* Odhner, the guide fossils of the Sammen Period, nor *Lamprotula* (*Cunolamprotula*) *kusabi* Otuka has been found, all of which are abundant in the Nihewan Area. *Lamprotula* (*Sinolamprotula*) *leai* (Gray) is found in northern Guangxi belonging to Holocene. The living samples of this species are extremely plentiful in the lower reaches of Changjiang River and in the lakes and streams in Guangdong and Guangxi. This species is also found in Mekong River of Vietnam. The Unionids in this province are of a transitionally mixed fauna between North and South China types. During Late Pleistocene of Quaternary, Changjiang River Delta Province was a transitional area for a mixed fresh-water lamellibranch fauna between North and South China types.

2. *Lenceolaria-Radioplicata-Parresia* province comprises southwestern Hunan and northern Guangxi; and the south-western parts of Hunan and Guilin and Liuzhou of northern Guangxi as well. In addition to *Lenceolaria* and *Radioplicata*, there are plenty of *Unio, Lamprotula, Pseudodon, Schistodesmus, Parraeysia, Cristaria Anodota, Corbicula,* etc.. Based upon their faunal characteristics, two types can be recognized: The first has a wide distrubution and a long range of period, including *Unio douglasiae* Griffitlh et Pidgeon, *Lamprotula* (*Sinolamprotula*) *leai* (Lea), and genus *Corbicula* spread worldwide is also better developed in this province. The second, showing strong local aspects and consisting of *Cuniopsis zhenpiyanensis* Huang, *Schistodesmus trapezoidus* Huang *Lamprotula* (*Cuneolamprotula*) *aculeats* Huang, *Radioplicata guangriensis* Huang, etc., all of which have not yet been found in other places. *Parreysia aurera* (Heude) is a living species, its fossil species was found for the first time, which is more primitive than its living samples. In Nihewan Area, Yu Xian County of Hebei, *Pseudodon* has also been found from the Lower Pleistocene but the species from the southwestern Hunan-northern Guangxi Provinces is different from that found in Nehewan, the former seems to be more highly developed and larger in size with a thicker shell-wall. In short, the fauna in this province bearing its own typical characteristics, is

140

quite different from those in North China.

3. The *Corbicula* province from the shore delta or embayment province includes Ganyu and Lianyunggang in Jiangsu, Nanhui in Shanghai, Hangzhou Bay in Zhejiang and Minhou areas in Fujian. The fauna in this province is mainly composed of *Corbicula* with *Corbicula fluminea* (Müller), *C. japonica samdiformis* Yokoyama and *Corbicula leana* Prime, etc., the leading forms of which, *Corbicula fluminea,* having a strong adaptability and a rapid reproduction, was found from the Quaternary up to the recent time in various kinds of water body, such as rivers, lakes, ponds, stream outlets and bays. And it is also a worldwide-spread species. *Corbicula fluminea* is rich in the strata of Early and Middle Holocene in Ganzhe and Baisha along the Minjiang River and in Wulong River basins as well. This genus commonly associated with *Ostrea (Crassostrea) pestigris* Honley, is found in the Holocene strata in Ganyu, Jiangsu and along the bays of Lianyungang. This shows that the fauna in this province is of the brackish water type in a transitional area between land and sea.

MARINE LAMELLIBRANCH PROVINCE

China's coast extends from the mouth of the Yalu River on the China-Korea border in the north to the mouth of the Peilun River on the China-Vietnam border in the south, about 14,000 km long. According to different assemblages, the marine lamellibranch fauna of East China may be divided into three provinces:

1. *Potamocorbula* province from the Huanghai Sea and the ancient Bohai Sea during the Quaternary: This province covers a vast area from Liaoning Province in the north to Hangzhou Bay of Zhejiang Province in the south. This fauna is rather few in species, with *Potamocorbula* as its dominant form. In addition, it contains in the province the following fossils: *Meretrix, Mactra, Siliqua, Dosinia, Anadara, Scapharca, Barbatia, Anomia,* together with *Lithophaga, Sphenia, Martesia, Poladidea. Potamocorbula,* the wide-spread species has been made known from Quaternary to recent in Hokkaido and northeastern Japan, Korea and the East China Sea. This species is living on sandy and muddy bottoms, about 20—60 meters deep, in the north temperate zone of Asia. *Siliqua minima* (Gmelin) is only distributed in the south of Zhejiang. A large number of *Ostrea (Crassostrea) gigas* (Thunberg), *O. rivularis* Gould and a small number of *Protothaca, Arca* form Cheniers made up of oyster shells along the western coast of the Bohai Gulf. This fauna is characterized by *Potamocorbula amurensis* (Schrenok) and the abundance of both temperate and eurythermal warm-water species. As a whole, it shows a temperate zone type having an affinity to the Quaternary fauna of Japan and Korea.

2. *Paphia (Paratapes) undulata* (Born)-*Timoclea minuta* (Yokoyama) province from the ancient East China Sea and the northern part of the ancient South

China Sea during Quaternary: This province covers an area from the Hangzhou Bay of Zhejiang Province in the north to the northern coast of Hainan Island in the south, expanding to the western coast of Taiwan and Penghu Islands in the east. In this faunal province, there are *Paphia (Paratapes) undulata* (Born) and *Timoclea minuta* (Yokoyama), often associated with *Sunetta, Cyclina, Chione (Clausinella), Pitar, Amusium, Chlamys, Placuna, Anadara, Barbatia, Striarca, Trisidos, Glycymeris*, etc. *Potamocorbula* is relatively uncommon, whereas *Corbula (Varicorbula), C. (Carycorbula), C. (Notocorbula)* are abundant. Two oyster shell-made cheniers lie in the mouth of the Jiulong River of Fujian, one of which is composed of *Ostrea (Crassostrea) pestigris* Hanley, while the other consists of *Ostrea (Crassostrea) gigas* (Thumberg), the largest one reaching 30 cm. in height. A large number of *Placuna placenta* (Linne) and a smaller number of *Anadara granosa* (Linne), *Polymesoda (Geloima) luodouensis* Lan, *Sinonovacula constricta* (Lamarck) are made up of the shell bed in the northern part of Hainan Island. *Placuna placenta* is living in warm current regions around Australia, the Phillippines, Indian Ocean and South China and East China Seas. This province is devoid of the characteristic speies of warm-water fauna living in the coral reefs of the tropical zone, for example *Tridacna* and *Codakia*. This fauna represents a mixed marine lamellibranch fauna between North and South China of subtropical type.

3. *Tridacna-Hippapus* province from the ancient South China Sea during Quaternary: This province includes southern Taiwan and Guangdong, the southern part of Hainan Island and South China Sea. This fauna is characterized by *Tridacna* and *Hippopus*, which are associated with *Timoclea, Chama, Anadara, Acar, Arcopsis, Tellina, Spondylus, Pitar*, etc.. It is characterized by a warm-water fauna living in the coral reefs of the tropical zone area. This province is quite different from the above-mentioned two faunal provinces.

THE LAMELLIBRANCH FAUNA CHARACTERISTICS OF EAST CHINA

1. Changjiang (Yangtze) River Delta Province (*Lamprotula, Cuneopsis-Arconaia*) is characterized by having a mixed fresh-water lamellibranch fauna between North and South China types.
2. The southwestern Hunan—northern Guangxi Province, (*Lanceolaria, Radioplicata-Parreysia*) bears a lamellibranch fauna, which shows features quite different from those of North China type.
3. The shore delta or embayment province (*Corbicula*) is a transitional area between sea water and land, close to the coast and embayment.
4. Ancient Huanghai Sea and the ancient Bohai Sea Province (*Potamocorbula*) are of North temperate zone type during Quaternary.
5. The ancient East China Sea and the northern part of the South China Sea Province (*Paphia (Paratapes) undulata* (Born)—*Timoclea minuta* (Yokoyama) during Quaternary represents a mixed marine lamellibranch fauna between

142

North and South China, belonging to the subtropical type.

6. The ancient South China Sea province (*Tridacna-Hippopus*) during Quaternary is typified by a warm-water fauna, living in the coral reefs of the tropical zone area.

REFERENCES

[1] Chow Min-chen, 1955, Pleistocene fresh-water pelecypods from Wuhohsien, Northern Anhwei. Acta. Palaeont. Sinica, Vol. 3, no. 1, pp. 73—82.

[2] Chow Min-chen, 1958, Description of molluscan shells. Inst. Vert. Palaeont. Acad. Sinica, Mon. No. 2, pp. 81—96.

[3] Gu Zhi-wei. Huang Bao-yu, Chen Chu-zhen et al., 1976, Fossil lamellibranchs of China, 522 pp., 150 pls. Science Press, Peking. (in Chinese)

[4] Haas, F., 1910—1920, Die Unioniden. In Martini und Chemnitz, Syst. Conch, Cab. Bd. 9, Abt. 2a-3.

[5] Heude, R. P., 1875—1886, Conchyliologia de province de Nanking et de la Chine centrale. Fasc. 1—10. Paris.

[6] Huang Bao-yu, 1981, The fresh-water Lamellibranchs from the site of the Zhenpiyan Cave in Guilin, Guangxi. Acta. Palaeont. Sinica. Vol. 20, No. 3.

[7] Huang Bao-yu, Guo Shy-yuan, 1982, Early Pleistoncene fresh-water Lamellibranchs fauna from Nihewan, Hebei. Bull. Nanjing Inst. Geol. & Palaeont. Acad. Sinica. No. 5.

[8] Lan Xiu et Wang Shu-mei, 1977, Cenozoic fossil Lamellibranchs from Jiangsu Province. Mem. Nanjing Inst. & Palaeont., Acad. Sinica, No. 8.

[9] Lan Xiu, 1981, Lamellibranchs (pp. 209—219). Fossils photographs of the continental shelf of South China Sea. Guangdong Sci. Techn. Press (in Chinese).

[10] Leroy, P., 1940, Late Conjoic Unionids of China. Bull. Geol. Soc. China. Vol. 19, No. 14.

[11] Nomura, S., 1933, Catalogue of the Tertiary and Quaternary Mollusca from the Island of Taiwan in the Institute of Geology and Palaeontology Tohoku Imp. Univ., Sendai, Japan pt. Pelecypods Sci. Rept. Tohoku Imp. Univ. Ser. 2, Vol. 16, No. 1.

[12] Odner, N., 1925, Shell from the Sanmen Series. Paleont. Sinica, Ser. B, Vol. 6, fasc. 1.

[13] Otuka, Y., 1942, Some new Unionidae from North China and Southern Mongolia. Proc. Imp. Acad. Tokyo, Vol. 5, No. 8.

[14] Tchang Si, 1959, Faune des Mollusques utiles et unisibes de la Mer Jaune et la Mer Est de la Chine. Ocean. Limno, Sinica, Vol. 2, No. 1.

[15] Tchang Si. Tsi Chung-yen, 1959, Fauna des Mollusques utiles et nuisibes de la Mer Sud de la Chine. Ocean. Limno. Sinica, Vol. 2, No. 4.

DISCOVERY OF CALCAREOUS NANNOFOSSILS IN BEIJING PLAIN AND ITS SIGNIFICANCE FOR STRATIGRAPHY, PALAEOGEOGRAPHY AND PALEOCLIMATOLOGY

WANG NAI–WEN AND HE XI–XIAN

(Institute of Geology, Chinese Academy of Geological Sciences)

Abstract—After a foraminiferal study on the Quaternary marine bed in Beijing Plain the authors have recently discovered a calcareous nannoplankton assemblage from the same bed, which is comprised of eight species in all, such as *Coccolithus pelagicus, Cyclococcolithus leptoporus, Gephyrocapsa oceanica, G. Protohuxleyi, Emiliania huxleyi, Pseudoemiliania* cf. *lacunosa* etc. The lower boundary and some problems on stratigraphy and palaeogeography are discussed. Based on the foraminiferal and nannofossil assemblages and magnetostratigraphical data, the authors prefer to accept the lower boundary of the Matuyama Epoch (2.43 to 2.48 m.y.) to be the lower boundary of the Quaternary.

In 1979 we reported an important discovery of a marine bed in Beijing Plain within a large continental sequence and a foraminiferal assemblage in the paper "Magnetostratigraphy of the core S5 and Beijing Transgression of the Early Matuyama Epoch". In later further study, which is in full swing until now, some new analyses have been made for extracting nannofossils. As a result, an excellent assemblage of calcareous nannofossils has been found from the same marine bed containing foraminifera. Though the number of species is not many, they are of significance for either palaeogeographical reconstruction or stratigraphical correlation.

Owing to this discovery we are able to make a new discussion on the stratigraphy, palaeogeography and palaeoclimate of the Early Quaternary based on the composition, and morphological characteristics of nannofossils, foraminiferal, palynological and lithological data in combination. Prior to the beginning of the study we realized that nannofossils could occur in the same marine bed containing foraminifera of stenohaline type. After a series of repeated treatments numerous specimens of calcareous nannofossils, just as we expected, were obtained from eight cores and, then, studied and identified by the scanning electronic microscope. The result shows that the present nannofossil assemblage has considerable significance both in age- and facies-indication. There are totally eight species belonging to seven genera of three families as follows:

144

Coccolithus pelagicus (Wallich) Schiller
Cyclococcolithus leptoporus (Murray and Blackman) Kamptner
Cricolithus sp.
Gephyrocapsa oceanica Kamptner
G. protohuxleyi McIntyre
Emiliania huxleyi (lohmann) Hay and Mohler
Pseudoemiliania lacunosa (Kamptner) Gartner
Syracosphaera sp.

Besides, a great many chrysophytes (*Cysta* spp.) appear to be the accompanying elements.

The reported nannofossil assemblage seems rather monotonous in composition which is corresponding to or, at least, not exceeding the recent population of the subarctic zone of the Pacific and Atlantic Oceans. Within the dominant forms of both Beijing fossil assemblage and the recent oceanic ones *Coccolithus pelagicus* and *Emiliania huxleyi* rank first in abundance of speciments. *C. pelagicus* is a stenothermous species, usually living in sea-water of 7°—14°C. This species, by the recent literature on the four biogeographic zones of the Atlantic, begins its existence in the transitional zone and becomes dominant only within the subarctic one. The described fossil specimens are not only abundant but also well-developed in morphology. *E. huxleyi*, on the contrary, is an eurythermous species, widely distributed at various latitudes in the recent oceans. But, fortunately, the morphological structures change in different environments: the central area is open and thirty to forty I-shaped crystallites are seen on its distal shield in the normal sea-water of the tropic and subtropic bands; but in cold water of the subarctic and transitional zones, in contrast to those mentioned above, the central area is usually closed and the number of I-shaped crystallites on its distal shield does not exceed twenty three to thirty-three. The majority of the specimens of the described assemblage has closed central areas and distal elements not more than thirty. There are, certainly, a few specimens with open central areas and distal elements exceeding thirty in number, but they will not affect our interpretation due to their little percentage. Hence, it can be seen that the described assemblage of calcareous nannofossils is characterized by a relatively monotonous composition or a low diversity and a cold-water palaeoecological aspect, mixed with a few warm-water elements. Obviously, there is a striking resemblance in indicating the climatic conditions between the nannofossils and foraminifers. Cold-water foraminifers, such as *Globigerina bulloides, G. pachyderma, Hyalinea baltica, Buccella frigida* and *Elphidiella arctica* are well-developed. All of the above-listed foraminifers are indicators for a cold-water environment. The accompanying cysts generally adapt to a relatively cold environment too. A climatic deterioration is well reflected on the palynological spectrum of the same marine bed, which is constituted of some herbaceous plants (Chenopodiaceae and *Artemisia*), coniferous trees (represented chiefly by *Pinus* accounting for 88% of woody plants) and a little percentage of broad-leaf trees e.g. *Betula* (1% of

145

woody plants). Besides, a few Tertiary relicts occur in the spectrum. All of these plants will serve to illustrate cold-trending climate. Such a climatic condition resembles with that of the recent subarctic zone of the Pacific at latitudes between Harbin and Bering Sea. On the cold-climatic background, however, some seasonal influence of warm-water currents like the recent Kuroshio can not be excluded. In a word, Beijing nannofossil assemblage must be regarded as a population of the subarctic bioprovince.

As a group of planktonic organisms, Coccoliths are adaptable to the open sea with normal salinity and, thereby, very rare in deltaic and off-shore regions, for example, in Changjiang (Yangtze) Estuary, Hangzhou Bay. Even in the East China shelf sea calcareous nannoplankton are not abundant until reaching the outer shelf and its continental slope. In the surface sediments of the East China Sea they are not well-distributed, but concentrated in the areas, crossed through by Kuroshio Current. No nannoplankton and planktonic foraminifers are found until now, even, such common benthic forms in the Quaternary in Beijing as *Cibicides, Hyalinea* and *Paromalina* were never recorded in the recent Bohai Gulf, where only *Nonion, Elphidium, Ammonia, Spiroloculina* and *Quinoueloculina* are common. Hence it can be said that the described nannofossil assemblage should be a normal marine population different from those of either gulf or off-shore ones.

Based on the discovery of nannofossils, we are able to go into the discussion on stratigraphy, palaeogeography and palaeoclimatology of Beijing Transgressional stage.

STRATIGRAPHY

By the discovery of nannofossils a new argument has been provided to identify the lower boundary of the Quaternary in Beijing. At present twenty-one calcareous nannofossil zones are accepted for the Neogene and Quaternary named as NNI to NN21 based on the materials from continent and, particularly, deep-sea drilling. The calcareous nannofossils of Beijing Plain comprise the components of NN19 to NN21, which involves all the three zones of the Quaternary. *Coccolithus pelagicus* and *Emiliania huxleyi* are always the dominant elements in samples of eight cores. *C. pelagicus* is ranging over a wide stratigraphical interval since the Miocene and, thus, has no significant role in stratigraphy. *E. huxleyi* occurs in the Late Pleistocene and marks NN21. *Pseudoemiliania lacunosa* appears within NN16 and disappears before the first occurence of *E. huxleyi*. The interval between the last occurences of *Discoaster brouweri* and *P. lacunosa* represents NN19. The zone between NN19 and NN21 is NN20, represented by *Gephyrocapsa oceanica*. The Beijing marine bed, as mentioned above, contains representatives of three Quaternary zones, so that this bed can reasonably be assigned to the Quaternary. The three-parting zonation of Quaternary nannofossils, however, would hardly be used to

146

Beijing Plain due to the simultaneous presence of all the three zonal fossils in a same bed or, even, in a same sample. The Beijing marine bed, dated magnet-ically as 2.26 m.y., is much older than the age given by most of the recent informations about the first occurence of *E. huxleyi*. However, Pyle (1968) also reported *E. Huxleyi* from the "Pliocene" of the coastal region of the Mexican Gulf. Therefore, the stratigraphical boundaries of *E. huxleyi* are needed to be restudied.

PALAEOGEOGRAPHY

The nannofossils and associated foraminifers of Beijing marine bed show a clear pelagic or stenohaline nature of the water mass, in which these organisms inhabited. Therefore, two palaeogeographical explanations for them may be proposed: Firstly, the ancient North China Plain was submerged by a rela-tively broad sea, the coastal line of which located, perhaps, along the Yanshan-Taihangshan Mts. and the salinity was nearly normal, like that of the recent East China Sea; secondarily, the sea water transgressed the coastal plain and penetrated deep into the continent along new active faults in trough or branch-shaped systems, like the recent Californian Gulf, which is not wide, but has a certain depth and normal salinity and keeps a more or less free cir-culation between its water mass and that of the open sea. Previously, we have suggested the first explanation and, now, we would like to provide the second one as an addition. From the paleontological view, both of those may be rea-lizable, but which one correct is in fact remains to be confirmed by further new information.

PALAEOCLIMATOLOGY

The nannofossil assemblage of Beijing Plain is characterized by the predomi-nance of cold-water forms over some warm-water ones and its low diversity, both of which indicate a cold climate, probably in subarctic environment. This conclusion accords with that based on foraminiferal evidence. *Globigerina pachyderma, G. bulloides, Hyalinea baltica, Buccella frigida* and *Elphidiella arctica*, etc. as well as nannofossils *Emiliania huxleyi* and *Coccolithus pelagicaus* with closed central area are main components of the recent northern waters. The latitudinal position of the present Beijing Plain essentially corresponds to those of the transitional zone of the Pacific. Therefore, we have considerable amount of evidence to decide that the southern boundary of the subarctic zone has shifted northwards for, at least, six degrees since 2.00 m.y. before present. Hisatake Okada (1970) referred the nannoplankton at N45°–50° of the recent Pacific to subarctic assemblage represented by predominance of *E. huxleyi*, which appears one of the two dominant forms in the Early Quaternary of Beijing Plain. According to Vincent (1975) and Bradshaw (1959), the southern boundary of the subarctic faunal zone of the Pacific conforms to the 15°C isotherm which is ranging within N45°–50° and trending south nearly to Los Angeles at the American side under the affecting of the southward cold Californian Current and deviates to the southern end of Sakhalin Island at the Asian side by the northward warm Kuroshio Current. The palaeoclimatic conditions of 2.00 m.y. before present seem to have been confirmed to coincide with the subarctic zone of the recent Pacific.

The Late Cenozoic, as we know, underwent two main climatic deteriorations, one of which took place in Miocene and marked by the formation of the antarctic icecover and the other — 2.40 m.y. ago, corresponding to the beginning of the Matuyama Epoch. The last interface is just recorded here in Beijing Plain and marked by the Beijing Transgression accompanied by a number of cold-water faunas and floras. Meanwhile, this is one of the most important arguments based on which we determine the Beijing marine bed as the lower boundary of the Quaternary in North China.

The authors are very grateful to Prof. Hao Yi-chun, who carefully reviewed this manuscript and made a number of important suggestions. We would particularly like to thank Prof. Huang Ji-qing, the head of the Geological Society of China. for his warm support and help for this study.

REFERENCES

[1] An Zi-sheng, Wang Nai-wen et al., 1979, Magnetostratigraphy of the core S5 and the transgression in the Beijing Area during the Matuyama Epoch. Geochemica, no. 4, p. 343—346.
[2] Bradshaw, J. S., 1959, Ecology of living foraminifera in the North and Equatorial Pacific. Contr. Cushman Found. Foram. Res., 10 (2), 1—25.
[3] Martini, E., Worsley, T., 1970, Standard Neogene calcareous nannoplankton zonation. Nature, vol. 225, p. 289—290.
[4] Müller, C., 1976, Tertiary and Quaternary calcareous nannoplankton in the Norwegian-Greenland Sea, Leg 38. DSDP Initial Reports, vol. 38, p. 823—838, pl. 1.
[5] Okada, H., 1970, Surface distribution of coccolithophore in the North and Equatorial (sic) Pacific. Geol. Soc. Japan, Jour., C. 76, p. 537—545, 1pl., 7 figs.
[6] Okada H, Hanjo, S., 1973, The distribution of oceanic coccolithophorids in the Pacific. Deep-Sea Res., vol. 20, no. 4, p. 355—374.
[7] Okada H, McIntyre, A., 1977, Modern coccolithophores of the Pacific and North Atlantic Oceans. Micropaleontology, vol. 23, no. 1, p. 1—55, pl. 1—13.
[8] Pyle, T. E., 1968, Late Tertiary history of Gulf of Mexico based on a core from Sigsbee Knolis. Amer. Ass. Petrol. Geol. Bull., vol. 52, p. 2242—2246, 2 figs.
[9] Robert, W. P., Hart, G. F., 1979, Phytoplankton of the Gulf of Mexico, Taxonomy of calcareous nannoplankton. Geoscience and Man, vol. XX.
[10] Vincent, E., 1975, Neogene planktonic foraminifera from the central North Pacific Leg 32. DSDP Initial Reports, vol. 32, p. 765—801.
[11] Wang Pin-xian, Min Qiu-bao, 1981, A preliminary study of calcareous nannoplankton in bottom sediments of the East China Sea. Acta Oceanographica, vol. 3, no. 1, p. 188 —192.

STUDIES ON VEGETATION AND PALAEOGEOGRAPHY FROM LATE TERTIARY TO EARLY QUATERNARY IN CHINA

LI WEN-YI

(Institute of Geography, Academia Sinica)

Abstract—From the viewpoint of palynology, the author intends to interpret the outline of the vegetation and palaeogeography from Tertiary to Early Quaternary in China.

I

Many studies has been made on the recent history of the physical geographic background in China.   The climatic changes from Late Tertiary to Early Quaternary led to the vegetational evolution, migration and reassemblage.   Based on the palynological records, the vegetational distribution in China during that period can be clearly outlined.   Professor Hsu Jen pointed out[1], in China the geomorphological and palaeogeograpical environments as well as the vegetational sapects of Quaternary differed slightly from those of Late Tertiary. So when discussing the characteristics of vegetation in that period, it is suitable to combine Pliocene with Early Pleistocene to interpret the vegetational situation and climatic changes.

This article mainly deals with the paleovegetation from Late Tertiary to Early Quaternary in China and some problems in its development and evolution from the viewpoint of palynology, and the results of pollen analytical studies as well. The climate in that time and the boundary between the Tertiary and the Quaternary are also disscused.

II

The outline of the pollen assemblages from Late Tertiary to Early Quaternary in some localities of China can be drawn as follows.

The pollen assemblage in the Sanjiang Plain of Northeast China*, in which tree

---

* Xia Yu-mei, 1978, Studies of the Late Tertiary-Quaternary sporo-pollen assemblage and palaeoclimate in the Sanjiang Plain.

pollen contained *Pinus, Picea, Carya, Alnus, Fagus, Quercus, Ulmus, Celtis,* etc. are present in the Late Tertiary. Herbs pollen grains are about 20%; among which Gramineae, Chenopodiaceae and Compositae are predominant. The assemblage reflects a warm temperate vegetation. Then the climate became cool as evidenced by the domination of *Pinus* and *Betula* and the subsidiary of *Picea, Abies,* etc. At this stage about a half of the assemblage was the herbs of *Artemisia* and Chenopodiaceae.

The Weihe Basin is situated in the centre of Shaanxi province, where pollen is abundant in the lacustrine sediments. Among them tree pollen occupy 50–60%, and some genera and species are referred to families of Pinaceae, Betulaceae, Fagaceae, Ulmaceae, Juglandaceae and Rosaceae. Polygonaceae, Compositae and Chenopodisceae are the main families of the herbs. The vegetation presents a semi-humid mesophilous needleleaf and deciduous mixed forest or a grassland landscape.

Along the coast of Bohai Gulf, the Pliocene pollen assemblage is found in the black shale[2]. The pollen elements are similar to those in Weihe Basin, but pollen frequencies are different that the percentage of tree pollen increases around 80—90%, in which the deciduous *Quercus* and *Ulmus* are the main components, and some thermophilous such as *Carya, Liquidambar, Corylopsis* are present as well. In the Early Pleistocene *Pinus* and *Picea* occur, the herbs rapidly increase more than 60%, which shows that the temperature and the moisture fluctuated in a wide range.

Pinus, Ulmus and *Quercus* are abundant in the Early Quaternary pollen assemblage from Shanghai[3]. Besides there are *Tsuga, Keteleeria* and *Taxodium* of the needleleaves. *Liquidambar* and *Carya* are the predominant elements within the broadleaf trees. Many genera of the Gramineae, Compositae are the main elements in herbs. In the later stage of the development, the appearance of *Picea* and *Abies* shows the influence of the cold climate.

In Anhui Province, there is no complete record of pollen sequences. The pollen assemblage comes from lignite deposits in southern Anhui. In the lower part there is a needleleaf and broadleaf mixed forest from subtropics to warm temperate. Among angiospermae, except Ulmaceae and Fagaceae, the subtropical elements occupy 20%. In the Quaternary, pollen flora pine pollen increases and broadleaf pollen decreases. The pollen assemblage can be correlated with the needle- and broadleaf mixed forest which lives in more humid condition nowadays.

In northern Hunan[4] a peat bed is interbedded in the thick bedded grey white clay formation. The vegetation of the Quaternary is a subtropical evergreen and deciduous forest, and the main elements are *Keteleeria, Quercus, Pterocarya, Liquidambar, Myrica* and *Ilex,* etc. *Myrtus* and *Laurea* are rare. The sequences of the pollen and spores are very similar to those of the present vegetation at the same local place. A trend of increase in *Pinus* and *Quercus* and a decresse of the subtropical elements is observed, reflecting that the climate tended to be milder and cooler than before.

150

In central Yunnan[5], the lignite deposits are found in the vicinity of Kunming, and the pollen flora has a very complex component. The distinct mixed pollen assemblages were formed in high altitude intermontane canyons of a peculiar geomorphological area. The pollen taxa contain tropical, subtropical, temperate and high mountain elements. Pinaceae and Fagaceae are predominant, as well as a lot of thermophilous ones such as *Podocarpus, Dacrydium, Carya, Ilex, Aralia, Platycarya, Liquidambar, Pittosporum, Caesalpinia, Sterculia*, etc. are found. In the upper part of the strata the assemblage shows some variation, and the increase of many conifers such as *Abies, Picea,* and *Cupressus*, the broadleaved trees such as *Alnus, Betula* and *Tilia* as well as some herbs are observed, which indicates that the climate became cooler.

Lacustrine and fluviolacustrine sediments are distributed in Mt. Xixiabangma and Jilong and Nieniexiongla Basins which lie to the west of Mt. Qomolangma. The pollen analysis shows that the vegetation growing there during Pliocene was predominantly *cedrus-Quercus-Pinus*, showing that warmer and humid climate coexisted on these regions. Later, the assemblage of *Pinus* and *Artemisia* rises to a higher percentage, showing a transition from temperate to cold. In the western Himalayas and western Gandise Range, lacustrine strata are distributed. The sporo-pollen assemblage is different from that nearby Mt. Qomolongma of South Xizang. It is predominanted by *Picea, Abies, Pinus* as well as herbs of Chenopodiaceae, Compositae and Cruciferae, which indicate a cold and dry climate.

In Northwest China, the dry-steppe had expanded since the end of Oligocene[6]. Then in Pliocene arbor was scarce and the vegetation was very poor. The climate became more and more arid in Pleistocene.

III

Based upon the above-mentioned, the flora of Late Tertiery and Early Quaternary in China show both spatial and temporal, changes in flora and the characteristics of the vegetation in East Asia during the period of global climatic fluctuations. The relationship between the vegetation and its environment can be analysed as follows:

1. According to the palynological studies on Late Tertiary and Early Quaternary of North China, the genera in the assemblage may be divided into two groups: developing and expanding temperate elements and gradually differentiating and retreating sub-tropical elements.

Picea and *Abies* act as indicators of cool climate for this region. In Early Pleistocene sediments they often reach as high as 60% of the total tree pollen. As for broadleaf trees *Quercus* act as an indicator of temperate climate of the region at that period. *Ulmus* has the similar evolution and feature, but represents drier environment.

As the climate became cool and dry, the terriberbosa rapidly developed and

151

expanded during this period, among which Compositae, Chenopodiaceae, Polygonaceae and Gramineae are predominant, and the development of Compositae was more obvious.

In the pollen assemblage the number of subtropic elements is not large. The predominant ones are *Podocarpus, Keteleeria, Tsuga, Pterocarya, Carya, Liquidambar*, etc., which retreated from warm-temperate zone in the north, but did not disappear in East Asia and still remained in the subtropical region in China.

As a whole, according to pollen analysis, the decrease of thermophilous elements and the increase of temperate elements during Quaternary in North China are clearly shown.

2. Since Late Tertiary the important factor influenced the vegetational development in South China was the monsoon. The southeastern and southwestern monsoons brought plenty of rain to land and therefore, plants could avoid being destroyed by the subtropical high pressure and preserved a large number of relics, e.g. Fagaceae, Proteaceae, Magnoliaceae, Hamanelidaceae, Aquifoliacaee, Araliaceae, Myrtaceae and Rutaceae, etc. They have been growing for a long time in many localities in South China since Tertiary till present.

3. In Northwest China, since Late Tertiary the dry climate with less rain formed due to the influence of the arid center in Central Asia and the barrier effect of Qinghai-Xizang Plateau. The aridity still continued when the climate became cooler. The characteristic of the pollen assemblage is that a great number of herbs pollen grains but poor in taxa remain. The common characteristic of the pollen assemblages in North China in Quaternary is that a lot of Chenopodiaceae and Compositae species remained throughout the period.

4. The development of the vegetation in Southwest China has been closely related to the uplift of the Himalayas since Late Tertiary. Because the time, speed and manner of uplifting of the Himalayas have been different, the influences of which on vegetation varied correspondingly.

During a period from Late Pliocene to Early Pleistocene three floristic regions can be divided:

1) Central Yunnan lies in the eastern side of the Himalayas. The assemblage was a subtropical evergreen and deciduous mixed forest. The vegetation was very similar to that of the present in the same region;

2) In the area near Mt. Qomolangma there was a warmer and more humid deciduous forest;

3) In the western segment of the Himalayas there was a needleleaf forest of cool climate;

5. Thus, it can be seen that, from Pre-Quaternary to Quaternary in China the regionalization of geographical distribution of the vegetation developed, and the outline of the zonation and regionalization was similar to that of the present.

At the begining of Quaternary, there was a stage of the universal development of vegetation, in which the amount of *Abies* and *Picea* showed an obvious increase. This characteristic reflects the cold climate from the north. But, no enough evidences are obtained to reflect the influence of ice sheet upon the vegetation at that time. In the glacial period, the change was only shown by lowering of the altitude of the vegetational zone in mountains and expanding of arid herbs in the north. As South China being a refuge where a lot of plants survived, a special flora was formed which bore distinguished features from those in other regions over the world.

IV

In China as well as over the world to place the boundary between the Tertiary and the Quaternary has been a subject of dispute for a long time. Based on the palynological data studied in China, if the boundary between Tertiary and Quaternary is placed in the geological stage which corresponding to the beginning of the cold climate, the evolutional stage of the pollen flora in North China may be expressed by the cold resistant elements such as *Picea* and *Pinus* of trees as well as a lot of Chenopodiaceae and *Artemisia* which were generally predominant in the assemblage. At the same time the temperate elements were only a few, *Tsuga* and *Pterocarya* were present occasionally. In the assemblage of the underlying sequence the thermophilous elements such as *Tsuga, Keteleeria, Carya, Pterocarya, Liquidambar,* etc. occupied about 5%, and the cold stage has never been discovered. According to the above-mentioned, it is possible that the boundary placed at the beginning of the cool climate could be correlated with the upper part of the Tiglian in Europe. Thus, the age should be 1.8—2.0 m.y.

REFERENCES

[1] Hsü Jen, 1978, Cenozoic flora of China in: Cenozoic plant of China. 186—212, Science Press, Peking.
[2] Li Wen-yi, Liang Yu-lian, 1981, The Pliocene sporo-pollen assemblage of Huanghua in Hebei Plain and its significance in palaeobotany and palaeogeography. Acta Botanica Sinica. 23 (6), 478—486.
[3] Liu Jing-ling, 1977, Studies on the Quaternary sporo-pollen assemblage from Shanghai and Zhejiang with reference to its stratigraphic and palaeoclimatic significance. Acta Palaeontologica Sinica. 16 (1), 1—10.
[4] Li Wen-yi, 1962, Sporo-pollen anslysis of the peat bed of the Tungting formation and its geological age and the problem of palaeogeography. Acta Geographica Sinica. 28 (1), 55—72.
[5] Li Wen-yi, 1978, A palynological investgation on the Late Tertiary and Early Quaternary and its significance in the palaeogeographical study in Central Yunnan. Acta Geographica Sinica. 33 (2), 142—155.
[6] Chow Ting-ju, 1965, Some palaeogeographical problems of Sinkiang in the late earth history. Acta Scientiarum Raturalium Scholarum Superiorum Sinensium. Pars Geologica, Geographica et Meterologica. 3, 230—244.

[7] Gao You-xi, 1962, Problem of monsoon. Some problems of East Asia monsoon. 12—27, Science Press, Peking.

[8] Hammer, T., Van der. Wijmstra, T. A., Zagwijn, W. H., 1971, The flora record of the Late Cenozoic of Europe. The Late Cenozoic glacial ages, 391—424, Yale Univ. Press.

[9] Leopold, E. B., 1969, Late Tertiary and Quaternary flora. Aspects of palynology. 377—438, Wiley Interscience Press.

QUATERNARY SPORO-POLLEN AND OSTRACOD ASSEMBLAGES FROM THE CONTINENTAL SEDIMENTS IN NORTH CHINA

TANG LING-YU AND HUANG BAO-REN

(Nanjing Institute of Geology and Palaeontology, Academia Sinica)

Abstract—This paper deals with the continental Quaternary sporo-pollen and ostracod assemblages from Beijing, Yangyuan Basin in Hebei, Datong Basin in Shanxi, Lantian in Shaanxi, Sanmenxia in Henan and Qinghai Lake area and Gonghe Basin in Qinghai of China, and the relevant vegetation and climate are also discussed.

LOWER PLEISTOCENE

The fossil sporo-pollen assemblage consist of plentiful pollen of dark coniferous forest, commonly herbaceous and a few Tertiary relics, mainly including *Pinus, Picea, Abies, Ulmus, Quercus, Betula, Artemisia,* Chenopodiaceae, Gramineae, Cruciferae and Polypodiaceae along with a few *Carya, Cedrus* and *Podocarpus.* The ostracods are composed of *Qinghaicyris crassa* and *Ilyocypris errabundis* in Gonghe Basin of Qinghai, and *Ilyocypris kaifengensis* and *Leucocythere mirabilis* in Shaanxi and Hebei.

The palynological assemblages from the known Lower Pleistocene in North China are listed in Table 1.

Table 1 Palynological assemblages from the known Lower Pleistocene in North China

Formation		Pollen assemblages	Vegetation	Climate
Nihewan Fm., Yangyuan Basin, Hebei[2]	I.	*Pinus, Picea, Abies, Tilia, Ulmus, Quercus, Artemsia,* Chenopodiaceae, Compositae, Gramineae, Polypodiaceae, *Podocarpus, Cedrus, Ephedra,* Cupressaceae.	Dark Coniferous forest	Humid-cold
	II.	*Artemisia,* Chenopodiaceae, Cruciferae, Gramineae, Compositae, *Betula, Quercus, Ulmus, Tilia, Celtis, Juglans, Acer, Salix, Carya.*	Park land	Dry-cool
	III.	Being similar to I.	Dark Coniferous forest	Humid-cold

(to be continued)

155

Table 1 *(continued)*

Formation	Pollen assemblages	Vegetation	Climate
Yiuhe Fm., Lantian Area, Shaanxi[1]	IV. *Pinus, Abies, Tsuga, Ulmus, Celtis, Juglans, Betula, Quercus, Tilia, Podocarpus, Larix, Cedrus,* and herbs (4—8%).	Piceetum	Humid-cold
	III. *Salix, Juglans, Aluns, Betula, Corylus, Carpinus, Quercus, Celtis.*	Needlebroad-leaved mixed forest	Humid-cool temperate
	II. Herbs 64% *Ulmus, Tilia, Celtis, Betula, Juglans, Carpinus, Ostrya, Quercus, Rosa, Pterocarya, Carya, Liquidambar.*	Broadleaf forest	Humid-warm temperate
	I. *Abies, Pinus, Tsuga, Picea,* Cupressaceae, *Ulmus, Celtis, Betula, Carpinus, Ostrya, Quercus, Tilia,* Rosaceae.	Needle-Broad leaved mixed forest	Cool temperate
Sanmen Fm., Sanmenxia Area, Henan[4]	*Pinus, Picea, Juniperus, Ulmus, Betula, Typha, Carpinus, Artemisia,* Chenoposiaceae, Gramineae, Liliaceae.	Forest steppe	Semi-arid
Zone III, Datong Bains, Shanxi	*Pinus, Abies, Picea, Tsuga, Cedrus, Larix, Ephedra, Artemisia,* Chenopodiaceae, Compositae, Leguminosae, Ranunculaceae, *Betula, Ulmus, Corylus, Tilia,* Polypodiaceae.	Forest steppe	Semi-arid

MIDDLE PLEISTOCENE

The important elements of this sporo-pollen assemblage are *Pinus, Celtis,* Rosaceae, Leguminosae in association with Gramineae, Cyperaceae and Polypodiaceae. Accordingly, the vegetation must have been the deciduous broadleaved forest in the temperate zone. The ostracods present at the same time are *Limnocythere dubiosa, L. sancti-patricii, L. binoda* and *Eucypris inflata,* which are distributed over the areas from Qinghai to Hebei.

The palynological assemblages from the known Middle Pleistocene in North China are listed in Table 2.

156

Table 2 Palynological assemblages from the known Middle Pleistocene in North China

Formation	Pollen assemblages	Vegetation	Climate
Zhoukoudian Fm., Beijing Area[3]	*Betula, Celtis, Ulmus,* Rosaceae, *Salix, Carpinus, Ostrya, Abies, Pinus,* Cupressaceae, *Caragana, Deutzia, Sparganium,* Rhamnaceae, Leguminosae, Gramineae, Cyperaceae, Liliaceae, *Potamogeton, Selaginella,* Polypodiaceae,	Deciduous broad-leaved forest or needlebroad-leaved forest	Humid-warm temperate
Xiehu Fm., Lantian Area, Shaanxi	*Artemisia,* Chenopodiaceae, Gramineae, *Pinus, Betula, Celtis, Carpinus, Borussonetia,*	Park land	Humid-cool temperate
Erlangjian Fm., Qinghai Lake Area	Trees 53%, Herbs 40%, Pteridophyta 6%. *Picea,* Pinaceae, Cupressaceae, *Ephedra, Populus, Juglans, Alnus, Betula, Allium,* Urtiaceae, Chenopodiaceae, Compositae, *Artemisia, Potamogeton.*	Conisilvae	Humid-cool temperate

UPPER PLEISTOCENE

The sporo-pollen assemblages contain abundant herbaceous elements and numerous pollen, such as *Pinus, Picea, Abies, Betula, Ulmus,* Leguminosae, Chenopodiaceae, Compositae, Gramineae and *Artemisia.* The common ostracods are *Ilyocypris bradyi* and *Candonialla albicans.*
The palynological assemblages from the known Upper Pleistocene in North China are listed in Table 3.

Table 3 Palynological assemblages from the known Upper Pleistocene in North China

Formation	Pollen assemblages	Vegetation	Climate
Qian Xian Fm., Lantian Area, Shaanxi[1]	V. *Pinus, Betula, Rosa.*	Forest steppe	Humid-warm
	IV. *Abies, Pinus, Carpinus, Celtis,* Leguminosae, Chenopodiaceae, Cyperaceae, Gramineae, *Artemisia,* Polypodiaceae,		
	III. *Pinus, Quercus, Juglans, Carpinus, Corylus, Ulmus, Celtis,* Compositae, Gramineae, Chenopodiaceae,	Needle broad-leaved mixed forest and forest steppe	Arid-cool
	II. *Pinus, Picea, Abies, Betula, Quercus, Carpinus, Ulmus, Celtis, Artemisia,* Rosaceae, Leguminosae, Gramineae, Cyperaceae, Cruciferae, Liliaceae, Polypodiaceae.	Piceetum and needle broad-leaved forest	Cool temperate

(to be becontinued)

Table 3 *(continued)*

Formation		Pollen assemblages	Vegetation	Climate
	I.	*Picea, Abies, Pinus,* Rosaceae, Leguminosae, Compositae, Polypodiaceae.	Picea-Abies forest	Cold
Malan Fm., Beijing Area		*Pinus,* Pinaceae, *Picea,* Cupressaceae, *Carpinus, Tilia, Juglans, Celtis,* Moraceae, Leguminosae, Chenopodiaceae, *Artemisia,* Gramineae, Polypodiaceae.	Steppe	Arid

HOLOCENE

The sporo-pollen assemblages are predominated by pollen of *Pinus, picea, Betula, Quercus, Carpinus, Ulmus, Artemisia,* among which *Pinus* and *Betula* occupy a dominant position in the early and late stages and more of broad-leaved trees in the middle stage. The vegetation reflected by those assemblages are almost similar to those of the locally living ones respectively. From the Holocene in the Qinghai Lake area, the ostracods *Limnocythere dubiosa* and *Eucypris inflata* have been found.

The palynological assemblages from the known Holocene in North China are listed in Table 4.

Table 4 Palynological assemblages from the known Holocene in North China

Formation			Pollen assemblages	Vegetation	Climate
Beijing Area[6]	Late	Liubin- tun Fm.	Pinus, Picea, Abies, Betula, Corylus, Carpinus	Pine forest	cold
	Middle	Yingge- zhuang Fm.	Pinus, Quercus, Ulmus, Celtis, Betula, Carpinus, Juglans, Corylus.	Broad-leaved forest or mixed forest	warm
	Early	Xiao- jiahe Fm.	Pinus, Picea, Carpinus, Betula.	Pine forest	cold
Buhahe Fm., Qinghai Lake[5]	Early Late		Gramineae, Allium, asparagus, Chenopodiaceae, Leguminosae, Compositae, Artemiasia, Salix.	Steppe	Arid-cool temperate
			Picea, Pinus, Salix, Populus, Chenopodiaceae, Leguminosae, Artemisia, Polypodiaceae.	Forest-steppe	Humid-cool temperate

REFEREACES

[1] Cenozoic Palynol. Gr. Inst. Bot., Acad. Sin. and Inst. Geol. & Min., 1966, Study on Cenozoic palaeontology from Lantian Area, Shaanxi, in: Inst. Verteb. Palaeont. & Palaeoanthrop., Acad. Sin., (ed.) "Symp. Cenozoic Lantian Field Meeting, Shaanxi," p. 157–182. Sci. Press, Beijing.

[2] Liu Jing-ling, 1980, Pollen analysis and geological age of the Nihewan Formation. kexue Tongbao, 25 (3–4), 584–587.

[3] Sun Meng-rong, 1965, Sporo-pollen assemblages from the Peking Man (*Sinanthropus Pekingensis* Black) bed at Zhoukoudian in the vicinity of Peking. Quat. Sin., 4 (1), 84–104.

[4] Song Zhi-chen, 1958, Plant fossils and sporo-pollen complex of the Sanmen Series. *Quat. Sin.*, 1 (1), 118–130.

[5] Lanzhou Inst. Geol. and Inst. Aquatic-Biology, Inst. Microbio. & Palaeont., Academia Sinica, 1979, A report on comprehensive expedition on Qinghai Lake, Sci. Press, Beijing.

[6] Liu Jing-ling, Li Wen-yi, Sun Meng-rong and Liu Mu-liang, 1965, Palynological assemblages from peat-bogs of southern Mt. Yan Shan in Hobei Province. *Quat. Sin.*, 4(1), 105–117.

[7] Liu Jing-ling, Tang Ling-yu, 1980, Pollen analysis of the Nihewan Formation. Paper for the 5th Interuational Palynological Conference.

QUATERNARY OSTRACOD BIOGEO-
GRAPHICAL PROVINCES IN CHINA

HUANG BAO-REN

(Nanjing Institute of Geology and Palaeontology, Academia Sinica)

Abstract—According to the ostracod ecology and stratigraphical and geographical distribution, the Quaternary ostracod biogeographical provinces show some characteristics succeeding to those of the Pliocene in addition to new features. It can be divided into six provinces as follows: North China, Xizang, Yunnan, Central China, Eastern Coastal Area and Southern Coastal Area.

NORTH CHINA PROVINCE

This province comprises Xinjiang, Qinghai, Shaanxi, Shanxi, Henan and western Hebei. It is characterized by the existence of euryhalinous species such as *Cyprideis torosa* (Jones) and *C. littoralis* (Brady), which were related to those of western ancient sea. This province can be subdivided into Xinjiang, Qinghai, Weisang and Yulu subprovinces. Xinjiang subprovince yields *Cyprideis torosa* (Jones) and *Candona nyensis* Gutentag et Benson. Qinghai subprovince bears *Cyprideis littoralis* (Brady), *Candona nyensis* Gutentag et Benson and *Microlimnocythere sinensis* Huang. In this subprovince the lower Pleistocene yields *Qinghaicypris crassa* Huang, the middle Pleistocene bears *Limnocythere dubiosa* Daday, *L. sancti-patricii* Brady et Robertson and *L. binoda* Huang and the Holocene contains *Limnocythere dubiosa* Daday and *Eucypris inflata* (Sars). Weisang subprovince, comprising Xhaanxi, Shanxi and western Hebei, yields *Cyprideis torosa* (Jones), *Limnocythere* and *Leucocythere*. In this subprovince the lower Pleistocene bears *Ilyocypris kaifengensis* Lee and the middle Pleistocene yields *Limnocythere sancti-patricii* Brady et Robertson. Yulu subprovince, comprising Henan and western Shandong, yields *Ilyocypris* and *Candoniella*. The lower Pleistocene in this subprovince commonly carries *Ilyocypris kaifengensis* Lee.

XIZANG PROVINCE

Many species of *Leucocythere* and *Leucocytherella* occur there. The former is also distributed in North China Province. In this province, the lower Pleistocene yields *Qinghaicypris* which is commonly known to occur in Qinghai subprovince of North China Province.

160

YUNNAN PROVINCE

This province is similar to Xizang Province in yielding *Leucocythere*, but the former commonly yields the forms of lake facies, such as *Neochinocythere* and *Candona* and the latter yields continental brackish water genus *Leucocytherella*.

CENTRAL CHINA PROVINCE

This province consists of Anhui, etc. *Ilyocypris* and *Candoniella* were found there.

EASTERN COASTAL AREA PROVINCE

This province comprises the coastal area from Liaoning through Hebei, Shandong, Jiangsu, to Zhejiang. In this province the lower Pleistocene, being of continental sediments, yields *Ilyocypris* and *Candoniella*. The Middle Pleistocene to the Holocene consists of the interbeds of marine and continental deposits. The marine beds mainly lie in the upper part, usually having three beds in total. The continental beds yields *Ilyocypris* and *Candoniella*, and the marine beds always carry *Sinocytheridea* and *Neomonoceratina*. In the northern coastal area, some of the Middle Pleistocene beds carry a mixed fauna consisting of marine fossils and northern inland continental brackish water species such as *Limnocythere dubiosa* Daday and *L. binoda* Huang.

SOUTHERN COASTAL AREA PROVINCE

It comprises coastal area of Guangdong, Guangxi and Taiwan. Many shallow sea genera such as *Neomonoceratina*, *Sinocythere Alocopocythere* and warm and highly saline water ostracods namely *Cytherelloidea*, *Bairdia* and *Uroleberis* were found over there.

From the geological ages, the typical ostracod-bearing Quaternary strata may be briefly described as follows:

1. The Lower Pleistocene. The lower Pleistocene Ayihai formation in Qinghai and the lower Pleistocene in Xizang yields *Qinghaicypris*. The lower Pleistocene Nihewan formation in Hebei contains *Ilyocypris kaifengensis* Lee.

2. The Middle Pleistocene. The Middle Pleistocene Erlangjian formation in Qinghai Lake area yields brackish water lake species *Candona neglecta* Sars, *Eucypris inflata* (Sars), *Limnocythere dubiosa* Daday, *L. sancti–patricii* Brady et Robertson and *L. binoda* Huang. The Middle Pleistocene Hutouliang formation in Hebei yields brackish water lake species *Limnocythere dubiosa* Daday, *L. sancti-patricii* Brady et Robertson, *L. binoda* Huang, *Cyprideis torosa*

161

(Jones) and *Cytherissa lacustris* Sars. The upper Middle Pleistocene in Bohai northern coastal area, being the first marine beds, yields brackish water lake species *Limnocythere dubiosa* Daday, *L. sancti-patricii* Brady et Robertson and *L. binoda* Huang and neritic facies species *Echinocy-thereis* sp. and Aurila sp.

3. The Upper Pleistocene. The upper Pleistocene in Bohai northern coastal area is composed of the interbeds of marine and continental sediments. Among them the continental beds yields fresh water pool and stream species *Ilyocypris bradyi* Sars and *Candoniella albicans* (Brady) and the marine beds contains neritic genera such as *Sinocythere*, *Sinocytheridae* and *Neomonoceratina*.

4. The Holocene. The Holocene Buhahe formation in Qinghai Lake area bears brackish water lake species such as *Limnocythere dubiosa* Daday and *Eucypris inflata* (Sars). In eastern coastal area the Holocene yields neritic species *Neomonoceratina*, *Sinocytheridae* and *Loxoconcha*.

The Quaternary continental ostracods *Ilyocypris*, *Candona*, *Candoniella* and *Cyprinotus* etc. originated from the Tertiary continental shallow water basin of China, and some genera of brackish to fresh water lake facies such as *Leucocythere*, *Leucocytherella* and *Limnocythere* originated from the Mesozoic and Cenozoic lakes in China. Few euryhalinous genera such as *cyprideis* originated from the Tertiary marine in Tarim Basin of China and even closely related to the Tertiary sea in Europe. The marine ostracods from the Miocene to the Holocene in eastern coastal area came from Pliocene and early Pleistocene seas of South China. Some marine genera like *Neomonoceratina* has been found from early Tertiary brackish water basin connected with ancient sea in eastern coastal area of China. The Quaternary ostracods in southern coastal area originated from the Teriary sea in this region and even from the Mesozoic to Eocene sea in southern Xizang in China.

REFERENCES

[1] Huang Bao-ren, 1964, Ostracoda fossils from Gan Sen District, Qaidam Basin. 12 (2), Acta Palaeontologia Sinica 12 (2).

[2] Huang Bao-ren, 1980, A preliminary study of Pleistocene Ostracoda from middle and lower Sanggan River valley. Kexue Tongbao, 25 (3).

[3] Lüttig, G., 1955, Die Ostrakoden des interglacials von Elz. Paleont. Z. 29 (3/4).

[4] Lanzhou Institute of geology, Academia Sinica et al., 1979, A report of comprehensive expedition on Qinghai Lake, Science Press.

MICROFOSSILS AND PALAEOGEOGRAPHY
OF THE NIHEWAN FORMATION

WANG QIANG

(Tianjin Institute of Geology and Mineral Resources Chinese Academy of
Geological Sciences)

AND WANG JING-ZHE

(Geological Bureau of Shanxi Province, Ministry of Geology)

ABSTRACT

This paper deals with the microfossils (ostracods, foraminifera, charophyta),
and fossil gastropods and pelecypods collected from the Nihewan formation
in Datong Basin (Shanxi Province) and Yangyun-Yuxian Basin (Hebei Province). In the formation four fossil assemblages have been found from the bottom upwards: *Candona-Candoniella-Ilyocypris*; *Limnocythere Limbosa*; *Limnocythere lacioidea — L. flexa*; *Limnocythere dubiosa*. The fossils reported
in this paper are useful palaeoenvironmental indicators, because the fossil
records reflect the physical environment. Based on the fossil assemblages,
reconstruction of the palaeogeography and the evolution of the mentioned
lacustrine basins in different stages of Nihewan—the Lower Pleistocene in North
China can be made.

THE OSTRACODA FAUNA OF LATE QUATERNARY AND PALEOENVIRONMENT FROM SHALAWUSU (SJARA OSSO-GOL) RIVER DISTRICT, INNER MONGOLIA AUTONOMOUS REGION

WANG QIANG

Tianjin Institute of Geology and Mineral Resources,
(Chinese Academy of Geological Sciences)

SHAO YA-JUN
(Lanzhou Institute of Desert, Academia Sinica)

AND ZHAO SHI-DE
(Geological Bureau of Inner Mongolia Autonomous Region, Ministry
of Geology)

ABSTRACT

This paper deals with the palaeoecology of Late Pleistocene-Holocene ostracod fauna collected from Shalawusu (Sjara Osso-Gol) River district, Inner Mongolia Autonomous Region.

The ostracod fauna of Shalawusu formation of early Late Pleistocene was divided into two assemblages. The lower assemblage includes *Candona* and *Candoniella,* suggesting swamp deposits; the upper assemblage includes *dubiosa* suggesting standing water deposits of lake with high salinity, and it is corresponding to the Last or Eemian interglacial ostracod assemblage.

According to palaeoecology of ostracods the authors infer that the Xisha formation consisting mainly of fluvial deposits can be referred to, late Late Pleistocene in age.

The Shalawusu Dagouwan consisting of deposits of slowly flowing water, lake-swamp and shallow lake can be assigned to Holocene in age.

164

THE DISTRIBUTION OF THE RECENT CORALS REEF OF CHINA AND ITS SIGNIFICANCE IN QUATERNARY STUDY

LIANG JING-FEN

(Department of Biology, South China Normal University)

Abstract — The author points out that the most important factor that influences the distribution of the reef corals is the water temperature.

According to the ecological environment of the reef corals, four regions may be divided in China as follows:

1. South China Sea Islands,
2. Taiwan Islands,
3. Hainan Islands,
4. Coast of Mainland.

The author enumerates in this paper 453 species in 72 genera and subgenera of the reef-corals which have a wide distribution in our country. 11 species of Hydrocorallinian are listed. Helioporian and Tubiporian are widely distributed in the Indo-Pacific regions, but are not abundant in the reef of our country.

The boundary of the corals distribution may be set on the north of the Taiwan Straits, the northern part of the Taiwan Island, and the Diaoyu Islands, which is also the northern boundary of the distribution belt of the reef-corals in the world. The boundary is coincident with the isotherm of 13°C of annaul winter water temperature.

The most important factor that influences the distribution of the reef-corals is the winter water temperature. The facts that the reef-corals are always absent in the river-mouths where cool water flows out to the open sea, may serve as a good example of the explanation. On the contrary, under the influence of the warm current, an entirely different phenomenon occurs, which may be found in the southern part of the Taiwan Island, where the fringing reefs are developed longer than 60 km along the coastline, uninterrupted by rivers where the warm water running out to the sea seems to have no influence on the fringing reefs.

There are more species of reef-corals near the warm current than away from it, so it may be noted that the further from the South China Sea warm current, the more species of the reef-corals decrease. The annual growth rate of the reef-corals is also controlled by the winter water temperature.

The rate of the reef-corals growth in the southern part of Taiwan Island is higher than that in the northern part of the Island, due to the water temperature of the former.

According to the ecological environment of the reef-corals, four regions may be divided from the coastal area in China: (1) The South China Sea Islands Region: (2) The Taiwan Islands Region: (3) The Hainan Islands Region and (4) The Mainland Coastal Region.

THE SOUTH CHINA SEA ISLANDS REGION

The author enumerates more than 160 species in 47 genera and subgenera of the reef-corals which have a wide distribution in this region.   The characteristics of the distribution of the reef-corals may be noted as follows: It is in the most prosperous region of the reef-corals of our country, and secondly, it covers a wide area, with its northern boundary set between on the south of the Taiwan Strait and on the south of the latitude near 3°52′. According to the ecological environment of the reef-corals, mainly to the factor of the water current, two regions may be divided from the South China Sea:

(1) Region of Dongsha and Xisha Islands.

(2) Region of Nansha and Zhongsha Islands.

The important factor that influences the water temperature of the region is the South China Sea warm current.   More species of the reef-corals are found near the water current than away from it, for example, 71 species are found in the Dongsha Islands, but 165 species in the Xisha Islands.

The Nansha Islands are also the prosperous region for reef-corals growth. In one of the atolls, Tizard Atoll, 106 species of reef-corals are collected.   In Zhongsha Islands, the drowned atoll, 30 species are found.   Some species are found in a depth of over 70 m, such as the *Pavona papyraces* (74 m), *Rhodarea largrenii* (74 m).

THE TAIWAN ISLANDS REGION

In general, the optimum temperatures for the growth of reef are not below 18°C during winter.   But in the Taiwan Strait some species can live between 13°C and 18°C.   An attempt was made to divide the water temperature zones of the reef-corals into the following zones:

Critical temperature	13°C
Growth temperature	13°C—18°C
Optimum temperature	18°C

In the northern part of Penghu Islands, the winter water temperature is always below 13°C, the reef-corals are killed by the cold water at the fringing reefs (such as in February, 1946).   So the water temperature is the important factor that influences the distribution of the reef-corals in the region.   From Table

166

1 it can be noted that more species of reef-corals are found near the Taiwan warm current than away from it.

Table 1 Relation between species number and water temperatures

Location	Lanyu	Eluanbi	Penghu	Aotou (Mainland coast)
Number of species	88	87	66	13
Winter water temperature	25.1°C	22°C	17.2°C	13°C

From Table 2, it may be noted that the further from the Taiwan warm current, the fewer species are found.

THE HAINAN ISLANDS REGION

The Hainan Islands Region consists of not only the main island but also islands or banks separated from the main island. The features of this region is very different from those of the South China Sea Islands and the Taiwan Islands. The islands are also an abundant zone of the reef-corals. The most luxuriant coral flat is observed in the southern part of the island, where about 87 species can be found at Luhuitou. Table 3 shows that more species of reef-corals are found near the warm current than away from it.

Table 2 Relation between species number and water temperatures
(Taiwan warm current)

Location	Lanyu	Suao	Jilong	Penghu
Number of species	88	82	66	66
Winter water temperature	25.1°C	22.7°C	19.9°C	17.2°C

Table 3 Relation between the number of species and the warm current

Location	Xisha Islands	Luhuitou	Shalao	Linchang
Landform	Atoll	Fringing reef		Barrier
Number of species	113	87	33	22
Winter water temperature	24.5°C	24°C	21°C	18°C

The reef-corals are also widespread in the lagoon along the coast of Hainan Island. The total number of the species of the reef-coral of the islands is not less than 168 of 43 genera and subgenera.

The Hainan Island belongs to the optimum temperature zone for the reef-corals, but the species are not as many as those of the South China Sea Islands Region, for example, there are 5 species of *Favites* in the South China Sea Islands but only 3 species in Hainan Islands. The following table would show this phenonmena (Table 4)

Table 4 Comparison of the genera between Hainan Islands and South China Sea Islands

Genera	South China Sea Islands	Hainan Islands
Favites	5	3
Favia	8	4
Fungia	9	5

The number of the same species both in the South China Sea Region and the Hainan Islands Region are 63, which indicates close relationship between the two regions.

THE COAST OF MAINLAND REGION

Along the mainland coast, there are about 45 species recorded. It is a sparse reef-corals zone. Since there is not enough clear water for photosynthesis, salinity is not constantly normal, fresh water from rivers or storm torrents easily damage the reef corals and the winter water temperature is always below 15°C. So the reef-corals are mainly distributed in the islands along the coast, such as Shangchuan Island and Xiachuan Island, etc.
It seems that the distribution areas of recent coral reefs have been enlarged after Pleistocene.

REFERENCES

[1] Ma Ting-ying, 1937, On the growth rate of reef corals and its relation to sea water temperature. Nat. Inst. Acad. Sinica. Zool. Ser, no. 1.
[2] Bassett-Smith, P.W., 1980, Report on the corals from Tizard and Macclesfield Banks Ann. Mag. Nat. Hist. 6th Ser, 6.
[3] Vaughan, T. W., Wells, H. W., 1943, Revision of the suborders, families and genera of the Scleractinia. Geol. Soc. Amer. Spec. Papers 44.
[4] Yabe, H., Sugiyama, T., Eguchi, M., 1936, Recent reef building corals from Japan and the South Sea Islands, Ibid. 1.
[5] Yabe, H., Sugiyama, T., 1941, Recent reef building corals from Japan and the South Sea Islands Sci. Rep. Tohoku Imp. Univ. 2nd Ser. (Geol.) Special 2.
[6] Siro Kawaguti, 1953, Coral fauna of the Island of Botel Tobago, Taiwan, Biology Journal Okayama University, 1 (3).

MIDDLE PLEISTOCENE MICROMAMMALS FROM HEXIAN MAN LOCALITY AND THEIR SIGNIFICANCE

Zheng Shao-hua

(Institute of Vertebrate Palaeontology and Palaeoanthropology, Academia Sinica)

ABSTRACT

A complete skull of fossil man was excavated from Hexian County, Anhui, in the winter of 1980. Associated with the skull are rich remains of mammals, of which 23 species and subspecies belonging to 11 families are micromammals. The micromammalian fauna is characterized by: 1. Its smaller percentage of fossil species and genera (21.7%) in the mammalian fauna; 2. the northern grassland animals and the western mountainous animals are mixed with the southern forest animals; 3. domination of the moist- and hydro-philous species. The fact mentioned seems to indicate that the living conditions of the ancient fauna are in disagreement with the present environment of the fossil man locality.

Based on the presence of the giant beaver, *Trogontherium cuvieri*, and the common occurence of 15 or 16 species of micromammals both in Hexian Man and Beijing Man localities, the two faunas are considered to be roughly the same age.

The Hexian Man-bearing bed which is quite homogeneous, about 0.5—1m thick, can only be corresponded to part of the thick sequence of Beijing Man locality, which is about 40 m thick and 11 layers are subdivided. The climatic changes inferred on the basis of the changes of the warm- and cold-loving micromammals distribution in different layers of Beijing Man locality, suggest that the cold-loving animals dominant in layers 5 and 4 are precisely the same as those in Hexian fauna. Though giant beaver is absent from layer 4, it is reasonable to correlate the bed of Hexian Man with layer 5, representing the deposit of one of the phases of the Dagu Glacial in China (as equivalent to Mindel Glacial of Europe).

QUATERNARY FAULTING IN CHINA

DING GUO-YU

(State Bureau of Seismology)

Abstract — Combining the studies of seismic risk zonation, earthquake prediction and engineering safety, a lot of observations and studies on some major Quaternary active fault zones in China have been made. The comprehensive studies have provided abundant data and some new ideas in understanding the features of Quaternary faulting. In this paper, certain information and some characteristics of the Quaternary faults in China are discussed.

THE DISTRIBUTION OF THE QUATERNARY FAULTS IN CHINA

The distribution of the Quaternary faults in China, in general, inherits mainly from the patterns of Mesozoic and Cenozoic faulting. It can be divided into eastern and western parts as demarcated by 105°E. In the eastern part, there are mainly tensional and tensional strike-slip faults with NE and NNE trending while in the west there are mainly strike-slip and thrust strike-slip faults[1].

In the western part, the Qinghai-Xizang (Tibet) Plateau is confined by a series of huge thrusts and strike-slip faults which have been strongly active since Quaternary. Within the plateau there is a set of NW and nearly EW trending huge strike-slip faults, which mainly belong to left lateral strike slip ones with some extent of compression during the Quaternary activity except some faultings in the southern and western parts. They offset geological terrains and distort a series of recent geomorphs such as drainage system. All these remarkably reflect in homogeneous lateral slips among the land-masses since the Quaternary or Himalayan episode in the Qinghai-Xizang Plateau. At the southeastern side of the plateau, a number of faulting belts confines a rhombic block-Sichuan-Yunnan rhombic block that slips and moves toward southeastern direction.

In the area to the north of Qinghai-Xizang Plateau, the Quaternary active faults are mainly distributed around the Tarim, Junggar and Alxa blocks and most of them are mainly characterized by compressional strike slip activities.

In East China, the Quaternary faults are mainly characterized by tension and tension-torque activities, such as the faults along the margins of the Ordos and a series of NE and NNE trending faults along the margins of basins in North China, Northeast China, and their lengths are relatively short.

In South China, the relatively active Quaternary faults are mainly distributed around southeastern coast, and their activities tend to be weaker from Taiwan toward inland.

170

DISPLACEMENT MAGNITUDE AND SLIP RATE OF THE ACTIVE FAULTS IN CHINA DURING THE QUATERNARY PERIOD

The total magnitude of vertical displacement along the faults since Quaternary is usually obtained through analysing the information about the altitude changes of denudation and terrace surfaces and the thickness of the Quaternary deposits on both sides of the faults. Accompanying by the thrustings of the faults since the Himalayan movement tremendously thick molasse deposits were formed in the piedmont depressions (in some regions named as Xiyu gravel group), such as in the Wuqia Area north of the Pamir piedmont, where the thickness of the molasse deposits reaches over 8,000 m, in Hotan-Yecheng Area of the Kunlun Mts. piedmont over 5,000 m thick, in the Artux Area of the Tianshan Mts. southern piedmont 5,000 m, in the Kuqa Area over 4,000 m, in Manas Area of the Tianshan Mts. northern piedmont 2,000 m and in the Baytik Mt. southern piedmont of Altay Mts. only less than 200 m. It can be seen that the magnitude of vertical dislocation of the Quaternary faults in West China reflected by the thickness of the deposits is gradually decreasing from south to north. It is analysed that if the deformation of the Early Quaternary Xiyu gravel group is taken as the total amplitude of tectonic movement since Mid-Pleistocene, the slip rate of faults at northern Pamir, western Kunlun Mts. piedmont, the southern piedmont of Tianshan Mts., the northern piedmont of Tianshan Mts. and Baytik Mt. is about 1.6 mm, 1 mm, 0.7 mm, 0.5 mm and 0.1 mm per year respectively. The vertical slip rate of the faults at Qilian Mt. piedmont in Quaternary was about 1 mm/year. In North and East China, the vertical slip rates of Quaternary faults along the margins of fault-bounded basins are greater only at Yinchuan Basin and Weihe Basin reaching over 1 mm/year (2.3 mm/year and 1.8 mm/year respectively) and those of the others are less than 1 mm/year and some of them even below 0.1 mm.

The most direct method for obtaining the total magnitude of horizontal displacement of Quaternary faults is to measure the torsion and offset distances by faulting on geological terrains, drainage systems and other geomorphs since Quaternary in field. In West China, due to the particularly favourable conditions of tectonics, morphology and climate, a variety of geological and geomorphological phenomena created by the horizontal displacement of many Quaternary active faults are displayed very clearly. Taking the Koktokay — Ertai fault along the northern margin of Junggar Basin in Xinjiang for an example, there are visible traces of fault horizontal displacement offsetting morphology and sediments of different periods, the valleys appeared in Upper Pleistocene have been displaced up to 700 m. There is a horizontal displacement about 2 km long in the river valley formed since Mid-Pleistocene. And there has been a right lateral displacement 26 km long on this fault since Neogene. A number of paleoearthquake traces are found along the fault indicating that there is a long history of fracturing activities in presentation of earthquakes along the fault.

The Altun fault located on the northwestern side of the Qinghai-Xizang Plateau is one of the largest strike slip active faults in China mainland. The Quaternary fan deposits and many river streams were deformed by the horizontal displacement along the fault. On the northeastern segment of the fault, the Shule river-bed that crossed the fault has been horizontally displaced more than 4 km. And moreover, some geological terrains and streams along the fault have been displaced from several tens of kilometers to even hundred kilometers.

In Quaternary, the thrusting of the Qilian Mts. fault located on the northern side of the Qinghai-Xizang Plateau was predominant, and the data obtained from geology, earthquake mechanism, earthquake faults, deformation measures, etc., obviously show this feature, and the horizontal displacement of this fault is also significant as well. Such as in the Changma fault there are a number of drainage systems which have been torsionally displaced. Extending southeast till the Liupan Mt., the horizontal slip movements along the West Huashan Mt. and South Huashan Mt. faults are the major form of recent faulting. In this region, the horizontal displacement at Haiyuan since Late Pleistocene reached up to 2—3 km. In Southwest China the Quaternary horizontal displacement along the Xianshuihe fault, Anninghe fault, Zemuhe fault and Xiaojiang fault in Sichuan-Yunnan Region left a large number of evidences on geological terrains and geomorphs. The southeastern segment on the Xianshuihe fault, the left lateral horizontal slip reached about 5 km during the Himalayan Episode. The phenomena of cutting drainage systems appeared here and there obviously. Since Late Pleistocene, on the northern segment, the torsional displacement reached about 1 km at Luhuo; on the middle segment, the torsional displacement reached 500—700 m at Daowu-Mazhi Zone; and on the southern segment, the drainage system nearby Kangding was left laterally torqued and the displacement reached 250—300 m. The horizontal slip rates of the fault derived from the torsional displacement of drainage systems reached 0.5—1 cm/year on the northern segment and 0.3 cm/year on the southern segment, which are similar to 0.9 cm/year of average slip rate along the fault obtained by the Brune[2] method on the basis of the data of strong earthquakes with magnitudes greater than 6 since 1725. Nearby Chechejie town along the Zemuhe River, a number of tributaries were left laterally torqued up to 1 km and more and some ancient tombs of Qin and Han Dynasties (about 2,000 years ago) along the Zemuhe fault were clearly deformed by horizontal left lateral torsions. In Dongchuan, Xundian and Yiliang regions of Yunnan Province, ^{14}C dates of the Holocene strata cut by the Xiaojiang fault were determined and the left lateral slip rates of the eastern and western branches of the Xiaojiang fault during Late Holocene were achieved, which are 4.8 ± 0.4 mm/year and 6.1 ± 0.4 mm/year respectively. ^{14}C date of strata nearby Dabaihe stream of Xundiangong mountain area is about $2,700 \pm 180$ years and the terrace was offset to 13m; ^{14}C date of strata nearby Yiliang-Maixi Area is about $1,322 \pm 85$ years and the layers of deposits were offset to 8m.

172

The Honghe (Red River) fault in Southwest China is a huge clear-cut active fault showed on landsat images, and the horizontal offset indicated by the cutting drainage systems reached up to more than 2 km and some segments show that the offset reached even as wide as 5—9 km.

The horizontal displacements of the Quaternary faults in the eastern part of China are not as intense as those in the West. However, the data obtained from all sources in recent years show that the ratio of horizontal and vertical displacements of the faults in North China since Quaternary or Pliocene is about 2:1. In general the total magnitude of horizontal displacement of the Quaternary faults is about several hundred metres to 1—2 km and the average slip rate during Quaternary was about several tenths of milimetres to 1 or 2 mm per year. But along tensional basin boundaries or in the transform junctions, the horizontal displacement is likely greater. For example, the Shilin fault of Huoxian in Linfen Basin left laterally offset Fenhe terrace of Mid-Pliocene up to 600 m and the average horizontal slip rate is 5 mm/year. Helan Mt. eastern piedmont fault in Ningxia right laterally offset the Great Wall of Ming Dynasty (1,506 A.D.) up to 1.45 m and the average slip rate is more than 3mm/year.

Based upon the data obtained from all possible sources within the country, the total horizontal displacement magnitude of the major active faults in China was about several km during the Quaternary period, and most of them reached 2—5 km. Along some most intensely active faults, especially those located on the margins of huge blocks, the total horizontal slipping magnitude might reach tens of kilometers during Quaternary.

Since Late Pleistocene, the horizontal displacement of the faults mostly reached tens to one hundred metres and some might reach several hundreds of metres. Since Holocene, the horizontal slip magnitude of the faults along some major active fault zones was mostly several to tens of metres.

In order to obtain actual data of the modern behaviour of the Quaternary faults, a large amount of deformation measuring work has been carried out in recent ten years or more, in which the short levelling and short base-line surveying crossing the faults shows a better observational effect. The general situation of recent activities of the faulting in different places are: the vertical dislocation rate for many faults in the North China Plain is about several hundredths to several tenths millimetres per year, and along some faults the vertical dislocation rate reaches up to several millimetres per year, and on the other hand the magnitude of horizontal displacement is as twice as that of vertical displacement. In Qilian Mt. and Liupan Mt. of Northwest China both base lines and triangulation survey demonstrate a shorterning in NE trending. In Southwest China, many surveying sites and triangulation neoworks crossing the faults are disposed along the Xianshuihe fault, Zemuhe fault, Xiaojiang fault, etc. Repeated surveying results show that these belts present a compressional and left lateral motion. The ratio of horizontal and vertical displacements along the Xianshuihe fault is about 6:1. In general, the data obtained from recent deformation surveying indicate that the pattern of these fault activities is consistent with that of faulting activities in a long history. However, the recent activity rate along the faults is higher than the average rate in a long time. In addition, there are many cases of earthquake in the historic past showing that some faults display significantly increased activity rates before a strong earthquake.

SOME CHARACTERISTICS OF QUATERNARY FAULTING ACTIVITIES IN CHINA

Sudden slipping occurred in an earthquake and slow creeping are the two basic forms of faulting activity[3]. In China, the seismic activity is intense and widely spread, which reflects directly that sudden slipping form is dominant in fault activities over China.

Based on the analyses of seismic activities in the long history of China, the seismic activities in China often bear the characteristic that strong earthquakes usually occur systematically and continuously along the fault zones. There are alternations of period of clustering activity and quiescent stage. Moreover, there is also feature that the seismic activities shift to certain surrounding seismic belts and regions of alternate and echo successively with each other and also a regularity that the equal distance is found in certain segments along which strong earthquakes frequently occur. All there seismic patterns essentially reflect a series of characteristics of faulting activities[4].

With respect to fault creep, in China there are no creeping faults found such as San Andreas fault in California of the United States. However, the phenomena of creeping in a shorter period before a large earthquake have aroused much attention in recent years. Some deformation data indicate that there were some creeping phenomena along the fault before the Tangshan earthquake. Along the Xianshuihe fault in Sichuan Province where a strong earthquake with magnitude 7.9 occurred in 1973, the fault creeping lasted for 3 years after the quake.

The inland-blocks are in mosaic with each other and the fault is the pivotal axis of the blocks movement. The faults are intersected, impeded and adjusted with each other, and with a characteristic of transforming one form into another during the movement. The ends of an active fault are bounded to transform into other forms in various patterns thus to regulate the tension, shortening and torsion of the earth crust generated by the blocks movements[5]. In China, many Quaternary tensional fault-bounded basins, compressional folding uplifts and magma activities as well as the scattered fault branches are considered as the results of the movement between blocks and the activities of the intermass faults. In recent years, thanks to the abundant data accumulated from geology, remote-sensing, seismology and crustal deformation observations, it is possible to draw a sketch of the relationship and transformation between the Quaternary major active faults within China mainland. And these also

174

shed some new light on the basic situation of intra-continental neotectonic deformation and stress field.

REFERENCES

[1] Institute of Geology, State Bureau of Seismology of China, 1979, Explanation on the seismotectonic map of the People's Republic of China. Cartographic Publishing House.

[2] Brune, J. N., 1968, Seismic moment, seismicity and rate of slip along major fault zones. J.G.R., 73, 2.

[3] Allen, C. R., 1975, Geological criteria for evaluating Seismicity. Geol. Soc. Amer. Bull., 86.

[4] Ding Guo-yu, Li Yong-shan, 1979, Seismicity and the recent fracturing pattern of the earth crust in China. Acta Geologica Sinica, vol. 53, no. 1.

[5] Tapponnier P., Molnar P., 1977, Active faulting and tectonics in China. J. G. R., vol. 82, no. 20.

175

CHANGES OF NATURAL ZONES IN CHINA
SINCE THE BEGINNING OF CENOZOIC ERA

ZHOU TING-RU

(Beijing Normal University)

Abstract — This paper deals with the paleogeography of China paying special attention to the shifting of the natural zones and notable climatic changes as well as to the general tendency of climatic development during the Cenozoic time. Such changes were considered mainly with regard to the intensive tectonic movement and the influence of the Quaternary glaciation. The author gives a brief picture for the zonality of Oligocene time of China as an example of Early Tertiary. Important changes are dealt with changes of environment in areal differentiation at various ages in China during Late Tertiary and Quaternary Time.

During Cretaceous and Early Tertiary, the tectonic movement in China was quite weak, and the mountains that were raised in the previous orogenic phase were levelled off by peneplanation. Most of Asia was surrounded by warm seas, containing warm marine fauna. Since there was no polar ice-caps, the oceanic and atmospheric circulation in the early stages was much weaker. The climatic zoning was less marked and the latitudinal temperature gradient probably gentler. The climate at this time was apparently extremely warm and moist, as is indicated partly by the distribution of a weathering crust on the earth's surface. The atmospheric circulation in the Early Tertiary should be considered in reference to the planetary system and normal pattern of natural zones.

The following table presents a brief picture of the environmental changes in Oligocene as an example:

In Late Pliocene and Early Quaternary, a vigorous uplift took place in the western part of China. It formed an immense area of high plateaux and lofty mountain ranges which descend gradually toward the eastern coastal plain. This remarkable change of relief caused a rearrangement of the natural zones, making them entirely different from those of earlier geological time. Notable changes occurred in the ancient physical environment as follows:

1. The extinction of the Tethys ocean and the unification of the Asian. European and Indian continents caused an increase of aridity in the interior part of the new, Euroasian continent and established the monsoon circulation system, which destroyed the original planetary system of the old Tertiary period. These large enclosed basins of the inner part of China were almost cut off from the moisture of the ocean.

2. The uplift of the Xizang (Tibetan) Plateau caused a distortion of the normal pattern by crowding extra tropical cyclonic tracks eastward, reinforcing the

dynamic activities of the monsoon circulation.

3.   At the end of Pliocene, the north polar region already had entered the pre-requisite stage of the Quaternary glaciation, and the resulting cold air mass penetrated Chinese territory in winters, the climatic zoning consequently became more marked, especially in eastern China, and the latitudinal temperature gradient undoubtedly was steeper.

The above-mentioned prominent changes of tectonic movements to a great extent explain the changes of climate and environment which occurred in three major zones of China.

EAST CHINA (NORTH, CENTRAL, AND SOUTH)

In East China, climate came to be influenced by a monsoon pattern of circulation, which caused seasonal changes in the landscape. This region possesses relatively low relief, from north to south, so the solar heat is unevenly distributed according to latitudinal position; thus, several sub-zones emerged, as follows:

Late Tertiary to Early Pleistocene

North China embraces the Northeast provinces and North China proper in a broad sense. The climate here became colder in Late Tertiary. This is indicated by plant fossils, such as in the "Wumiji" formation, which show a gradual decline of thermophilic elements in vegetation that was predominant in Early Tertiary, as well as the appearance of phytocoenosis, evidence of moderate or even cold winter climate. It should be noted that this cold spell of Early Pleistocene produced a Taiga forest, with a permafrost formation, in the northern portion of Da Hinggan (Greater Khingan) Mts.

In North China, "the Shan-Wang" flora of the Miocene age[3] is evidence of a summer-green forest. At the time of Pliocene and Early Pleistocene, fossils of the "Po-te" and "Sanmen"[4] beds show the winter to have been dry, and the summer to have been humid, herbivorous animals wandered about on the open grassland of the great plain. Thus North China could be divided into two zones: a) the Northeast Provinces, characterized by a moderate, warm climate; with a mixed forest composed of coniferous and broad-leafed deciduous trees; and, b) North China, with a summer-green forest and grassland.

Central China involves a broad belt of subtropical climate. After the establishment of the monsoon circulation pattern from Late Cenozoic, the area was influenced by the summer monsoon rains and continental cyclonic storms which mitigated the effects of the dry, subtropical high pressure cells of the northern hemisphere. Thus, a landscape of subtropical evergreen and deciduous mixed forest formed, which can be identified by pollen analysis from the "Gutagtang"

177

peat beds near Dongting Lake and some other localities[5]. But some high pressure cells still occurred in areas of the middle and lower Changjiang River basins, where the weather was hot and dry during the summer, so that the chemical weathering process in this section was significant. Some seams of gypsum in the lacustrine beds are still observed.

In South China, sediments containing pollen of subtropical evergreen trees as well as some deciduous trees are still found in fossil assemblages. In the most southern part, the fossil remains of the basin deposits reveal the presence of a tropical rain forest and tropical animals.

Middle and Late Pleistocene

In North China conspicuous climatic changes occurred in the Middle and Late Pleistocene periods. An alternation of relatively cold and warm periods, related to the regression and transgression of the sea, appeared more pronounced than during the previous time. During the warm stage, caused by the inward flow of warm, humid air from the seas, the zonal pattern was probably quite similar to that of the present time. The well-known "Peking Man" and the animals associated that were found in Zhoukoudian Cave deposits represent this period of the early Middle Pleistocene. In the cold stage, the polar air mass pushed southward to the northern foot of the Qinling Mountains (Lishi stage), or even as far as the northern side of Taihu Lake, south of the lower Changjiang River (Malan stage). This can be determined by the southern limit of the early loess distribution, where the polar front frequently visited over a certain period of time. During the cold stage, the climate became quite severe in the most northern part of East China. Development of permafrost on uplands and deposits in basins were observed. Sediments usually have no appreciable traces of chemical weathering, but contain a good deal of humus and a "chernozem" soil layer. A great part of the great plains was occupied by low lying marshes as well as extensive herbs, scrubs and trees.

In North China, plateaux and rolling hills were covered by a veneer of loess and sediments of different origins. During the age of the last glaciation, the loess steppe extended 4 degrees of latitude further southward than at present. The mammoth and *Coelodonta antiquitatis* fauna are noted as mammals of this stage.

Owing to the insufficient moisture in the cold air mass, no glaciation exited in East China except on those high peaks of Taibai Mt. in the Qin Ling Mts. and Mt. Yu Shan in the Central Range of Taiwan, both of which rose above the ancient snowline. Here some traces of Quaternary glaciation have been preserved.

During Middle Pleistocene, the loess deposits in North China manifested the intercalationa of paleosoils and horizons of weathering to indicate frequent fluctuations of the climatic conditions. In the warm stage, the weathering of

178

the loess of the earlier stage into orange colour was taking place. Active wash-out of the loess from uplands occurred and was redeposited on the low plains to form alluvial and lacustrine beds which retained plentiful lime concretions. In the southern part of East China, the temperature fluctuations did not seem to be significant; this may be ascertained by the character of fossil animals which possessed a thermophilic or tropical habitat, and also by the lithological character of the sediments which show significant traces of active weathering. Consequently, an increase of humidity and temperature caused the formation of iron oxide concretions and plinthilic horizons. The occurrence of the latter was confined to an area between the Lower Changjiang Valley and the Nan Ling Mountains and formed the index soil of the Middle Pleistocene. Moreover, a profound weathering process contributed to the formation of an iron oxide crust, with laterites in alluviums and on rocks surfaces as is noted in South China. This erosion and corrosion produced high terraces and a magnificent karst landscape. The area's alluvial and cave deposits contain tin and zinc minerals, as well as mammal fossils, such as the well-known *Ailuropoda-Stegodon* associates.

Holocene

After the last Quaternary glaciation, the polar ice caps diminished, causing a climatic change and a rise in the level of the sea. This transgression of the sea reduced the width of the coastal plains and created shallow continental shelves. Higher temperatures shifted the natural zones toward the northern pole, which resulted in the migration of subtropical plants and animals to the north. As a result, a degradation of permafrost and solifluction occurred in the most northern part of East China.

In North China, an outwash of loessal material reoccurred. This climatic change is mostly determined from the changes observable in the peat beds. Three or four sub-stages can be divided through such techniques as pollen analysis. The time of this so called "climatic optimum" has been variously estimated, ranging from 6,000—8,000 years BP By the process of transgression of the sea, it would seem that by about 6,000 years ago, the modern sea levels were approximated. This period of time of near stabilization dates the beginning of the present detailed outline of the coast, the cut of present sea cliffs, the building of sand spits, lagoons, modern beaches and bars, and the formation of salt marshes and new deltas.

THE ARID REGION OF NORTHWEST CHINA

This arid region stretches from Nei Monggol (Inner Mongolia) to Xinjiang (Sinkiang). Humidity decreases gradually westward, along with changes of landscape, as one moves across the steppe into a desolate desert. Two sections can be differentiated as follows:

The Steppe and Steppe-forest of Nei Monggol (Inner Mongolia).

Because of the uplift of marginal mountain barriers into the Nei Monggol plateau, there was a decrease of oceanic influence, resulting in an increase in the aridity of the climate and creating the prevalent steppe and the forest stepps pattern of the region.

In Pliocene and early Pleistocene, steppe grasslands dominated the area at the north bounded by the Yin Shan Mt., which functioned as a division between the interior drainage system in the north and the exterior drainage system in the south. In the north, the vegetation was comparatively scanty. The arid denudation process and short distance transportation of detrital material were all significant features. Variegated formations filled up the depressions, which also contain carbonates, salts and gypsums. The Pliocene deposits show clear evidence of weathering, and the presence of secondary clayey minerals, especially with beidellite and illite, indicates the semi-humid to semi-arid climate.

South of Yin Shan Mt., a relatively more humid climate prevailed in summer, which provided better conditions for plant growth. The landscape was grass-land, with patches of trees. Sediments were formed as they were brought from higher places by the torrential waters of storms. The clayey materials from the basalt plateau retained its neutral ratio of weathered crust minerals. In Late Pliocene, the northern temperate steppe became very dry. Winds swept over the soft sediments on the land surface, and fine sands and silts drifted into sanddunes and loess deposits.

In the south, the region was relatively desolate. Steppe grass, swamps, shallow lakes and saline patches were scattered all over the open land. Sanddunes and loess deposits existed extensively along the southern margin of this region. At the end of Pleistocene, the rock sea (Felsmeer) and permafrost were still found in higher regions.

During Holocene, the climate became warm and less arid. The steppe grass grew abundantly, and forest growth occurred on mountain slopes. The chestnut soil developed on the grassland. The water level of the lakes rose, creating an environment that supported distinctive fish and birds and producing a favorable setting for neolithic man. The sanddunes were stablized by a dense growth of scrubs. But overgrazing of the steppe grass by domestic herds and reclamation of the land by farmers in historical time, removed much of the vegetation cover and revived the activity of sanddunes and the shift of sand deposits.

Proceeding westward from the steppe grassland, a far drier climate resulted in desert conditions. In contrast to the eastern area, the western zone is less favorable to all forms of life. During the late Cenozoic era, the vigorous uplift of mountains in the western part resulted in increased aridity for these large, enclosed basins. They were characterized by an ineffective drainage and

180

a deep detrital cover that could not be transported out of the basins by streams. Although the vegetation was sparse, yet zonality can still be identified by plant growth. In the Junggar Basin, for example, the temperate steppe is dominated by a dense growth of *Artemisia*.

The flora of the Tarim Basin is represented by the "Kucha" formation of the Miocene age, when plants seemed to be related to the supply of ground water, but fossils show xerophytic or succulent characteristics and indicate that the climate was drier here than in the north. From the pollen data of Turpan, Jiuquan,[6] and Qaidam[7] we may conclude that the area was subtropical and semi-desert during the Pliocene.

During the Pleistocene, the Tarim Basin became more arid, and sanddune accumulation became more pronounced. The whole basin was treeless, except for some scrubs that were scattered on the open desert. Only *Populus* and tamarisks usually gathered in formations along rivers.

The landscape changed markedly from the north to the south, with a warm, temperate semi-desert in northern Junggar Basin, with a warm semi-desert in southern Junggar Basin, and with a subtropical desert in the Tarim Basin. In Quaternary, most parts of Xinjiang were affected by the Arctic air masses, that originated at the polar ice cap, and glaciation occurred above the snow-line on high mountains. During the Great Ice Age, there were about 4—5 occurrences of glaciation, of which the earliest were the largest. It seems quite possible that the greatest glaciation occurred in the early Pleistocene because of the creations due to the uplifting of the peneplain to the zone of maximum snowfall, of more flat space for the accumulation of snow. Later on, as peneplains raised to a height above the zone of maximum snowfall, they received less and less precipitation, so the extent of glaciation was gradually reduced.

It has been argued by some authors that the dense network of river systems in the Tarim Basin indicates that a humid climate existed during Quaternary, showing that the desiccation of the Taklimakan Desert must have occurred by the beginning of Holocene, but I believe that this desiccation in the basins of Xinjiang dates from very remote geological past. New evidence shows that no special climatic changes occurred in the Quaternary and historical periods, with the exception that a few layers of deposits might indicate very short pluvial phases within the general background of an arid climate.

THE QINGHAI-XIZANG (TIBETAN) PLATEAU AND AFFILIATED AREAS

The great uplift of the Qinghai-Xizang Plateau occurred at the end of Pliocene and the beginning of Quaternary, when the Indian Plateau drifted northward against the Qinghai-Xizang Plateau. In the southern portion of the plateau the Himalayas arose, highest in its middle section but with an assortment of

181

high, rugged peaks elsewhere. During Quaternary, as the global climate turned cold and the uplift of mountains occurred, four periods of glaciation developed, according to the pollen data and ages of moraines from Mount Qomolangma[8] (Mt. Everest) and other localities. In basins and valleys, Quaternary beds of lacustrine, fluvial, and glacio-fluvial deposits were built up to a great thickness. Variegated molasse beds of several hundred meters indicate different ages within Cenozoic time. The *Hipparion* fossils found in the ancient lacustrine deposits[9] remind us of the Pliocene forest or forest glassland that was replaced in the vertical zonal section by the present Alpine meadow, indicating rapid change from warm-humid to cold conditions that can be accounted for velocity by the young upheaval movement here during the Late Cenozoic Period.

The vast expanse of the Qinghai-Xizang Plateau became much colder and drier because the marginal high mountains prevented oceanic moist air from penetrating the interior. Thus in areas where glaciation had diminished, the lakes shrank quickly as is indicated by many shore terraces that rise above the present water level. Many lakes filled with saline water and salt deposits. The well-developed karst topography of Tertiary time, suffering from congelifraction and destruction by weathering and eolian action, was worn down somewhat like ancient ruins. Since its uplift to the present level, the plateau has been covered with scanty grass and rosette plants. The herbivorous animals of the region adjusted to this severe environment by developing thick fur and strong jaws and teeth in order to adapt to such a severe environment and the consequent shortage of food.

Grabau held the view that Qinghai-Xizang was the original[10] habitat of man. Recently neolithic remains have been found in Hoh Xil (Kokoshili) and Heihe (Black River) on the plateau. The level of skill in tool manufacturing which is indicated in these remains is quite similar to that of the upper Huang-he River basin. In the areas adjacent to Qinghai-Xizang, such as in the frontal hills of the Himalayas, Ladak, and the Kashmir Valleys, many Neolithic human skeletons, tools and fractions of animal bones have been found in different localities. It is worth noting that the upheaval of the plateau and the radical change in the unfavorable environment to human habitat on the plateau area like today seems to have occurred in the very late stage of the Quaternary.

REFERENCES

[1] Liu Hong-Yun, 1955, Developments of geological stratigraphy of Northeast China. Institute Geol. Pub. no. 1.

[2] Hong You-chong et al., 1974, Stratigraphy and palaeontology of Fushun coal field Liaoning Province. Acta Geologica Sinica v. 2, 113–158.

[3] Chaney, Hu Xuan-xiu, 1940, A Miocene flora from Shantung Province, China. Acta Palaeont, Sinica. 112.

[4] Song Zhi-chen, 1958, Plant fossils and spore-pollen complex of the Sanmen Series.

Quaternaria Sinica, v. 1, n. 1. 118–130.

[5] Sze H. C., Hommer Lee, 1944, A Late Tertiary flora from Hunan. (abstract.. Acta Palaeont. Sinica. v. 2, n. 2.

[6] Sung Tze-chen, 1958, Tertiary spore and pollen complexes from the Red Beds of Chuichuan, Kansu and their geological and botanical significance. Acta Palaeont. Sinica v. 6, n. 2.

[7] Hsu Jen et al., 1958, Sporo-pollen assemblages from the Tertiary deposits of the Tzadam Basin and their significance. Acta Palaeont. Sinica. v. 6, n. 4, 429–440.

[8] Huang Wan-po, Ji Hong-xiang, 1979, The fi!st discovery of Xizang *Hipparion* fauna and its significance to the plateau rise, Kexue Tongbao. v. 9, 885–888.

[9] Xizang Scientific Investigation Team, Academia Sinica, Report on the scientific investigation (1966–1968) in Mt. Qomolangma district-Quaternary geology, Science Press.

[10] Grabau, W. 1936, Tibet and the Origin of man; Hyllnisskrift Tillognod, Svan Heden po Hans 70 Arstag den 14 Febr.

Table 1 Zonality of Oligocene in China   *(to be continued)*

Climatic belt	Area	Major sedimentation represented	Landscape
1. Temperate, warm and humid; with rainfall all year round	North-east provinces; Northeast part of Nei Monggol (Inner Mongolia)	"Fu Shun" coal series	Tall trees consist mostly of angiospermae, Pteridophata; lakes and swamps with plentiful insects; some tree leaves show special draining point and stomata, supposed to be due to cyclonic westeriles.
2. Warm, humid; with a slight increase of summer temperature	Noth China & most part of Nei Monggol	Mainly variegated beds, with gray-green and dark formations	Mixed forest; an increase of subtropical elements
3. Warm and semi-arid area in the interior part	Gansu corridor; Junggar Basin	Red Beds of little carbon content; fluvial, pluvial and lacustrine deposits alternated	Northern and eastern portions, forest and herbivorous mammals predominant; in the western portion, grassland and rodents prevailed.
4. Hot and semi-humid (northern subtropical belt)	Huaiyang Mt., Qin Ling Mts. Liupan Mts. highland of Ordos and Alxa	Variegated formations, with coal seams and gypsum intercalations	Mainly subtropical forest, mixed with elements of temperate belt; large fossii deposits of mammals and rodents; zonal position shifted north and south from tim to time.

Table 1 *(continued)*

Climatic belt	Area	Major sedimentation represented	Landscape
5. Hot and dry, savanna type (middle sub-tropical belt)	Drainage basin of Changjiang River; Yunnan and Guizhou Plateaus; Qaidam and Tarim Basins.	Red beds, of con-glomerate and sandstones; lacus-trine deposits, usually containing salts and gypsum of great thickness	Ephedra dominant, scattered gingko Juglans, Fagus, Zelcova; also, gullery formations, animal life noted for *Tetrobelodon, Crocodilus, Trigont,* etc.
6. Hot and humid, drier in winter (southern tropical belt)	Taiwan, south of pressent 24°N, down to the southern border of China	Variagated with red and purple layers; dark deposits, usually containing asphalt, coal and oil shale seams; gypsum also exists, especially in western part.	Mainly subtropical evergreen leaf trees, mixed with some tropical elements; also a few deciduous trees; fossil animals affiliated with "Pongtang" fauna of Burma; zonal position shifted to the north or to the south from time to time.

THE PALEOENVIRONMENT OF THE SHUIDONGGOU SITE OF ANCIENT CULTURAL REMAINS IN LINGWU COUNTY, NINGXIA

Zhou Kun-shu

(Institute of Geology, Academia Sinica)

and Hu Ji-lan

(The Laboratory of Geologial Bureau of Nei Monggol)

Abstract

The Shuidonggou site of ancient cultural remains is located 30 km and more north of Lingwu County of Ningxia, at 106°29'E by 38°19'N.

Shuidonggou is a valley more than 40 km long along a small tributary in the middle reaches of Huanghe River. The ancient cultural site lies in a small basin in the valley about 11 km from the Huanghe River in the west. Since Late Tertiary the area has been uplifting very slowly, and the Early and Middle Pleistocene strata are not developed.

Overlying the Late Tertiary Red Clay Late Pleistocene river and lake deposits over 10 m thick were accumulated. There is a layer of sand and gravels at the bottom, a layer of dark gray lens-like mud in the middle overlying a layer of clayey sand that contains a layer of nearly horizontal wavy curly bedding which is probably periglacial involution. The top layer of the deposits is composed of grayish yellow silt. In 1920s abundant paleolithic stone implements and some animal fossils were discovered in the upper part of the deposits. In Holocene, a new period of erosion occurred, and an erosion break surface, is found upon which Holocene river and lake deposits of 8 m thick were accumulated. The lower part of these deposits is composed of gravels, in which some paleolithic and neolithic stone implements are buried. The middle part of the layer consists of swamp clay deposits, upon which is a layer of black loam.

The results of the sporo-pollen analysis of these deposits show that the main contents are plants of steppes and deserts, such as *Artemisia*, *Ephedra*, Chenopodiaceae, Compositae, Graminia, *Nitraria*, etc. But in the black mud of the late Pleistocene some woody pollen was discovered such as *Pinus*, *Picea*, etc. In the swamp soil of the Middle Holocene some herbaceous pollen was also discovered.

The Shuidonggou cultural strata can be correlated with the late stage deposits to the Salawusu strata, and their age is from late stage of the Late Pleistocene to Holocene.

185

DEVELOPMENT OF THE PEKING MAN'S CAVE IN RELATION TO EARLY MAN AT ZHOUKOUDIAN, BEIJING*

REN MEI-E

(Department of Geography, Nanjing University)

AND LIU ZE-CHUN

(Nanjing Institute of Geography, Academia Sinica)

Abstract — Peking Man's Cave, Zhoukoudian, Beijing, is a world-wide known site where fossils of Peking Man have been discovered. The evolution of Peking Man's Cave may be divided into 5 stages: stage of deep burial, stage when an East entrance was opened, stage when Peking Man mainly dwelled in eastern part of the cave, stage of the collapse of cave roof of Pigeon's Chamber and westward migration of Peking Man's habitat and stage when the cave was filled up and Peking Man left the cave.

Zhoukoudian, about 50 km southwest of Beijing, is the site where fossils of the famous "Peking Man" (*Homo erectus pekinensis*) were discovered. As Peking Man dwelled in limestone caves, detailed study of the evolution and filling of limestone caves in relation to early man may throw light on the life of Peking Man.

I

In Zhoukoudian area there are four caves with evidences of activities of early men, i.e. the 1st site or Peking Man's Cave, the 15th site, the 4th site or New Cave and Upper Cave. These caves, developed along surfaces of the strata or joints, all have considerable space for habitat of early men who left their fossil ash layers or stone tools in the caves.

According to cave morphology, these caves may be grouped under two categories, i.e. The vertical type and the horizontal type. Peking Man's Cave, more than 40 m high and 107 m long, may be cited as an example of vertical cave. The cave trends N60—70°W, parallel to the strike of the strata. Its walls are very steep, with angle about 60—70°, also corresponding to the dip surface of the limestone. The northern fissure and southern fissure in this cave are eroded along vertical joints. The New Cave is an example of hori-

---

* Wang Xue-yu, Peng Bu-zhuo and Wang Zong-han have done some work in the laboratory.

zontal caves; here, limestone is gently inclined, with dip angle only 10—20°. Developed along the surface of the strata, the cave has even floor and roof and its spacious room is not filled with cave deposits. As the cave is generally more stable, its cave deposits consist chiefly of sandy loam and stalactite with only few breccias. Upper cave, at 125 m a.s.l., is the highest cave in Longgu Hill. Its western part is mainly developed along the surface of the strata (dip 20—30°) and has a morphology quite like horizontal cave. But the northern part of the cave assumes a knee-like form as the dip of the strata steepens to 50—60°. Moreover, the presence of shear joint causes corrosion penetrating deeply downward, forming a vertical shaft which had once connected (Upper Cave) with Peking Man's Cave.

Although these caves are horizontal ones, their elevations and age of cave deposits can in no way be correlated with terraces outside the cave, owing to the fact that they had developed under the influence of structures. The principle of correlation of caves with terraces is also not applicable to Peking Man's Cave, a typical vertical one. As the elevation of the floor of Peking Man's Cave has already reached the present bed of the Zhoukou River and the lowest part (below the 14th layer) of cave deposits in Peking Man's Cave is older than 700,000 yrs BP, the formation of the cave probably dates back to more than 1 million yrs BP, whereas the lower gravels on terraces of the Zhoukou River are younger than 700,000 yrs BP . Evidently, the ages of Paking Man's Cave and terraces of the Zhoukou River are not the same and therefore, they can not be correlated. Moreover, the elevation of Peking Man's Cave is 80—128 m a.s. l., while the highest terrace of the Zhoukou River is about 90—100 m a.s.l. (Table 1)

II

The unusually thick cave deposits in Peking Man's Cave (more than 40 m thick) are seldom seen in other caves of China. These deposits form scientific basis for the reconstruction of evolutionary stages of the cave. According to data of recent prospecting wells and stratigraphic section of the Zhoukoudian Research Group, Academia Sinica, the cave deposits may be divided into 17 layers, i.e. in addition to former 13 layers the 14th—17th layers can be distinguished. They may be grouped into four genetic types:

1. Collapse and slide deposits

These deposits, chiefly limestone breccia, can be found from the 13th layer upward, especially in the 9th, 6th and 3rd layers. Some breccia layers contain huge limestone blocks, with only little fine sediments. About 50% of these breccias is more than 12 cm long, and a few reach 2—3m long. They are

Table 1 Sedimentological characteristics of cave deposits in Peking Man's Cave and sediments outside the cave

Stratigraphic layer	Heavy mineral assemblage	Clay mineral* association	Other lithological characteristics	Chemical composition in quartz grains
1. Cave deposits (1) 17th layer	sphene-magnetite-limonite-hornblende-apatite	kaolinite, glagerite, illite, vermiculite, epidote	with crossbedding, rounded small gravels (metamorphic rocks), much mica in sand	major: Si trace: Al, Zn
(2) 14th layer (formerly basal gravels)	sphene-limo-nite-zircon-magnetite	illite, kaolinite, vermiculite, epidote	lithological composition of gravels similar to upper gravels	major: Si trace: Fe, Al, Zn Nb, Ta
(3) 12th–13th layer	magnetite-limonite-apatite-sphene-zircon	illite, kaolinite, vermiculite, epidote	red-brown clay breccia with coarse sand layers, locally clay forms aggregates	major: Si trace: Ba, Ta
2. Sediments outside the cave (1) Upper gravels	sphene-limonite-zircon	kaolinite, glazerite, illite vermiculite, epidote	sandstone, schist, slate, phyllite and quartzite	
(2) Gravels at East Cave entrance	sphene-magnetite-limonite-hornblende	kaolinite glagerite illite, vermiculite, epidote	lithological composition of gravels similar to lower gravels, but their diameters are smaller	major: Si trace: Al, Zn Nb, Ta
(3) Lower gravels	sphene-magnetite-hornblende-zircon-limonite-apatite	illite, kaolinite, vermiculite, epidote	siltstone, slate coarse and fine sandstones	

* We are indebted to Prof. Xiong Yi, Xu Ji-quan and Yang De-yong of Institute of pedology, Academia Sinica for their help in the analysis and explanation of clay minerals.

mainly the results of collapse of cave roof. In Pigeon's Chamber of Peking Man's Cave, breccias of the 6th layer are largely deposits of this origin. Therefore, it is inferred that in this part of the cave, collapse of cave roof occurred during the 6th layer period.

2. Fluvial sediments brought into the cave by the old Zhoukou River

In Peking Man's Cave, there are several layers of sand, silt and gravels. According to heavy mineral assemblage, surface texture of quartz grains and trace elements in sand layers, it seems evident that sand of the 17th and 12th layers are similar both to fluvial deposits of the old Zhoukou River, i.e. gravels in East Cave entrance and lower gravels and to sand of the present Zhoukou River. Their heavy mineral assemblage is chiefly aspidelite-magnetite-hornblende. Surface of quartz grains are also rather similar, mostly subangular, with various types of impact scars, formed by impacts between sand grains in mountain streams. Trace elements in semi-opaque quartz grains are also about the same, mainly Al, Zn, etc. (See Table 1).

In the correlation of stratigraphic layers, although sand of the 17th layer in Peking Man's Cave have the same provenance as the lower gravels, but according to paleomagnetic measurement, cave deposits below the 14th layer belong to Matuyama's magnetic reversal period and those above the 13th layer belong to Brunhes magnetic period, while the lower gravels are magnetic normal. Therefore, in the time scale, only cave deposits from the 13th layer upward can be correlated with the lower gravels. The clay mineral association of the 17th layer is same as gravels at the East Cave entrance, both having kaolinite and illite as chief clay minerals. On account of these data, it may be concluded that the 17th layer in Peking Man's Cave probably corresponds to gravels at the East Cave entrance, both being deposited by the old Zhoukou River before the deposition of the lower gravels.

3. Deposits brought into the cave by sheet flow

This kind of cave deposits consists of clay silt, sand and gravels. Red clay and larger gravels in the 14th layer came from the upper gravels. This can be proved by the fact that lithological composition of gravels, heavy mineral assemblage and surface texture of quartz grains and trace elements in sands are about the same. The heavy mineral assemblage is aspidelite-limonite-zircon-magnetite, noted by the absence of hornblende. Trace elements are mainly Ba.

Some limestone breccias in the 3rd layer are somewhat rounded, indicating that they have been corroded by rainwash and consequently their surfaces are scattered with numerous corrosion marks. It can be reasonably inferred that during this period, mountain slopes were covered with large amounts of weathered debris which had been brought into Peking Man's Cave by sheet flow. But on the bottom of the 3rd layer, piles of ash, burned bones and stones were discovered, indicating that Peking Man had lived on large stone slabs fallen from cave roof.

189

4. Ash layers are rather common in Peking Man's Cave

They first appear in the 10th layer and is the thickest in the 4th layer, attaining a maximum thickness of 6m. The distribution of ash layers in the cave provides valuable key for the reconstruction of living activities of early men in the cave.

III

According to principles of development of limestone caves, comparison between cave deposits and sediments outside the cave, data of isotopic dating, former records of excavation of Peking Man's Cave and valuable suggestions from colleagues*, developemt of Peking Man's Cave may be divided into five stages which are closely related to the livelihood of early men.

1. Stage of deep burial (about $1,000 \times 10^3$ yrs ago)

During Pliocene when a widespread peneplain of the Tangxian stage was developed, the climate of Zhoukoudian Area was moist-subtropical, favourable to karstification which penetrated deep into the rock massif along steeply dipping surface of limestone strata, gradually forming vertical caves. This is the infant stage of the present Peking Man's Cave.

2. Stage when on East entrance was opened (about 700×10^3 yrs ago)

In early Pleistocene, with the uplift of the Tangxian surface and downcutting of the river, Peking Man's Cave was further corroded and enlarged. The Zhoukou River flowed close to the eastern slope of the Longgu Hill. Fluvial sand-gravels of the old Zhoukou River have been discovered near the East Cave entrance at about 93 m a.s.l. In Early Pleistocene, truncation of eastern slope of Longgu Hill by lateral erosion of the Zhoukou River destroyed the eastern wall of the cave, forming a small cave entrance which is about the site of the present East Cave entrance. Excavation of the two recent prospecting wells reveals that the floor of Peking Man's Cave was originally very rugged at that time, and water of the old Zhoukou River slowly flowed into the cave through the small cave entrance, depositing the 15th—17th layers of cave deposits which contain some pollen of wood and grasses. This indicates that during lower Pleistocene when the 15th—17th layers were deposited, Peking

---

* Oral communication from Prof. Chia Lan-po and Yuan Zhen-xin, Institute of Vertebrate Paleontology and Paleoanthropology, Academia Sinica.

190

Man's Cave already had some connection with outside. But as cave floor was still rugged and cave entrance probably rather small, the early men were unable to enter the cave, and live there. The upper part of the 14th layer contains abundant pollen, more than 85% of which are *Pteris*, growing on the limestone hills near the cave. It seems that during deposition of the 14th layer, connection between Peking Man's Cave and outside was still rather limited.

3. Stage when Peking Man mainly dwelled in eastern part of the cave (about 500×10^3 yrs ago)

With continuous filling of the cave with fluvial deposits of Zhoukou River and sheet flow deposits, the floor of Peking Man's Cave was gradually flattened and a considerable tract of flat floor was formed. At the same time, the East Cave entrance was gradually enlarged by corrosion and erosion. Therefore, after deposition of the 13th layer, Peking Man came into the cave from the East Cave entrance and stayed there for a long time. Stone implements have been discovered from the 13th layer upward and ash layers were not seen until the 10th layer. From distribution of ash layers and stone implements, it seems evident that in the early stage, Peking Man mainly dwelled in the eastern part of the cave where more than 10,000 pieces of stone tool have been discovered and ash lagers are thicker than those of the western part of the cave. The eastern part of the cave had obvious advantage for the habitat of early men. It lies close to the East Cave entrance, near the Zhoukou River and the roof of this part of the cave was at that time well preserved, forming a perfect shelter for early men. But thin ash layers have also been found in the western part of the cave. This is because the roof of the cave was not continuous and there was an opening through the top of the cave connecting it with outside so that the western part of the cave still had enough air and light. The existence of a vertical opening is also proved by the fact that the 14th layer mainly came from the upper gravels on the top of the cave as has already been elucidated in the preceding section.

4. Stage of collapse of cave roof of Pigeon's Chamber and westward migration of Peking Man's habitat (about 300×10^3 yrs ago)

During the period when the 6th layer was deposited, large-scale collapse of cave roof of Pigeon's Chamber blocked the East Cave entrance. As a result, after deposition of the 6th layer, activities of Peking Man were more concentrated in the western part of the cave where ash layer of the 4th layer reached a maximum thickness of 6m, but the ash layer gradually thinned out towards Pigeon's Chamber. The thick ash layer gives vivid witness to the activities of

Peking Man in this period. If factor of composition is taken into account, the maximum thickness of ash layer of the 4th layer must have exceeded 10 m, indicating that Peking Man had stayed in this part of the cave for a long period and his hunting products were quite plentiful.

5. Stage when the cave was filled up and Peking Man left the cave (about 230×10^3 yrs ago)

With the deposition of the 3rd-1st layer, Peking Man's Cave was almost completely filled up at about 230,000 yrs BP, Peking Man left the cave because its favourable condition for habitation had already been lost.

From the above brief account, it can be seen that Peking Man had lived in the cave at least from 500,000 yrs BP (i.e. the 10th layer), until 230,000 yrs BP. In other words, Peking Man had lived in the cave for about 300,000 yrs without interruption. Evidently, this is due to particular favourable environment of Peking Man's Cave as a babitat for early men. In the first place, the location of the cave near the Zhoukou River gives early men an easy access to water supply. Secondly, the cave, in addition to East Cave entrance, had a vertical opening on the top, and consequently there were enough air and light inside the cave. Furthermore, dense temperate forest and wooded steppe in that time yielded abundant wild animals. Therefore, it is not a mere accident that Peking Man had lived in Peking Man's Cave for such a long time and had left abundant valuable materials for paleoanthropology. Evidently, the activities of early men were closely related to the development and filling of the cave and paleogeographical environment. The present paper elucidates the life of Peking Man as related to evolution of Peking Man's Cave—his habitat and thereby throws some new light on the knowledge of life of Peking Man.

REFERENCES

[1] Chinese Academy of Geological Sciences, The Geotectonic Map of China (1: 1,000,000), Geology Press, Beijing, 1962

[2] Pei Wen-chung, 1934, Note on excavation of the caves at Choukoudien, Special geological report (B), No. 5.

[3] Pei Wen-chung, 1960, The living envirnoment of Chinese primitive man Resume Vert. Palas. 4(1), 40–44.

[4] Chia Lan-po, 1964, *Sinanthropus pekinensis* and their stone artifacts, 11–14. Zhonghua Shuju.

[5] Chia Lan-po, 1951, Upper Cave Man (*Homo sapiens*), 20–21, 75–76, Longmen Shuju.

[6] Woo Ju-kang, 1956, Human fossils found in China and their significance in human evolution, Scientia Sinica, 5(2), 389–397.

[7] Chow Min-chen, 1955, *Sinanthropus* living environment inferred from vertebrate fossils, Kexue Tongbao, 1, 15–22.

[8] Hsu, Jen. 1965, The climatic condition in North China during the time of *Sinanthropus*,

Scientia Sinica, 15, 410–414.

[9] Huang Wan-po, 1960, Restudy of the CKT *Sinanthropus* deposits, Vert. Palas., 4(1), 83–95.

[10] Chiu Zhong-lang, 1973, Newly discovered *Sinanthropus* remains and stone artifacts at Choukoudien, Vert. Palas., 11(2), 109–131.

[11] Qian Fang et al., 1980, Magnetostratigraphic study on the cave deposits containing fossil Peking Man at Zhoukoudian, Kexuo Tongbao, 25(4), 359.

[12] Zhao Shu-sen et al, 1980, Uranium-series dating of Peking Man, Kexue Tongbao, 25(5), 447.

[13] Guo Shi-lun et al, 1980, Age determination of Peking Man by fission track dating, Kexue Tongbao, 25 (6), 595.

CLIMATIC CHANGES DURING PEKING MAN'S TIME

XU QIN–QI AND OUYANG LIAN
(Institute of Vertebrate Palaeontology and Palaeoanthropology,
Academia Sinica)

ABSTRACT

The fossiliferous deposits at Zhoukoudian Loc. 1, where Peking Man lived, may be divided into 13 layers. According to the dates, Layer 3 can be correlated with ^{18}O stage 8 and Layer 10 is equivalent to stage 12. In order to correlate the Zhoukoudian sequence with the ^{18}O records, we have used cluster analysis and other mathematic techniques. As a consequnce, the mammalian faunas in Layers 3—4, 6—7, 10 indicate three cold stages which are correlated with ^{18}O stages 8, 10, 12 respectively; while the faunas in Layers 5, 8—9 represent two warm stages which correspond to ^{18}O stages 9, 11 separately.

FOSSIL SOIL LAYER IN PLEISTOCENE
SEDIMENT OF HUANGHAI SEA*

Liu Min–hou

(First Institute of Oceanography, National Bureau of Oceanography)

Abstract — The preliminary physico-chemical analyses of so called "hard mud layer" or "hard clay layer" in cores of Late Pleistocene in Huanghai Sea (Yellow Sea) indicate that they actually represent buried fossil layers in three different periods. Their common characteristic is that none of them has formed inherent stratified units of typical fossil soil section, but only stopped at the early period of soil development, similar with buried weathering layer in loess. But according to the analysis of organic materials there exists another possibility, i.e. owing to the later transgression, the most part of the well developed fossil soil layer at that time had been washed away, and only the relict part and its parent material layer left there.

I

The results we have gained in recent years from the study of Late Pleistocene sediment in Huanghai Sea shows that it may be divided into three layers of continental phase i.e. lower layer, middle layer and upper layer. Their ages of formation are possibly around 72,000—69,000 years, 60,000—42,000 years and 30,000—12,000 years BP,** which reflects that three land-formed periods in different scales did occur in Huanghai Sea shelf since Late Pleistocene. The land-form extent of the latest period was the greatest and the whole Huanghai Sea shelf was exposed as land[1,2]. Peat layers and humus soil layers of different developed scales and thickness formed on land surfaces in basins and depressed regions of these three periods, but so–called "hard mud layer" or "hard clay layer" formed in the other regions. These dense "hard mud layers" with pelitic silt indicate some characteristics occurred in the development of fossil soil layer. It is worth noticing and studying these not well developed buried fossil soil layers, which is of great significance to study the fossil soil layers in Quaternary sediments. It is not only because the nature and pattern of them could reflect certain temporal and spatial physiographical conditions and climatic changes, but also provide one of the valuable indicators for division and correlation of stratigraphy. Therefore, from stratigraphic point of view, the fossil soil layer is used to be called the special "index fossil" in geological history.

Remarkable results have been gained from the study of fossil soil layer in

---

* Data in this paper were provided by labs. of the Institute.

** The ages of continental phase layer should be open to discuss further.

Quaternary sediments, especially those buried in loess on our land[3]. But unfortunately there is few report on the fossil soil layer in sediments of our coastal seabed, in other words, attention has not been paid enough yet to that so far. The report of National Bureau of Oceanography on the marine geological comprehensive investigation in East China Sea area once mentioned that yellow brown "hard clay" with some soil characteristics*** existed in certain cores, but did not go into detail. The author tries to present this viewpoint here based on the occupied data and expects to have further discussion and study in the future.

II

The six-metre cores, the longest, taken by vibrating piston-corer show that buried fossil soil layers exist in core sections of sts. H_{19}, H_{23} in northern Huanghai Sea and H_{13}, H_{17} in southern Huang Hai Sea, etc. They belong to the lower layer (H_{17}), the middle layer (H_{19}), and the upper layer (H_{13}, H_{23}), respectively. The water depths of these stations range from 55 m (H_{23}) to 75 m (H_{17}). Sometimes, obvious erosional surfaces occur between fossil soil layers and overlying marine layers. Sometimes the erosional surfaces are not clear. Most parent soil materials are pelitic silt. The fossil soil layer is 48—80 cm thick in general and 120 cm in maximum. The distribution condition of buried fossil soil layer in Huanghai Sea area is not yet clear so far. It seems that it has nothing to do with the water depth, but relates to the paleogeomorphological position of sea bed. Based on the estimation of occupied data, it may be closely related with the ancient Huanghe River deltas, terraces and adjacent areas of lake and depression.

The fossil soil layers in three different periods mentioned above are quite similar from the viewpoint of morphology and results from analyses in lab. None of them developed into inherent stratified units of typical soil section, but some characteristics in pedogenic process still could be seen. The fossil soil layer is yellow brown in colour. The lower part, sometimes is a bit light, i.e. dark grey, sometimes dark brown due to the dissemination of oxides or spots. The fossil soil layers are dense, harder with great viscosity. The phenomenon of $CaCO_3$ vertical eluviation can be seen in some parts. The plant roots cross each other, which made fossil soil into porphyritic mass. Sometimes bright brown peal appears on the surface of the structure body. Clay particles fill along the root pores and worm holes channels. A large quantity of $CaCO_3$ spotted materials and cementation materials can be seen under microscope, and quartz and feldspar and other mineral particles expose

---

*** Report on marine geological comprehensive investigation in the East China Sea area, 1978 by the First and Second Institutes of N.B.O.

196

out after adding hydrochloric acid. According to the results of grain size analysis, the average clay particle content in fossil soil layer varies from 29.3% to 43.4% and it has an increase trend from top to bottom. For example, the clay particle content in the fossil soil layer in st. H_{17} increase downward from 25.7% at the surface to 31.3% at the bottom, in H_{23} from 36.3% to 45.2%. According to a rough statistics, the porosity of fossil soil layer decrease downward from about 40% at the top to 5% at the bottom.

The results of chemical analysis of the buried fossil soil layers in four stations (H_{13}, H_{17}, H_{19}, H_{23}) show that approximation exists among their chemical component contents and the average content (%) are as follows SiO_2—53.34, 57.32, 58.20, 50.68; Al_2O_3—11.60, 11.06, 11.56, 10.60; Fe_2O_3—5.22, 4.55, 6.43, 5.21; MnO—0.102, 0.088, 0.072, 0.098; MgO—3.08, 2.81 2.64, 3.04; CaO—5.29, 6.16, 5.17, 8.13; Na_2O—2.07, 2.17, 2.01, 1.98; K_2O—2.09, 1.93, 1.93, 1.96; TiO_2—0.16, 0.56, 0.58, 0.50; P_2O_5—0.208, 0.179, 0.99, 0.197. The average content (%) of $CaCO_3$ is higher. They are: 9.89, 10.47, 9.14, 13.29 respectively. Organic materials (%): 0.22—0.37 (H_{19}), 0.28—0.39 (H_{23}). pH values varies from 8.30 to 8.48 (H_{19}), 8.20 to 8.45 (H_{23}). The ratio of SO_2/Al_2O_3 varies from 3.90 to 5.71, Fe^{2+}/Fe^{3+}—0.41 to 1.00. The variation of chemical component in fossil soil section is not obvious, but the increase trend from top to bottom in contents of $CaCO_3$ and Fe_2O_3 can be seen. They increase from 9.76% and 5.69% at the surface to 10.96% and 7.11% at the bottom.

The results of clay mineral analysis indicate that the clay mineral in buried fossil soil contains illite, kaolinite, chlorite and montmorillonite. The relative content of illite is high. The average contents (%) of various clay minerals in fossil soil layers in three stations (H_{17}, H_{19}, H_{23}) are: illite, 51.9, 64.6, 53.6; kaolinite, 26.6, 17.1, 21.6; chlorite, 11.0, 14.3, 17.4; montmorillonite, 10.5, 4.0, 7.2. The variation of content of clay mineral in fossil soil section is relatively obvious. e.g. in st. H_{19}, illite at the top of the section is 55.4% and increases downward to 75.9% at the bottom, while kaolinite and montmorillonite decrease from 19.9% and 7.7% to 9.5% and 2.1%, respectively.

The results of heavy mineral analysis indicate that buried fossil soil layer contains mainly amphibole and epidote and secondarily garnet, zircon, tourmaline and titanite and metallic minerals, etc. The results of analysis of three stations indicate that the average content (%) of amphibole is: 43.2, 40.7 and 55.2 and epidote 16.0, 34.6, 17.6 respectively. The variation of amphibole content in fossil soil section is most obvious. It increases from 29.4 at the top to 51.5 at the bottom (H_{19}).

III

From the data mentioned above we can see that the fossil soil layers in three different periods in Late Pleistocene sediments in Huanghai Sea are quite similar either in morphological structure or in chemical and mineral components, which shows that pedogenesis did occur in layers of three different periods but did not form inherent stratified units of typical soil section. This is firstly shown in the content variation of clay particles and mineral component in these sections. This variation reflects the result of clay grouting caused by biogenic activity and chemical weathering. And secondarily it is probably just because of the clay grouting which made the fossil soil itself too hard to be penetrated when we took core from the sea. The variation of $CaCO_3$ content in the section also shows that eluviation indicating pedogenic process once occurred but was quite weak. In other words, it was just in the early period of development similar to the buried weathering layer in loess.

According to the results of sporo-pollen analysis, herbaceous pollen is in superiority in the latest fossil soil section. Pollen assemblage shows that at that time steppe was dense with sparse woods. marshes and depressions everywhere. The climate was cold and wet[2]. Such climate and physiogeographical condition is suitable for the formation of humus soil. But the result of organic matter analysis shows that the content of organic matter in the section is less, vertical change is not great and no indication of obviously humus soil layer. This phenomenon is consistent with above inference, however, this does not exclude the other possibility i.e. the humus soil layer was formed at that time and even a great part of fossil soil had been washed away and what we observe is only the relict part and its parent material layer.

IV

According to the preliminary observation and analysis about buried fossil soil layers in three different periods of columnar cores of Huanghai Sea in Late Pleistocene the author has learnt that so called "hard mud layer" or "hard clay layer" actually represents three different pedogenic periods and reflects certain biogenetic climatic condition and natural environment. Their common characteristic is that none of them formed inherent stratified units of typical fossil soil section, but only stopped at the early period of soil development, similar with buried weathering layer in loess. But according to the analysis of organic matters, there exists another possibility, i.e. owing to the later transgression, the most part of the well developed fossil soil layer at that time had been washed away. So, only the relict part and its parent material layer left there.

The buried fossil soil may be extensively distributed in Quaternary sediments of Huangnai Sea. This is a new subject that attention should be paid to. This paper only mentions some general morphological observations and preliminary studies on the physical and chemical properties about buried fossil soil layers in some cores. So final conclusion has not been obtained, and in the meanwhile possible well developed buried marsh soil in this sea area has

198

not been involved yet. These problems should be certainly solved in further studies in the future.

REFERENCES

[1] Lab. No. 3, First Insititute of Oceanography, National Bureau of Oceanography. 1979, The evolution of paleogeographical environment of the Yellow Sea since Late Pleistocene. Kexue Tongbao, no. 12.

[2] Xiu Ja-sheng, 1981, The Yellow Sea in the Latest Ice Age—The acquisition and study of some new data about paleogeography of the Yellow Sea. no. 5. Scientia Sinica.

[3] Zhu Xian-mo, 1965, Fossil soil in loess deposit in China. Quaternaria Sinica. vol. 4. No. 1.

[4] Liu Tung-sheng, 1965, Loess Accumulation in China. Science Press.

TRANSGRESSIONS AND SEA-LEVEL CHANGES IN THE EASTERN COASTAL REGION OF CHINA THE LAST 300,000 YEARS

ZHAO SONG–LING AND CHIN YUN–SHAN
(Institute of Oceanology, Academia Sinica)

Abstract –– Three marine transgressions invaded the western coast of Bohai Gulf in the last 100,000 years while the western coast of the Huanghai (Yellow) Sea was subjected to 5 marine transgressions in the last 300,000 years.

INTRODUCTION

The distributive regularity of the buried marine formations found in drilling cores in the eastern coastal region of China are of great importance for the study of the marine paieoenvironmental changes in the shelf sea of China during the Quaternary period, the eustatic movements of the world sea-level, the climatic changes between glacial and interglacial ages as well as the regional tectonic movements.

The discovery of three cheniers along the western coast of Bohai Gulf in the sixties of the 20th century pushed on the investigation of marine regression. A new core-drilling study in the seventies reached a greater depth of about 500 m, and abundant micropaleontologic, molluscan ^{14}C dating and paleomagnetic measurement data were accumulated.

Many authors have discussed from different viewpoints that three transgressions have occurred and three marine formations have buried in a wide range of area since 100,000 yrs BP. At the same time in the coastal plain of the southern Huanghai (Yellow) Sea and the East China Sea three equivalent marine formations have also been discovered. In recent years micropaleontologic analyses and paleomagnetic measurements from drilling cores of Yancheng, Qinggang, Lianyun Harbour, Gk8, Ph4 and Fangyang Harbour along the coastal region of the Huanghai Sea and the East China Sea as well as Dc1 and Dc2 cores in a water depth of about 30 m in the East China Sea were made by Institute of Oceanology, Academia Sinica. We also made radiocarbon dating with some layers adaptable. So far, plenty of bio- and chrono-stratigraphic data are obtained. In addition, many micropaleaontological analyses in this region were also made by Nanjing Institute of Geology and Paleontology, Academia Sinica, Tong-Ji University of Shanghai and

Jiangsu Hydrogeological Brigade.

From all the data mentioned above, we are able to extend the study of marine formation in the eastern plain of China to 300,000 yrs ago. This article will give a preliminary discussion on the transgressive cycle, extent and age, coast line and sea-level changes of the eastern coastal plain of China in the past 300,000 yrs.

THE SUBDIVISION OF MARINE AND CONTINENTAL FACIES FORMATIONS

In subdividing marine and continental facies formations, foraminifera, ostracod and mollusc faunas are important fossil indicators of facies. Collections of these specimens were usually made in equal distance, ca. one meter or so. On the western coast of Bohai Gulf, 7 cores (1,200 samples) were taken and analysed, while 10 cores (ca. 1,000 samples) were analysed from the coastal plain of the Huanghai Sea and the East China Sea. In analysing the samples, "pollution" factors were eliminated as far as possible. Then the horizons containing rich marine faunas were combined as marine facies formation, and the thickness was known. The so-called continental facies formations include beds containing only fresh water ostracods, molluscs and no fossils. Obviously marine and continental facies formations were alternatively formed clearly showing the circles of transgressions and regressions as well as the periods of sea-level changes. The dominant species (foraminifera, ostracod and mollusc) from ca. 20 cores are listed in Tables 1 and 2.

Table 1 The three transgressions in the western coast region of Bohai Gulf

Sea area	Marine Formation	1st marine formation dominant species	2nd marine formation dominant species	3rd marine formation dominant species
Bohai Sea Area	Foraminifera	Ammonia tepida Pseudononionella variabilis Stomoloculina multangula Pseudoeponides andersomi Elphidium magellanicum	Ammonia tepida Ammonia annectens Elphidium advenum Quinqueloculina akneriana rotunda Spiroloculina laevigata	Ammonia annectens Pseudorotalia Schroeteriana Elphidium advenum Quinqueloculina seminula Spiroloculina laevigata
	Ostracod	Sinocytheridea sinensis Sinocytheridea latiovata Neomonoceratina dongtaiensis Basquetina sinucostata Albilegens sinensis	Sinocytheridea sinensis Albileberis sinensis Boswquetina sinucostata Lequminocythereis hodgi	Sinocytheridea sinensis Leguminocythereis hodgi Neomonoceratina dongtaiensis Munseyella japonica
	Mollusc	Mactra quadrangularis Arca suberenata Cyclina sinensis	Arca subcrenata Nassarius sp. Chione isabelina Solen sp. Chlamys sp.	Chione isabelina Murex sp. Mitra sp. Oliva ornata Arca subcrenata Ostrea sp. Nassarius.

Table 2 The five transgressions in the coastal plain regions of Huanghai Sea and East China Sea

Sea area / Marine fm.	1st formation dominant species	2nd formation dominant sp.	3rd formation dominant sp.	4th formation dominant sp.	5th formation dominant sp.
The Huanghai Sea and the East China Sea Areas — Foraminifera	Ammonia tepida Ammonia annectens Elphidium advenum Quinqueloculina seminula Pseudorotalia gaimardii Pseudoeponides anderseni	Ammonia tepida Elphidium advenum Pseudoeponides anderseni Cribrononion incertum	Ammonia tepida Cribrononion incertum	Ammonia tepida Cribrononion incertum	Ammonia tepida Cribrononion incertum
Ostracod	Sinocytheridea sinensis Neomonoceratina dongtaiensis Bosquetina sinucostata Leptocythere sp.	Sinocytheridea sinensis Neomonoceratina dongtaiensis	Sinocytheridea sinensis Neomonoceratina dongtaiensis	Sinocytheridea sinensis	Sinocytheridea sinensis
Mollusc	Arca subcrenata Solen sp.		Arca subcrenata		Ostrea sp.

Table 1 shows that in the western coast of Bohai Gulf three marine transgressions formed three marine formations. In the second and third marine formations (from top to bottom) warm water foraminifera *Pseudorotalia schroeteriana* and mollusc *Chione isabelina, Murex* sp. *Oliva ornata*, etc. are found. These warm water fauna now only live near the southern coastal region of Jiangsu and Zhejiang Provinces where the annual mean water temperature is 18° C–20°C, while the annual mean water temperature is only 12°C now in the present western coast of Bohai Gulf, a difference of 5.5°C—7.5°C. So it is concluded that the climatic characteristic during the second and third marine transgressions were similar to that along the coastal region of Fujian Province at present.

Table 2 shows that the coastal plain regions along the Huanghai Sea and the East China Sea went through 5 marine transgressions. Yet the number of species and genus in these regions is less than that in the western coast of Bohai Gulf. The reason for this phenomenon awaits further investigation. The distributive characteristics of foraminifera, ostracod and mollusc in the marine formations along the eastern coastal region of China are shown in Fig. 1.

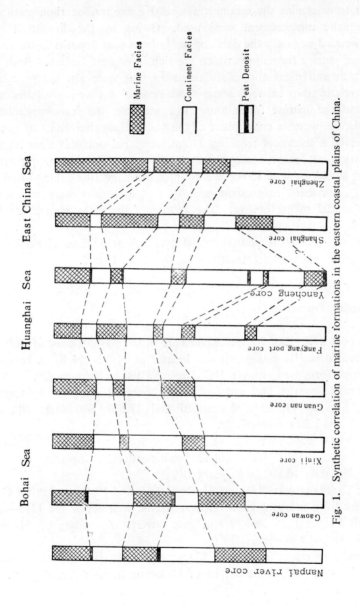

Fig. 1. Synthetic correlation of marine formations in the eastern coastal plains of China.

It is very interesting to note that peat deposits occur in many marine formations, among them the Nanpai river core and the Yancheng core being the most typical ones (Fig. 1). We hold that the Quaternary peat deposits in the coastal plain of China could be a forerunner of both the onset of warm paleoclimate and the marine formation.

DETERMINATION OF MARINE FORMATION AGE

In order to determine the marine transgressive age, radiocarbon method and paleomagnetic measurement were used. Owing to the limitation of ^{14}C dating, we analysed only the dates of the bottom peat deposits in the first and the second marine formations which are rich in organic carbon. Both paleomagnetic measurement and microfaunas analysis were also adopted to find out the relationship between some short reversal events in Brunhes normal epoch and the marine formations. For example, the Yancheng core was located in the western coast plain of the south Huanghai Sea, 60 km from coastline, of a thickness reaching 110 m. and paleomagnetic measurements were carried out on vertically oriented specimens. An alternating magnetic field with a peak value of 150 Oe was adopted for magnetic cleaning to eliminate secondary remanent magnetization. All the specimens were measured by Lam-24 model magnetometer. The results of the measurement are shown in Fig. 2.

It is obvious that these short reversal events bear a close relationship with transgressions or marine formations as shown in Fig. 2.

First Marine Formation

1. The peat deposit of Nanpai River core in the western coast of Bohai Gulf was measured in ^{14}C dating. It was located at 14.72—14.87 m. beneath the first marine formation, with a ^{14}C date of $8,590\pm170$ years BP.
2. The mud deposits, Dc1 and Dc2 cores, beneath the top marine formation, their ^{14}C ages are $11,510\pm570$ years BP and $11,520\pm690$ years BP, depths 19.0 m and 15.2 m respectively.
3. The peat deposit of Yancheng core in the western coast plain of the South Huang Hai Sea. It was located at 13.5 m beneath the first marine transgressive deposit, ^{14}C date $10,800\pm140$ years BP.
4. The Yancheng core, Lianyun Harbour core and Dc1 core exhibit Gothenburg event from paleaomagnetic measurements, their depths are 13.5—14.0m, 10.0—10.5 m and 18.86—19.4 m respectively. According to Mörner, its age is 12,350—13,750 years BP.

As a result, we consider that the first marine formation could take place ca. 10,000 years BP, which should be of Holocene transgressive deposit.

Second Marine Formation

1. The peat deposit of Nanpai River core in the western coast of Bohai Gulf was located at a depth of 40.5—41.5 m, ^{14}C date 32,000 years BP.
2. Mungo event was recorded in Yancheng core and Dc2 core from paleo-

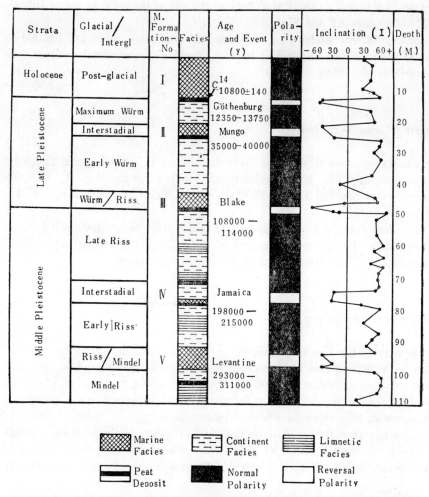

Strata	Glacial/Intergl	M. Forma-tion-No	Facies	Age and Event (y)	Pola-rity	Inclination (I) −60 30 0 30 60+	Depth (M)
Holocene	Post-glacial	I		C14 10800±140			10
Late Pleistocene	Maximum Würm	II		Göthenburg 12350—13750			20
	Interstadial			Mungo 35000—40000			30
	Early Würm						40
	Würm/Riss	III		Blake			50
Middle Pleistocene	Late Riss			108000 — 114000			60
							70
	Interstadial	IV		Jamaica			80
	Early Riss			198000 — 215000			90
	Riss/Mindel	V		Levantine			100
	Mindel			293000 — 311000			110

Marine Facies Continent Facies Limnetic Facies
Peat Deposit Normal Polarity Reversal Polarity

Fig. 2 Correlation between paleomagnetic polarity change and the distribution of marine formation in Yancheng core.

magnetic measurements, with depths 22—25 m and 29.99—30.4 m respectively, located beneath the second marine formation, ages either ca. 30,000—40,000 years or 35,000—36,000 years BP.

It is evident that the second marine formation corresponds well to the interstadial deposit of Würm glaciation.

Third Marine Formation

Blake event was recorded in the silty clay deposit of Xinji core located at a depth of 60—65 m in the western coast of Bohai Gulf. The equivalent clay

205

deposit of Yancheng core is at 47—49 m, the age of the event is about 108,000 —114,000 years BP see Fig. 2.

According to our study, this transgression took place from ca. 100,000 to 70,000 years BP, corresponding to the interglacial transgressive deposit of Würm-Riss glaciation.

Fourth Marine Formation

This is transgressive deposits of short duration, it is located in the Yancheng core at a depth of 77—78 m, beneath which there is a thin peat deposit. Jamaica reversal event (ca. 198,000—215,000 years) was found at a core depth of 74—78 m.

It is evident that the sea regression had taken place before the short reversal event ended. Its age may be estimated at ca. 200,000 years BP and should be assigned to be interstadial deposits of Riss glaciation.

Fifth Marine Formation

As seen from the results of paleomagnetic measurement, the short reversal event in Yancheng core at a depth of 97—99 m. may correspond to Levantine event, with an age of ca. 293,000—311,000 years and a thickness of 92—99 m. We consider that the fifth marine formation may correspond to the interglacial transgressive deposit in Riss-Mindel glaciation.

To sum up, we may say that the variation of geomagnetic field, the climatic warming up, the formation of peat bed, sea level changes, the variation of sedimentary facies and the evolution of fauna are closely related (see Fig. 2).

PALEOCOASTLINE CHANGES

Paleocoastline Changes

Based on the analysed data from 76 cores in the western coastal region of Bohai Gulf, we compiled three paleocoastline maps from 100,000 years BP on. According to analysed data in the western coast of southern Huanghai Sea, three Paleocoastlines from Late Pleistocene on* and 5 paleocoastlines from 300,000 years BP on were compiled.

---

* Xia Dong-xing and Wang Yong-jie. On the coastline changes of the Huanghai Sea and Bohai Sea from the Late Pleistocene (unpublished).

The Sea Level Changes

The alternative occurrence of marine and continental formations in the eastern coastal region of China in the last 300,000 years provided reliable evidences for sea-level changes of the shelf sea in China and those of the world. In fact, the extents of the interglacial transgression and the glacial regression were dependent on the earth's climate changes as well as on the tectonic factors in that particular area and period. Consequently, those marine formations as shown in drilling cores of the coastal plain in China reflect only the highest level the sea water can reach in the interglacial period; but the lowest sea-level in glacial period can only be found in the shelf sea far away from the coastal region. Based on our study the sea-level of the shelf sea in the East China Sea was ca. 130 m lower at its maximum 17,000 years age. Dc2 core taken from the East China Sea, water depth 29 m, with a core deep of 29,99—44.00 m shows a sedimentary environment of seashore-lagoon facies providing rich evidences that the sea-level was ca. 70 m lower in early Würm glaciation than that at present. In addition, Dc2 core at 60 m below is of sequential marine formation, the horizon at 74 m may correspond to Blake reversal event in Brunhes normal epoch, which means the lowest sea-level in Riss glaciation could be higher than that of Würm glaciation. According to Fairbridge's estimation, the world sea-level in Riss glaciation was only 55 m lower than that at present. Based on the above data, a sea-level change in the shelf sea of China in the last 300,000 years is known.

To be sure, the eustatic line of sea-level reflects only essential characteristic of the world sea-level changes during the last 300,000 years. The high peak should correspond to that in transgressive stage while the low curve should correspond to that of sea regression. A more detailed curve reflecting the shelf-sea changes is to be shown after further study.

CONCLUSIONS

1. Three marine transgressions occurred on the western coast of Bohai Gulf in the last 100,000 years while the western coast of the Huanghai Sea has subjected to 5 marine transgressions in the last 300,000 years.

2. Except the fourth marine transgression, each marine transgression became greater in extent and more rapid in frequency from bottom up in the western coast of the south Huanghai Sea and the East China Sea. But in the western coast of Bohai Gulf, the second marine transgression was the greatest.

3. In the second and the third marine formations in the western coast of Bohai Gulf warm water foraminifera such as *Pseudorotalia schroeteriana* and subtropic molluscs such as *Chione isabelina*, *Murex* sp., *Oliva ornata* and others are found, which means that the sea water temperature at that time was 6°C higher than that at present. The change of water temperature has not been

found in the western coast of the Huanghai Sea and the East China Sea.

4. When the third marine transgression took place, fluctuations of sea-level of short duration occurred.

In the third marine formation in the western coast of Bohai Gulf a thin bed of land deposit is found, so that the eustatic line of sea-level shows "double peak".

5. There are interstadial glacial marine transgressive deposits both in Würm and Riss glaciations, which constitute the second and the fourth marine formations.

6. As seen from Yancheng core, each short reversal event in the Brunhes normal epoch seems to correspond to the climatic warming, the formation of peat bed, the sea-level change, the variation of sedimentary facies and the evolution of fauna.

7. We consider that peat deposit in Quaternary in the coastal plain of China may mark the turning warm of the paleoclimate and the beginning of marine formation.

8. Based on the results of our paleomagnetic research, we may also conclude that the Laschamp event should consist of two short reversal events, one is called the Gothenburg event, and the other the Mungo event.

Acknowledgement — The authors are greatly indebted to Mr. Gao Liang, Sun Wei-ming, Zhang Hong-cai and Cang Shu-xi for their valuable suggestions in preparing this paper, to Mr. Li Ben-zhao who helped making paleomagnetic measurements of the samples and to Ms. Li Qing and Mr. Jiang Meng-rong for drawing the figures.

REFERENCES

[1] Li Shi-yu, 1962, Investigation of relics of paleocoastline in Tientsin region. Hebi Daily, April 30.

[2] Zhao Song-ling, Xia Dong-xing, et al., 1978, On the marine stratigraphy and coast-lines of the western coast of the Bohai gulf. Oceanologia et Limnologia Sinica, 9(1): 15–25.

[3] Cang Shu-xi, et al., 1979, Middle Pleistocene paleoecology, paleoclimatology and paleo-geography of the western coast of Bohai gulf. Acta Paleontologica Sinica, vol. 18 (6), 579–591.

[4] Zhao Song-lin, Huang Qing-fu, 1978, Sea-level changes in the world in the last 100,000 years. Journal of Marine Science, 1, 46–49.

[5] Wang Shao-hong, Han You-song, 1980, Transgressive study in the southern coast of Haizhou Gulf in the Quaternary period, Journal of Marine Science, 2, 19–23.

[6] Wang Pin-xian, et al., 1981. Strata of Quaternary transgressions in the East China Sea: a preliminary study, Acta Geologica Sinica, vol. 55 (1), 1–13.

[7] Chin Yun-san, et al., 1981, Synthetic study of Dcl core in the East China Sea. Proceedings of the 3rd Conference of Oceanologia et Limnologia Sinica, (in press).

[8] Zhang Hong-cai, et al., 1978, Paleomagnetic study of two sediment cores from the eastern coastal region of China, Oceanologia et Limnologia Sinica, 9 (2), 183–193.

[9] Zhao Song-ling, et al., 1981, Sea-level changes of the East China Sea since Late Pleisto-

cene. Geology of the Yellow Sea and the East China Sea. (in press)

[10] Mörner, N. A., 1977, The Gothenburg magnetic excursion. Quaternary Research. (7): 413–427.

[11] Barbetti McElhinny, 1972, Evidence of geomagnetic excursion, 30,000 years BP, Nature 239 (5371), 327.

[12] Smith, T. D. Foster, J. H., 1969, Geomagnetic reversal in Brunhes normal polarity epoch, Science, 163, p. 565–567.

[13] Martine Rossignol-Strick, 1973, Palynologic de sapropeles méditerranéens du villa-franchien A L' Holocene. in Les Méthodes Quantitatives d'Etude des variations du climat au cours du pléistocène Editions du centre national de la recherche scientifique Paris 1974.

[14] Fairbridge, R. W., 1961, Eustatic changes in sea level p. 99–185. in Physics and Chemistry of the Earth. vol. 4, New York. Pergamon Press.

BASIC CHARACTERISTICS OF THE HOLOCENE SEA LEVEL CHANGES ALONG THE COASTAL AREAS IN CHINA

ZHAO XI-TAO
(Institute of Geology, Academia Sinica)

AND ZHANG JING-WEN
(Institute of Geology, State Bureau of Seismology)

Abstract—The present situation of studies on the Holocene sea level changes, and the geomorphological, sedimentological and biological indicators to restore the ages and positions of the ancient sea levels are reviewed in this paper. Based on the available data along the coastal areas in China, we hold that: 1. The Holocene sea level changes can be divided into three periods, i.e. the rapid rise before 6,000 BP; the highest sea level between 6,000—5,000 BP; and the relative stability or slight descent for the past 5,000 years; and 2. The Holocene sea level has been fluctuated, and there were six peaks of the sea level fluctuation, i.e. 8,500—7,800, 7,300—6,700, 6,000—5,000, 4,600—4,000, 3,800—3,100 and 2,500—1,500 BP, the latter four of which were the Holocene sea levels higher than that of present.

DISPUTE OF TWO DIFFERENT OPINIONS

Since the radiocarbon method has been applied to the study of sea level changes, some unanimous views about the sea level movement of Early Holocene has been agreed, however, currently a dispute is still going on about the sea level changes during the past 6,000 years. One opinion holds that the ancient sea level at the end of the Atlantic stage rose rapidly up to 3 m above the present sea level, and since then the fluctuations of sea level have had amplitude of 6 m[1]. Another opinion emphasizes that the Holocene sea level has been continuously rising, denying that any Holocene sea level was higher than that of present[2]. These two views have been conflicted for a long time. In addition, there have been a series of other views which are either intermediate between the two opinions or different from them. In China, the situation of the studies on the Holocene sea level changes is also the same.

In recent years, owing to the considerable development of the geology, geophysics, mathematics and mechanics and their infiltration into each other, a series of new ideas and models emerged. Formerly, when the problem of sea level changes was discussed, it was held that the sea level changes caused by the advances and retreats of ice sheet were global, so the changes were called the eustatic sea level, while the crustal isostasy resulted from the loading and unloading of the ice sheet was local, which only influenced the glaciated and their surrounding areas. The main cause for the variation of sea level curves shown is the tectonic movement.

Nowadays, from the view of the earth rheology, it is considered that the earth is a viscoelastic body under the force effect in a time unit of one thousand years, the glacio-isostasy resulted from the loading and unloading of ice sheet is global. In addition, there is a hydro-isostasy resulted from the rise and descent of sea level. According to J. A. Clark's numerical model of the global sea level changes, the relative sea level changes on a viscoelastic earth's surface can be divided into six zones. Zone I is the glaciated areas of the late glaciation; Zone II is North America and Northwest Europe, the surrounding areas of the glaciated areas; Zone III is the central part of the Atlantic Ocean and the northern part of the Pacific Ocean; Zone IV is the central part of the Atlantic Ocean and the North Pacific Ocean; Zone V is throughout much of the southern oceans and Zone VI is all the continental margins exclusive of those adjacent to Zone II, the zone of considerable foreburg submergence. In four (I, III, V, VI) of these zones, the Holocene sea levels higher than present are predicted and confirmed by observations from various areas on the earth's surface[3]. This also provides a new approach for resolving the above-mentioned dispute about the Holocene sea level changes, because the studied areas where these different curves of the Holocene sea level changes obtained belong to the different zones proposed by Clark. Therefore, the concept that the sea level changes are globally coincident is worthy of careful consideration. It should be considered that they change from place to place. We should establish the curve of the Holocene sea level changes along the coastal areas in China.

ANCIENT SEA LEVEL INDICATORS

In order to restore the actual features of the Holocene sea level changes along the coastal areas in China, it is necessary to find the geomorphological, sedimentological and biological indicators so that ages and positions of sea level at that time can be determined. According to the present data and the authors' studies, (4) the ancient sea level indicators are mainly as follows:

Marine Layers

Under the surface of the coastal plains in China, one (along the coasts of South China) or several marine layers (along the coasts of North and East China) or the marine-continental transition layers have been found. A number of marine molluscs, foraminifers, marine ostracods and diatoms are found in the marine layers. Based on the ^{14}C dating, the first marine layer was formed in different phases of Holocene. Because the marine layers deposited in different

211

localities and water depths have different biological assemblages and sedimentological features, we can deduce the ancient sea level positions during their deposition[4,5].

Coral Reefs

In South China Sea Islands, and along the coasts of the northern South China Sea and Taiwan, the Holocene coral reefs are well developed. Based upon the stratigraphical studies and ^{14}C dates, their ages of formation can be divided into seven phases. Because reef corals cannot survive from sea water and are also difficult to live on the sea floor below a depth of a few tens meters, and since the reef flat takes the low tide level during the stable phase of sea level as its upper limit of development, the ancient sea level positions can be deduced from the biological assemblages and flats of coral reefs[6].

Cheniers

Along the coasts of silty and muddy plains of North and East China, cheniers are well developed. Based on ^{14}C dating, the cheniers on the upper part of the plains have been formed since the 6,000 BP. Because the cheniers are products of breaker near the high tide level, their bases are generally equivalent to the high tide level at that time[7,9].

Beachrocks

In South China Sea Islands and along the coasts of the northern South China Sea, the Holocene beachrocks are widely distributed. Based on ^{14}C dating, beachrocks exposed on the surface are formed during the recent 6,000 years, and they can be divided into four phases of formation, while a recent beachrock is still forming now. Because the beachrock is a kind of sediments deposited and cemented in the intertidal and spray zones, the beachrock of beach type should represent the ancient sea level position then, and the beachrock of beach ridge type is several meters or more higher than the ancient sea level at that time[10,11].

Ostracean Reefs

Ostracean reef is a kind of deposits of thick growing ostracea in the littoral zone near the mouths of small rivers where silt is less abundant. At present,

212

the ostracean reefs have mainly been discovered along the western coast of the Bohai Gulf and the southern coast of the Laizhou Bay, and their ages of formation are younger than 6,000 years. Besides, in silty and muddy littoral plains in China, a large amount of ostracean deposits have been found, which partly may be ostracean reefs, and partly may be yielded in bay or lagoon. Because the ostracean reefs are formed in the river mouths near shore, their tops are often equivalent to the low tide level, and the relationship between the ostracean deposits of bay or lagoonal facies and sea level is similar to that between ostracean reef, and the sea level[4].

Peat Beds

A few meters or 20 to 30 meters under the surface of the coasts and littoral shelves in China, peat or mud beds of the Holocene lagoon or lake and swamp facies related to sea water are often found. They are distributed immediately beneath and above the first marine layer. The upper peat bed and the lower peat bed are gradually merged landward. The peat bed of lake or swamp facies indicates that the ancient sea level lay below, whereas, the lagoonal peat bed indicates that the ancient sea level lay close to or slightly higher than it[4,5].

In addition, there are also other marine and marine-continental transitional deposits and abrasion landforms, etc., which not only contain samples that can be determined by ^{14}C dating, but also can reveal ancient sea level positions. The authors will not introduce here in detail.

CHARACTERISTICS OF THE SEA LEVEL CHANGES

During the recent ten years, the authors have investigated the relics of the Holocene sea level changes along the coastal areas in China, collecting and dating more than 200 samples closely related to sea level changes. Other researchers also have obtained ^{14}C dates as many as ours. Based on the analyses and correlations, most of the ^{14}C dates are reliable, the sampling heights are accurate, and the horizons, preserved positions and the relationships with ancient sea levels are clear, which provides a sound bases for the discussion on the regularity of the Holocene sea level changes.

Because there is no place on the earth where the crust is absolutely stable, and the coastlines in China are very long and traverse several different geotectonic units. Therefore, impacts of crustal movements with different characteristics and extent superimposed on the eustatic and hydro-isostatic sea level changes were considerable, which makes the study all the more complicated. So our studies are mainly based on the surveyed results of long-term subsident coastal plains of the Bohai Sea, the Huanghai Sea, the East China Sea and the South China Sea Islands, and of the bedrock coasts of the mainland which have been slightly or moderately uplifting. As to the Taiwan Island situated at the boundary of two large plates, the Eurasia and the Pacific, especially along the East Coastal Range and in the Hengchun Peninsula of Taiwan Island, the impacts of crustal movements, although strong, are not intense enough to cover the relics of the sea level changes, so the coral reefs and marine layers may also be correlated well with other stable or subsident areas in China. Therefore, they still may be regarded as important evidences for the study on the sea level changes.

Now, the processes, evidences and characteristics of the Holocene sea level changes along the coasts in China can be described as follows:

So far as the position of the sea level at the lower boundary of Holocene, about 10,000 BP is concerned, the available data are not enough. Based on the tendency in sea level changes during the past 20,000 years[4], the lower boundary is deduced to be placed at about 35—40 m below the present sea level. Based on the ^{14}C date of muddy clay of 5—6 m in depth, the top of the first continental layer, from a drill hole with a water depth of 25 m off the Luanhe River mouth is 9,165±110 BP, which shows that the ancient sea level 9,000 years ago did not reach −30 m yet. The peat and mud beds beneath the first marine layer in drill holes of Sixin, Chentanzhuang of Tianjin and Nanpaihe of Huanghua County were formed in the phase between 10,000—8,000 BP, and their altitudes are mostly situated at −10 to −15 m at present, which suggests that the Holocene transgression did not reach the present coastline of the Bohai Gulf[4,5]. On the bedrock coasts, relics of ancient sea level are situated higher. For instance, during 8,500—7,800 BP when ancient sea level was at a higher position, it probably reached a height of about −10 m. For example, the ostracean deposits of Qianjiang of Fenghua County is at −15 m, a marine mud bed of Qiangyang of Donggou County is at 3 m s.a.l.[12], the coral reefs of Xiaodonghai and Ximao Island of Yaxian County are exposed in intertidal zone, the Longgong formation of Miaoli County, Taiwan, outcrops above the high tide level[13], the Yieliu raised coral reef of Taibei County is at 5 m a.s.l.[14], and the Tanzi raised coral reef of the Hengchun Peninsula reaches up to 20 m in altitude[15].

During 7,300—6,700 BP, the ancient sea level was at another relative peak, which was close to or slightly lower than the present sea level. For example, the marine shell of Huangcaotuo of Tianjin is 2—3 m lower than the present sea level[8], the primary coral reef of Dengloujiao of the Leizhou Peninsula is exposed at 1.5—2 m above the low tide level, the Agongdian coral reef of Tainan is at 10 m a.s.l.[15], and the Shiniuqiao raised coral reef of Hengchun Peninsula reaches 15 m in altitude[16].

The sea level during 6,000—5,000 BP was 2—4 m higher than present, at the same time, the Holocene transgression also reached its maximum. This transgression had effects on the Ninghe, Baodi, Tianjin, Qingxian, Wenan, Cangxian and Haixin areas of the western coast of the Bohai Gulf, the Lixiahe drainage

214

of the western coast of the Huanghai (Yellow) Sea, the Taihu Lake drainage and the Hangzhou-Jiaxing-Huzhou, Xiaoshan-Shaoxing and Wenzhou-Ruian plains, etc. of the western coast of the East China Sea, and invaded into most of the bays and river mouths of the bedrock coasts, separating many islands from the mainland. In the plain areas, the marine shells, such as Dongditou of Tianjin and Maomaojian of Ninghe, the ostracean reefs and other ostracean deposits of Lutaizakou and Dongzhuangzi of Ninghe, Xinhe of Pingdu and Linjiacao of Qingpu whose altitudes of tops are close to or slightly higher than the present sea level[4,5,7,8]. The bases of cheniers of this phase, such as in Maqiao and Zelin of Shanghai and Guojinzi of Shouguang County, are close to or slightly higher than the present high tide level. Along the coasts of Southeast China, the coral reefs and beachrocks are well developed, for example, the Luhuitou raised coral reef of Yaxian County[6], the Guoxingpu and Gengziliao raised coral reefs on northern Taiwan[15], the raised coral reef of the northern coast of the Diaoyu Island, whose altitudes are all at 3—5 m a.s.l.[17]; the raised coral reefs of Fenggang and other sites in Hengchun Peninsula and of Houzishan in Taidong are at 15—20 m a.s.l.[16], the beachrocks south of the Huanglong Village of Raoping County and on the first terrace of Shidao Island of the Xisha Islands are at about 5 m a.s.l.

During 4,600—4,000 BP, the sea level was 1—2 m higher than present, which is mainly shown by the Tongju-Miaozhuang chenier whose base has the same altitude as the present high tide level[7], the chenier of Liucatuozi of Donggou County with altitude of 5—15 m[12], the beachrock exposed at the intertidal zone in Yinggehai of the Hainan Island[18], the ancient coral reef flats of Beidao Island and Dongdao Island of the Xisha Islands, and the Kending formation with the altitude of 14 m in Hengchun Peninsula[16] etc.

The sea level between 3,800 and 3,100 BP was 1—3 m higher than present. It is mainly reflected by Zhangguizhuang-Chanzhuang chenier of the western coast of the Bohai Gulf[7], the Zhanglin-Limei chenier of the Shantou Area and the Meilong chenier of Haifeng County, whose bases are slightly higher than the present high tide level, the secondary coral reefs in Luhuitou and Dongmao Island of Yaxian County[6], the raised coral reefs along the eastern and western coasts of Diaoyu Island of 2—3 m in altitude[17], the Yagouhai coral reef of Hengchun Peninsula with altitude of 10 m and the Milun coral reef in Hualian County with altitude of 14—20 m[14] etc.

During 2,500—1,500 BP, the sea level was 1—3 m higher than present. It is mainly shown by the Nigu-Qikou chenier of the western coast of the Bohai Gulf, whose base is same as the present high tide level[7], the raised beach deposits and beachrocks in Yandun of Wenchang County and Qingzhou Island of Dianbai County of 3—5 m in altitude[11], the Guoxingpu formation of the northern Taiwan[14], the coral reef flat of Yongxing Island of the Xisha Islands, the coral reefs of the western and southern coasts of Diaoyu Island[17], the Hongchaikeng raised coral reef in the Hengchun Peninsula and the Hualian coral reef of the Taidong coastal range[19] etc.

Up to now, a Holocene curve of sea level changes along the coasts in China can be roughly given (Fig. 1), and the following conclusions are reached:
1. The Holocene sea level changes can be divided into three periods, i.e. the rapid rise before 6,000 BP, the highest sea level between 6,000 and 5,000 BP and the relative stability or slight descent during the past 5,000 years.
2. The sea level in Holocene has been fluctuated, and six peaks, i.e. 8,500—7,800, 7,300—6,700, 6,000—5,000 4,600—4,000, 3,800—3,100 and 2,500—1,500 BP, were found, the latter four of which were the Holocene sea levels higher than the present.

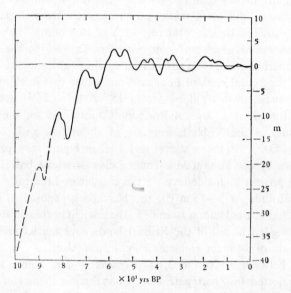

Fig. 1. The Holocene curve of sea level changes along the coastal areas in China.

REFERENCES

[1] Fairbridge, R. W., 1961, Eustatic changes in sealevel. Physics and Chemistry of Earth, 4, p. 99—185.
[2] Shepard, F. P., 1964, Sea level changes in past 6,000 years; possible archaeological significance. Science, vol. 143, p. 574—576.
[3] Clark, J. A., 1980, A numerical model of worldwide sea level changes on a viscoelastic earth, in: Earth rheology, isostasy and eustasy, John Wiley and Sons, Ltd. p. 525—534.
[4] Zhao Xi-tao et al., 1979, Sea level changes of eastern China during the past 20,000 years, Acta Oceanologia Sinica, vol. 1, no. 2, p. 269—281.
[5] Zhao Xi-tao, 1980, The Holocene coastline shift of the western coast of the Bohai Bay, in: Formation and development of the North China fault block region, p. 302—309, Science Press, Beijing.
[6] Zhao Xi-tao, 1979, Ages of formation of the Luhuitou coral reefs, Hainan Island, and their effects on shoreline changes, Kexue Tongbao, vol. 24, no. 21, p. 995—998.
[7] Zhao Xi-tao et al., 1980, Chenier ridges on the western coast of the Bohai Bay, Kexue

216

Tongbao, vol. 25, no. 3, p. 243—247.

[8] Groups of Quaternary geology and ^{14}C dating, Institute of Geochemistry, Academia Sinica, 1980, Paleocoastline shifts of the western coast of Bohai Gulf in Holocene, Quaternaria Sinica, vol. 5, no. 1, p. 64—69.

[9] Zhang Jing-wen et al., 1979, Radiocarbon dating of the chenier ridge in Maqiao and Zhelin, Shanghai, Seismology and Geology, vol. 1, no. 4, p. 10.

[10] Zhao Xi-tao et al., 1978, Holocene beachrocks at Hainan Island, Scientia Geologia Sinica, no. 2, p. 163—173.

[11] Zhao Xi-tao et al., 1979, A preliminary study of Holocene stratigraphy and sea level changes along the coast of Hainan Island, Scientia Geologia Sinica, no. 4, p. 350—358.

[12] Lab. of Quaternary palynology and Lab. of radiocarbon dating, Kweiyang Institute of Geochemistry, Academia Sinica, 1978, Development of natural environment in the southern part of Liaoning Province during the last 10,000 years, Scientia Sinica, vol. 21, no. 4, p. 516—532.

[13] Hashimoto, W. et al., 1970, Studies on the younger Cenozoic deposits in Taiwan: Part I. The younger Cenozoic deposits of the middle part of West Taiwan: Geol. Paleont. Southeast Asia, vol. 8, p. 237—252.

[14] Lin C. C., 1969, Holocene Geology of Taiwan: Acta Geologica Taiwanica, no. 13, p. 83—126.

[15] Hashimoto, K., Taira, K., 1974, Radiocarbon dating records of some calcareous fossils from the younger Cenozoic deposits of Taiwan and problems brought about by the dating: Geol. Paleont. Southeast Asia, vol. 14, p. 117—134.

[16] Konishi, K. et al., 1968, U^{234}–Th^{230} dating of some Late Quaternary coralline limestones from southern Taiwan: Geol. Paleont. Southeast Asia, vol. 5, p. 211—224.

[17] Konishi, K. et al., 1979, Holocene raised coral reef on Senkaku Islands,* An active remnant arc, Proc. Japan Acad. v. 55, ser. B, no. 7, p. 335—340.

[18] Zhang Jing-wen et al., 1981, The age of formation of the marine shells and beachrocks in the Yinggehai Region, Hainan Island, Seismology and Geology, vol. 3, no. 1, p. 66.

[19] Peng Tsung-hung et al., 1977, Tectonic uplift rates of the Taiwan Island since the Early Holocene, Memoir of Geological Society of China, no. 2, p. 57—69.

---

* The authors' note: Diaoyu Islands of China.

217

GLACIAL VARIATION SINCE LATE PLEISTOCENE ON THE QINGHAI-XIZANG (TIBET) PLATEAU OF CHINA

ZHENG BEN–XING AND SHI YA–FENG

(Lanzhou Institute of Glaciology and Cryopedology, Academia Sinica)

Abstract—The climate of the plateau became more and more cold and dry because of the strong uplifting of Qinghai-Xizang Plateau. As a result, the development of glaciers has been seriously restricted since Late Pleistocene.

The fluctuation of the Holocene glaciation may be divided into three major stages: slow regression stage at Early Holocene (10,000—8,000 BP), qucik regression stage at Middle Holocene (8,000—3,000 BP) and frequent advances and regressions of glaciers at Late Holocene i.e. the Neoglaciation (3,000—0 BP).

During Neoglaciation there were three great glacial fluctuations i.e. Xuedan glacial advance (^{14}C 2,980 ± 150 BP) or Rongbude glacial advance, Ruoguo glacial advance (^{14}C 1,920 ± 110 BP) or Tongzhulin glacial advance and little ice age (300—0 BP). In recent 10 years, the temperature fall has been accompanied by a little increase of precipitation in West China. The majority of glaciers are still in recession although their rates are obviously reduced and some are stable. A small proportion of glaciers are clearly advancing and even a few surging glaciers appear.

Qinghai-Xizang (Tibet) Plateau with its surrounding high mountains, including the Himalayas, Karakorum, Pamirs, Kunlun, Qilian and Hengduan Mts., may be the most developed area of mountain glaciers in the world. Since 1958 glacial expeditions have been carried out continuously by Chinese scientists, which enables us to reconstruct more definitely the history of variation of glaciers since Late Pleistocene on Qinghai-Xizang Plateau and to inquire their regional peculiarities.

LAST GLACIATION ON QINGHAI-XIZANG PLATEAU

The spectacular elevation, topography and particular climate of Qinghai-Xizang Plateau in China exert great influence on the development of glaciers. The present glacierized area on the plateau within Chinese territory reaches about 46,640 km², 83% of the total glacierized area in China[1] distributed mainly in western Kunlun (8,438 km²) Nyainqentanglha (7,536 km²), Himalayas (11,055 km²) and Karakorum (3,265 km²). Many large glaciers extend their length over 20—30 km. The Yinsugaiti glacier in the northwest of K₂ in Karakorum with a length of 41.5 km and an area of 329 km² is the largest one known in China[2]. The heights of snowline range from 4,400 m in East Qilian Mt. and Southeast Xizang to 6,000—6,200 m in South and West Xizang and the isolines are some what like irregular concentric circles with their centre in Southwest Xizang. This peculiar form of variation of snowline is mainly owing to the rapid decreasing of precipitation and intensification of plateau thermal effect from some outer mountains to the inner part of the plateau. At least four Quaternary glaciations could be distinguished. The greatest glaciation occurred in Middle Pleistocene, the scattered piedmont glaciers covered the plateau but no great ice sheet was formed at that time. After Middle Pleistocene, the successive uplift of the plateau and its surrounding mountains rendered their elevation reaching 4,000 m in the inner part of the plateau and 5,500 m in the main ridges of the Himalayas, and played the role of a substantial barrier to the South Asia Monsoon. The climate of the plateau became more and more cold and dry in Late Pleistocene. As a result, the magnitude of glaciers was seriously restricted and they became separated valley glacier. The length of the largest valley glacier reached 100 km and more in Southeast Xizang, about 50 km at the northern slope of West Kunlun and about 30 km at the northern slope of Central Himalayas. A few glaciers extended out of valley with wide snouts or combined together forming piedmont glaciers. In the inner part of the Qinghai-Xizang Plateau, where the remnant of Tertiary or Early Pleistocene peneplain rose above the snowline, there appeared some small ice caps or flat-topped glaciers. In the southeastern part of the plateau, small ice caps were discovered at Midika-Sangba Basin (about 31°N, 93°E) to the northeast of Lhasa and on the Haize Mt. of Daocheng (about 29°N, 100°E) each with an area of 2,000—3,000 km². The lowering value of the snowlines during the last glaciation as compared with present snowlines was from 1,000—2,000 m in west Pamir, south of Mt. Qomolangma, and southeastern Hengduan Mts., from 600—800 m in Qilian Mt. 500 m on the inner Qinghai-Xizang Plateau and only 300 m at the northern slope of Mt. Qomolangma. The present preserved altitudes of paleosnowlines are clearly influenced by neotectonic movement, where the uplifting is more intensive and the lowering value is smaller. It also shows that the decrease of precipitation from the marginal mountains of Central Asia towards the inner plateau was more severe in Late Pleistocene than that at present. A continuous permafrost zone extended to about 30°N in the plateau, more than 2 degrees of latitude lower than that today. At the ^{14}C date 14,780±700 years BP, the coastline of the East China Sea was located at the outer margin of the continental shelf, 600 km east of Shanghai, where the present sea water depth is 155 m[3] and this might have been one of the reasons why the climate of China during the Last Glaciation was drier than that of present.

GLACIAL VARIATION DURING EARLY HOLOCENE

Since the general tendency of climate in the last glaciation was turning warm, glaciers started to retreat. By the results of sporo-pollen analysis in South and Central Xizang, three plant associations reflecting the climatic changes during Holocene can be divided: 1) The Early Holocene ranges from 7,500—13,000 BP with wide distribution of sparse shrub steppe on the plateau; 2) The Middle Holocene from 3,000—7,500 years BP. It was warm humid in South Xizang with forest-shrubsteppe land-scape, and cold dry in North Xizang with sparse shrub-steppe; 3) The Late Holocene is 0—3,000 years BP. The landscape was typical of small shrub-steppe or steppe similar to the present one[4].

In Early Holocene, the glacial variation showed a characteristic of slow recession and alternative stagnancy. For example, in front of the Yebokanjiale Glacier at the northern slope of Mt. Xixabangma in southern Xizang, there are two end moraines, indicators of the Last Glaciation, at the elevations of 5,080 m and 5,100 m[5]. At that time, this glacier connected with another ancient Daqu Glacier at their termini and formed a small piedmont glacier. After that, while receding, the Yebokanjiale Glacier separated from the other and formed a moraine terrace above 5,200 m a.s.l. and an end moraine at about 5,280 m a.s.l. Though lacking data of isotopic chronology, it might be taken as the indicators to show the glacial recession of Early Holocene (Table 1, Fig. 1). In southeastern Xizang, there are many examples of glacier recession, e.g. in Puqu Valley, to the south of Arza Lake, Lhari County, in East Xizang a series of small terminal and lateral moraines at the back of two end moraines of Baiyu Glaciation corresponding to Würm at about 4,600 m and 4,660 m a.s.l. are

Table 1 **Since Qomolangma Glaciation, changes of glaciers in Lakdaula River at the northern slope of Mt. Xixabangma**

Age			Glacial and interglacial	Length (km.)	Area (km.)	Height of ice tongue (m)	Depression of snow line (m)	Height of pale-osnow-line (m)	Type of glaciers
Holocene	Late	Neoglaciation	Modern little ice age	14	20	5,530	−87	5,931	valley
			Tongzhulin ice advance	16	40	5,400	−198	5,810	valley
			Rongbude ice advance	17.8	49	5,300	−260	5,740	valley
	Early Middle	Yali climatic optimum	Quick regression	10	10	5,800	+300	6,300	valley or cirque
			Slow regression	19	53	5,280	−275	5,725	complex
				22	58	5,200	−342	5,658	valley
Late Pleistocene	Late	Qomolangma	Late Baiyu (Rongpu stadial)	23	64	5,100	−397	5,603	piedmont
			Early Baiyu	24	70	5,080	−422	5,578	piedmont
	Early		Last interglacial (interstadial)						
			Guxiang (Jilong stadial)	31	100	4,937	−546	5,436	piedmont

220

found, respectively.　The former gives the evidence of glacier recession in Early Holocene.

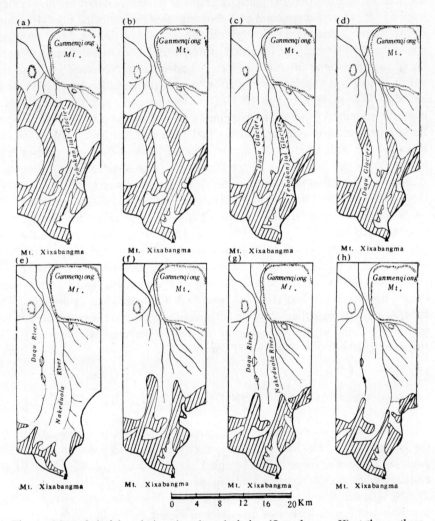

Fig. 1.　Map of glacial evolution since last glaciation (Qomolangma II) at the northern slope of Mt. Xixabangma.
(a) and (b) showing the first stage and second stage of Late Qomolangma II in the Late Pleistocene; (c) and (d) showing the first stage and second stage of Early Holocene; (e) showing the Yali Climatic Optimum of Middle Holocene; (f), (g) and (h) showing the Neoglacial-1, 2 and 3 of Late Holocene.

GLACIAL RECESSION DURING MIDDLE HOLOCENE

Middle Holocene was a stage warmer than the other stages in Holocene and the

221

glaciers were collapsing quickly, only some small ones remained in higher alpine mountains and those in lower alpine mountains all disappeared. According to the upper limit of lateral moraine with a great amount of round gravels, it can be recognized that the glacier termini of Rongpu Glacier at the northern slope of Mt. Qomolangma and of Yebokangjiale Glacier at the northern slope of Mt. Xixabangma were at a position above the elevation of 5,800 m which is about 300—600 m higher than that of present. In Southeast Xizang, according to the upper limit of trimline of glacial advance at Neoglaciation, the terminus of Ruoguo Glacier, Yi'ong Area, was at the elevation of 4,300 m or so and 600—700 m higher than that of existing glacier. On the lower alpine mountains, such as those near Changma of Dingri and near the Dingka Pass of Burang in Southwest Xizang, glaciers had been extinguished. This stage can be called the Yali Climatic Optimum because in the calcareous tufa of Yali (4,300 m) to the north of Central Himalayas, fossils of *Rhododendron* and *Willow*, etc. show that the mean annual air temperature was 3—5°C higher than that at present [6] and these kinds of plants can only live at the elevation of 3,400 —3,800 m on Central Himalayas nowadays. It is also called the Guangchang Hypsithermal Interval in southeastern Xizang. For example near the Ruoguo Glacier, at the elevation of 3,800—4,000 m, the pollen analysis shows that there was a forest of mixed coniferous and broadleaved trees, which corresponds the subtropical vegetation at about 2,500—3,000 m nowadays, and that the mean annual air temperature at that time was 5—6°C higher than at present. The increase of *Quercus* pollen to some extent in the lacustrine deposits at 4,750 m in Nariyon Lake of South Xizang also proves a warmer paleoclimate within ^{14}C date 6,380±100 — 3,625±100 years BP[4].

GLACIAL VARIATION DURING LATE HOLOCENE

The hypsithermal interval mentioned above was followed by a new stage called the Neoglaciation, e.g. in the Indian peaks area of Rocky Mountain, North America, it began at the so called Temple Lake Ice Advance (4,000—2,550 BP)[7] and this might be earlier than those in Southeast Xizang.

The Neoglaciation in Southeast Xizang can be divided into three stages. The first ice advance can be shown obvioulsy in Arza Glacier, in Southeast Xizang where a decayed wood buried in the lateral moraine proves that the forest was destroyed by ice advance at ^{14}C 2,980±150 BP[8]. At that time, the end moraine was 2 km beyond the present terminus and the lateral moraine 100 m higher than the present ice surface.

The second cold period as shown clearly in Ruoguo Glacier, where the middle lateral moraine was formed at ^{14}C 1,920±110—1,540±85 BP after which there was an obvious warm interval, causing a large scale recession of the glaciers.

The third cold period clearly coincided with the Little Ice Age, and caused a new advance of glaciers in West China. According to the studies of Zhu Ke-zhen (1973), the lowest temperature in East China occurred at about 17th century. Generally, 3 parallel ridges of fresh end moraine were formed in front of modern glaciers. From the growth rate of *Rhizocarpon geographicum* growing on the moraines, Li Ji-jun (1975)[9] estimated that 3 moraine ridges at Woguo-dulin Glacier near Yadong, South Xizang, were formed in 1818, 1871 and 1885, respectively. Due to the fact that the advance of large glaciers may be several decades behind the onset of a cold period, the Batura Glacier, in the Karako-rum in Pakistan, and the Arza Glacier in Southeast Xizang were still advancing in the 1920s.

RECENT VARIATIONS OF GLACIERS

The majority of glaciers began their recession from the end of the last century. The rise in temperature was so clear as during the warmest 5 years in 1940s which was 0.5—1.0°C higher than the average value recorded in 1901—1957 in East China[10]. The termini of larger glaciers in West China retreated be-tween several hundred metres and several kilometres. The magnitude of glacial recession was generally greater in the humid marginal mountains and less in the arid, interior mountains of the plateau. For example the Arza Glacier in south-eastern Xizang retreated 700 m between 1933 and 1973 (a rate of 17.5 m a^{-1}), and the Skyang Glacier northeast of Mt. Qogir (K_2) in Karakorum retreat-ed 5.25 km between 1937 and 1968 (a rate of 169 m a^{-1}). (2) However, the retreat rates of glaciers in the interior of the plateau and in the western Qilian Mt. are generally less than 10 m a^{-1}. Some glacial termini, with the protection of a thick debris cover were stable during the recession period. For example, the terminus of the Rongpu Glacier at the northern slops of Mt. Qomolangma showed no change according to three surveys made in 1921, 1959 and 1966, respectively[11].

The prominent temperature fall between the 1950s and the present has been accompanided by a little increase of precipitation in West China. In the north-eastern margin of the plateau i.e. in Qilian Mt., the average temperature at an elevation about 4,200 m (600 mb) between 1967 and 1976 fell 0.8—1.3°C as compared with the decade before 1967, and the precipitation at 3 stations above 3,000 m a.s.l. increased 3% during the same time (from 345 to 355 mm year^{-1}). The mean temperature of 1960s in Xizang area decreased by 0.7°C as compared with that of 1950s and the precipitation increased 5—27% in the same period. From 1974, at 4 glaciers in Qilian Mt. where mass balances were observed, positive values appeared successively for several years, and the mean height of snowline was 100—200 m lower than that of 1950s. The majority of glaciers are still in recession although their rates are obviously reduced. A small proportion of glaciers are clearly advancing. Comparison of various Landsat images of 1970s and the actual observations from 1973 to 1981 with the aerial photo maps of 1950s and 1960s shows that advancing glaciers appeared in many mountains, such as Qilian, Kunlun, Karakorum and the Himalayas. Of the total of 116 glaciers with the information on terminus advance or recession, about 53.4% (i.e. 62 glaciers) were in retreat, 30.1% (i.e. 35 glaciers) were clearly in advance and 16.5% (i.e. 19 glaciers) were stable or without prominent change[1]. The most prominent glacier advance probably occurred in the Karakorum at Sino-Pakistan border and A'nyemaqen Mts. at the eastern end of Kunlun. Five out of eleven glaciers along Sino-Pakistan highway advanced prominently including three of surging type[2]. According to the latest information fifteen out of nineteen glaciers in A'nyemaqen Mts. were in advancing with the extension of 50—790 m from 1966—1981[12]. According to the climatic fluctuations shown by tree rings and other data, such as the record of 935 years from *Sabina* in Qilian Mt. the downward tendency of temperature will possibly continue to the end of this century. The amount of advancing glaciers would increase in this and next decade.

REGIONAL PECULIARITIES

The strong uplift of the Qinghai-Xizang Plateau had great effects on climate and glaciation. The magnitude of uplifting of the plateau is estimated at about 3,000 m in Quaternary and about 1,000 m from Late Pleistocene to Holocene. Owing to the great height of the Himalayas, Karakorum, Kunlun and other marginal mountains as well as plateau itself they act as substantial barriers to atmospheric circulation which has impacted the climatic condition of the inner part, especially the western and the northern parts of the plateau to become drier and intensified the monsoon. As a result summer humid weather has prevailed in the southern and the eastern marginal mountains, such as at the southern slope of the Himalayas, southeastern Xizang and Hengduan Mts. Also, the development of glaciers was seriously restrictred in the greater part of the plateau. No great ice sheet has been discovered even during the maximum glaciation in Middle Pleistocene. The length of several large valley glaciers in the Last Glaciation was only 1.6—5.9 times of those at present. The lowering value of snowlines during Last Glaciation as compared with present snowlines ranges only from 300—500 m on the inner plateau. Since the Last Glaciation, the warmer tendency has been much reduced by the incessant and rapid rising of the plateau. The scale of glacial variation is as great as that in Alps. The rate of variation also decreases from marginal mountains towards the interior.

REFERENCES

[1] Shi Ya-feng, Li Ji-jun, 1981, Glaciological research of the Qinghai-Xizang Plateau in China, Geological and ecological studies of Qinghai-Xizang Plateau. II, p. 1589—1597,

Science Press, Beijing.

[2] Zhang Xiang-song, 1980, Recent variations of the Yınsukati Glacier and adjacent glaciers in the Karakorum Mountains, J. Glaciology and Cryopedology 2. 3. p. 12—6 (in Chinese).

[3] Shi Ya-feng, Wang Jing-tai, 1979, The fluctuations of climate, glaciers and sea-level since Late Pleistocene in China, in: sea level, ice, and climatic change (Proceedings of the Canberra Symposium, December 1979). IAHS Publ. no. 131.

[4] Huang Ci-xuan et al., 1981, A pollen study to discuss the natural environment of the Central and Southern Qinghai-Xizang Plateau of Holocene, in: Geological and ecological studies of Qinghai-Xizang 1. p. 215—224, Science Press, Beijing.

[5] Zheng Ben-xing, Shi Ya-feng, 1976, Quaternary glaciation on the Mount Qomolangma Area, Report on the Scientific Expedition (1966—1968) in the Mount Qomolangma Area Quaternary Geology, 30—63 Science Press, Beijing (in Chinese).

[6] Hsu Ren et al., 1976, Quaternary palaeobotanical studies on the Mount Jomolangma Region, Report on the Scientific Expedition (1966—1968) in Mount Jomolangma Area, Quaternary Geology, Science Press, Beijing (in Chinese).

[7] Richmond G. M., 1972, Appraisal of the future climate of the Holocene in the Rocky Mountains, Quaternary Research 2 (3).

[8] Zheng Ben-xing, Li Ji-jun, 1981, Quaternary glaciation of the Qinghai-Xizang Plateau, in: Geological and ecological studies of Qinghai-Xizang Plateau, 2, p. 1631—1640, Science Press, Beijing.

[9] Li Ji-jun, 1975, Recent study on glaciers in Southeast Xizang, J. Lanzhou Univ. no. 2. (in Chinese)

[10] Zhu Ke-zhen, 1973, A preliminary study on the climate fluctuations during the last 5,000 years in China. Scientia Sinica 16 (2).

[11] Zheng Ben-xing, Shi Ya-feng, 1975, Variation of glaciers on the Mount Qomolangma Area, Report on the Scientific Expedition (1966—1968) in the Mt. Qomolangma Area Glaciology and Geomorphology, p. 92—110. (in Chinese)

[12] Wang Wen-ying, 1982, Variations of glaciers in A'nyemaqen Mts. (in Press)

ON LATE PLEISTOCENE PERIGLACIAL ENVIRONMENTS IN THE NORTHERN PART OF CHINA

Cui Zhi-Jiu

(Department of Geography, Beijing University)

and Xie You-Yu

(Institute of Geography, Academia Sinica)

Abstract—Judging from chronological data and stratigraphical correlation, in addition to records of periglacial phenomena, results of palynological analysis and studies on fauna and sealevel changes, the periglacial environments in the northern part of China during the later period of Late Pleistocene are summarized.

PERIGLACIAL PHENOMENA AND CHRONOLOGY

Based on chronological data, the geomorphological positions and the equivalent stratigraphic horizons, it is suitable to divide the formation of the paleoperiglacial phenomena in North and Northeast China since Late Pleistocene into three stages:

— The first frigid period of the later period of Late Pleistocene (equivalent to Würm I), being called Late Pleistocene frigid "I" for short, is >26,000–35,000 yrs BP.

(The relative warm period is 23,000—26,000 yrs BP.)

— The second frigid period of the later period of Late Pleistocene (equivaient to Würm 11), being called Late Pleistocene frigid "II" for short, is 12,000—23,000 yrs BP.

(Turning into the warm period of Early and Middle Holocene being 3,000—12,000 yrs BP.)

— The frigid period of neo-ice age of Middle Holocene, being about 3,000 yrs BP.

Evidences Based on Isotopic Chronology

The Late Pleistocene frigid "I" is 26,000—35,000 yrs BP, such as the calcareous sinter in involutions at Hutouliang, Yangyuan County, Hepei Province is

27,675±745 yrs BP, and the black, muddy layer is >41,600 BP*, the involutions in the lower layer at Shalawusu is 27,940±200 yrs BP.** In the Late Pleistocene frigid "II" such as at Zhalainuoer, Nei Monggol, fossils of mammoth are found in the upper layer while involutions are found in the middle and lower layers. ^{14}C dating is 11,460±230 yrs BP[1]. Zhou Ting-ru and Li Hua-zhang (1981) reported that two layers of involutions existed in the Gonggouyan profile, at Daihai of Nei Monggol (41°N, 1,240 m), the mud in the involutions of the upper layer being 8,005±100 yrs BP. A spacing is found 3 metres below and another layer of involutions occurs beneath it which is considered to be no later than 10,000 yrs BP, by the same two scholars, and is equivalent to Late Pleistocene frigid "II". The example of neo-ice age is the involutions in rather loosen structure developed in black soil layer in Northeast China and Nei Monggol region. The black soil layer is dated to be 7,500 yrs BP at the Zhonghe Commune, Butha Qi, Da Hinggan Mts[2]. Three layers of involutions are found in the Holocene black soil in Zhalainuoer, Nei Monggol, the upper layer (0.8 m deep) being 1,450±140 yrs BP; the middle (2.2—2.3 m deep), 3,010±80 yrs BP; and the lower (2.6—2.8 m deep), 3,770±70 yrs BP. They are equivalent to the frigid period of Zhou and Han Dynasties beginning 3,000 yrs BP followed the Yangshao warm period (i.e. the optimum period) of China[3]. Two layers of involution are also found in the Dongheyan profile, nearby Dai Hai Lake. The mud in the lower layer is 6,039±90 yrs BP, and that in the upper layer, 5,085±85 yrs BP. Thus, the frigid period also belongs to Holocene. The age of the frigid period represented by involutions should be younger than that of the black soil layer and mud layer discussed above. The ages of the two stages of Late Pleistocene frigid "I" and "II" discussed here coincide basically with the relevant data of Japan and North America[4,5,6].

Evidences of Geomorphological Positions and Stratigraphic Horizons

The periglacial phenomenon of Late Pleistocene "I" of North China is mainly developed in the sand-gravel bed and brecciated layers, i.e., the lower layer of so called "doubled layer structure". (the upper layer is Malan Loess). For instance, the ice wedge developed in the lacustrine facies atrata north of Datong, Shanxi overlain by Malan Loess; the rock sea on top of the Baihua Mt. overlain by Malan Loess and the similar phenomena found at the foot of the Changbai Mts.; the accumulation of solifluction in the Huanggongliang Area at the southern end of Da Hinggan Mts. overlying the unsymmetrical valley; involutions in the fine sand gravel bed of Tianzhen, Shanxi and fauna in the overlying loess layer[7]; the typical ice wedge in sand-gravel bed of Handaqi, Xiao Hinggan Mts. and the involutions of the later period of Late Pleistocene in Nansanqi,

---

* According to Zhou Ting-ru and Li Hua-zhang (1981).
** According to Zhou Kun-shu (1982).

Da Hinggan Mts.; Jingbian of Shaanxi; Baiyinchang of Gansu; Zhongning and Zhongwei; the eastern margin of Yinchuan Basin and Wuyuan of Nei Monggol are all the examples. Involutions equivalent to Late Pleistocene frigid "I" ($27,940 \pm 200$ yrs BP). "II" and frigid period of neo-ice age (3,200—5,070 yrs BP) are also found in Shalawusu horizon.*

PERIOD AND CHARACTERISTICS OF PERIGLACIAL VEGE- TATION AND FAUNA AND THEIR INTERRELATIONSHIPS

Sporo-pollen analysis and fauna studies of the later period of Late Pleistocene (30,000—40,000 yrs BP) of North and Northeast China indicate that firstly vegetation experienced three stages i.e. the stage of coniferous forest of frigid temperate zone and coniferous and broadleaved mixed forest of temperate zone; the stage of steppe; the stage of coniferous forest and coniferous and broad-leaved mixed forest of frigid-temperate zone in North China. The stage of tundra and forest-tundra—the stage of coniferous and birch forest—the stage of desert steppe or tundra—steppe are found in the central and the northern parts of Northeast China. Secondly; dark coniferous forest belt with dragon spruce and fir in predominance has not only once expanded toward middle latitude regions starting from high latitude mountains to hills or piedmonts. It is thought the southward migration may reach 16° latitude while the downward may reach 1,500—2,000 m or even greater. Thirdly, "Coelodonta— Mammuthus fauna" was quite active in Northeast and North China[8,9,10,11]. But, the authors think according to ecological characteristics, source of origin and migrational scopes of Coelodonta antiquitatis and Mammuthus primigeneus, though being named as one fauna jointly, these two kinds of animals are found either in association or in separation yet. As for Northeast China, 144 sites are found to have mammoth fossils 73 sites are found to yield Coelodonta antiquitatis and 47 sites are found to yield both, which occupies 1/4 of the total of sites. As for Jilin Province, Mammuthus primigeneus makes up the highest amount of Late Pleistocene mammoth fossils unearthed, being 27.8% of the to-tal amount, while Coelodonta antiquitatis, only 16.6%[12]. Mammoth came from the polar area and needed a frigid environment. The southern boundary of its distribution lies further north than that of the Coelodonta an-tiquitatis. The scope of its activity is mainly confined within tundra and forest — tundra zones. Hence, it is more significant to take mammoth as an indica-tor of frigid periglacial climate. Concerning in Northeast China, north of the 42°N is a concentrated distribution area while north of 41° is a sparse distri-bution area in terms of fossil distribution density. But, actually, either in Eur-asia or North America, the scope of mammoth distribution is equivalent basi-cally to that of latitudinal permafrost. But as for Coelodonta antiquitatis, it could transverse coniferous forest southward and departed from the perma-

* Followed Zhou Kun-shu (1982).

228

frost zone. Therefore, it is unsuitable to take the individual occurrence of *Coelodonta antiquitatis* as an indicator of periglacial climate. This particular type of fauna coexisted with typical polar fauna such as arctic fox and reindeer. However, it coexisted with primitive ox, water buffalo, etc. in the northern part of North China, which indicates that it was adaptable to middle latitude temperate environment. So it should be admitted that zonal difference also exists in regard to climatic indicative role of this particular fauna.

There are two horizons bearing *Coelodonta antiquitatis—Mammuthus* fauna fossils in Northeast China. The relative wide distribution is found to be located in the lower part of so called "double layered structure" of Late Pleistocene period, that is, fine sand-gravel bed (at Zhalainuoer, Yakeshi, Tieli, etc.)*, and the fossils are also found in the upper part of "double layered structure", of loess silt and sandy clay, and loess sediments such as in the upper part of first terrace of the Nenjiang River Besin,** and Linkou County, in Northeast China. According to the available data, the latter is mainly confined in an area north of 45°N. Sporo-pollen analysis of fossil-bearing horizon in the lower part indicate that the fauna was principally in an environment with the decrease of wood pollen (*Pinus, Picea, Abies, Betula* predominant) and the increase of herbaceous pollen.*** In combination with other data of pollen analysis, we think that among the three stages of Late Pleistocene vegetation in Northeast China, the first and third tunda-meadow stages are the major active periods of *Coelodonta antiquitatis—Mammuthus primigeneus* fauna. Because mammoth was a grass-eating animal living around tundra and meadow instead of forest animal if ecological characteristics and distribution regularity are taken into account. Hence, the authors hold that it is possible to distinguish the time and space of activity of *Coelodonta antiquitatis—Mammuthus primigeneus* fauna from those of coniferous development. In Northeast China according to fossil distribution and horizons in combination with part of the isotopic chronological data, it seems that two periods of comparatively full activity of *Coelodonta—Mammuthus* fauna in Northeast China can be distinguished,**** one is 26,000—35,000 yrs BP***** which is equivalent to those in Hokkaido, Japan (32,000 yrs BP) and those in the southern part of Siberia of USSR (39,000—44,000 yrs BP) where this fauna was widely spread. A.E. Heintz suggested (1965) that 35,000—40,000 yrs BP was one of the high-tide periods of activity of mammoth, and the other is 23,000—12,000 yrs BP.****** A.E. Heintz also sug-

---

* Followed Yang Da-shan (1974).
** Followed Zhang Ai-xing (1979).
*** Followed Qiu Shan-wen et al. (1979).
**** Similar views are held by Qiu Shan-wen (1979).
***** For example, the ages of the teeth of mammoth found in Mudanjiang, Heilongjiang are 20,900±1,000 and 21,540 yrs BP and the ages of mammoth fossils in Zhaoyuan are 21,200±600 and 20,600±600 yrs BP
****** For instance, the ages of the teeth of mammoth and *Coelodonta antiquitatis* in Mingyuegou, Antu, Jilin is 26,560±550 and 28,720±750 yrs BP respectively. The ages of *Coelodonta* fossil in Zhoujiayoufang, Yushu, Jilin is 31,800±900 and 35,370±1,800 yrs BP and incisor of mammoth in Yuquanzi Commune, Yongji is 25,900±500 yrs BP (After Qiu Shan-wen 1981)

229

gested, that 11,700—11,200 yrs BP saw a destroyed period of mammoth since individuals became small. These two periods of activity precisely correspond to the two stages of tunda-meadow prior to and followed the dark coniferous-birch forest stage in Northeast China. A.E. Heintz and V.E. Gerutt also pointed out that mammoth dramatically decreased 25,000 yrs BP was due to climatic changes[13] (23,000—30,000 yrs BP)*. Since Late Pleistocene frigid "II" is rather short which though was the most frigid period since Late Pleistocene as people think, mammoth was not widely distributed, even failed to reach the southern boundary of period "I" which also shows it might be related to the short frigid period.

RECONSTRUCTION OF ENVIRONMENT

Similar to those in Eurasia, Britain and North America[4,6], relics of polygonal ice wedge are mostly distributed, in Northeast and North China. Ice wedges are mainly concentrated in an area north of 45°N in Northeast China with greater depth and width (2.0—2.5 m deep, 1.0 m wide). The large patches of ice wedge in modern world are distributed simply in tundra zone which is equivalent to continuous permafrost[4]. Ice wedge differs from other periglacial phenomena in permafrost area for it needs more frigid climatic conditions and can grow after repeated freezing and cracking of ice bodies. Hence, we hold that areas around 45°N should be the southern boundary of tundra zone or forest tundra zone in the later period of Late Pleistocene period. Areas around 42°—45°N were the transitional zone between forest tundra zone and northern coniferous forest zone. Besides, areas north of 42°N were also where ice wedge and periglacial loess were comparatively well developed and mammoth mostly concentrated. The sporadically distributed ice wedges south of the boundary are seen in Jilin (43°N), western Liaoning (42°N) and Datong (41°N), Shanxi Province. A number of ice wedges occur in the same profile of the lacustrine strata (1,000 m) north of Datong. The top width is 30—50 cm, and 1—1.3 m deep, spacing 1—2 m or 3—5 m**. The surface is overlain by Malan Loess of 20 cm thick. However, the depth and width are both shallower and narrower than those found in the northern part of Northeast China, showing they were weaker in frigid degree and smaller in thawing frequency than those in the Northeast. According to A.L. Washburn's (1980) opinion, the annual average temperature indicated by ice wedge may be estimated as −5°C (only lower rather than higher)[6], or according to T.L. Pewe (1966) may be estimated as −6°—

---

* For instance, the stem of *Picea* in Huangshan Mt. of Harbin is 28,300±940 yrs BP in age. And the stem of that on the western slope of the main branch of the same mountain is 23,680±50 yrs BP in age. The deciduous pine stem in the second terrace of Zhoujiayoufang, Yushu, Jilin is 26,100±850 and 27,530±500 yrs BP in age (After Sun Jian-zhong et al., 1981).

** Followed Yang Jing-chun (1980).

-8°C[4]. The area along 45°N in the Northeast is basically equivalent to +3°C averaged annually. It dropped 8°C or 10°C as compared with that presently.* Area along 42°N is basically equivalent to 6°C averaged annually, being 11°C lower than the present. This line should be taken as the southern boundary of sparse steppe then, permitting *Coelodonta—Mammuthus* fauna (especially mammoth) could exist to the north of the line. The ice wedges sporadically scattered from south of 42°N to 39°N and north of 40°N represent a discontinuous permafrost zone or insular permafrost zone. The main feature north of the line is the extensively developed area of involution. Besides, periglacial loess, rock sea, wide valley with flat bottom unsymmetrical valley, solifluction depression and wind erosive depression, etc. are widely distributed. Most of them indicate the existence of permafrost, which is probably similar to the present case of discontinuous permafrost in Canada[15].

This area covers Liaodong Gulf and Changshan Islands in the southern part as a consequence of sea regression[14]. In addition, there is also a report on mammoth activity reaching even to the northern part of present Bohai Gulf (121°20′E, 38°40′N). Along the western section of the southern boundary from northwestern Hebei via northern Shanxi and northern Shaanxi to Liupan Mt., larger patches of insular permafrost may be developed with the declining of permafrost of Late Pleistocene frigid "I" due to a series of 2,000—3,000 m high mountains. The lower limit may reach about 1,000 m. Insular permafrost is also found to be developed in Baihuashan, West Hill, Beijing. Based on the above-mentioned data we suggest that the southern boundary of the northern coniferous forest could reach the area around 39°—40°N with the southward migration of tundra and forest tundra zones shifted to 45°N during the frigid period of the later period of Late Pleistocene. It is estimated, that once annual average temperature drops 1°C, climatic zone would migrate southward 100 km[15], the southern boundary of the northern coniferous forest could migrate from the present southern boundary of 50°N to 39°—40°N if temperature dropped 11°C in the later period of Late Pleistocene period.

As for temperature variations, the southern boundary of present insular permafrost (48°N) in Northeast China corresponds to an annual average of 0°C. The southern boundary of insular permafrost at that time (40°N) corresponds basically to the present isothermal line of 10—11°C, which also fits well to a temperature drop of 10—11°C then. It is estimated that it was 10—11°C lower in the southern part of Northeast China, Nei Monggol, northwestern Hebei, and northern Shanxi than in the present. This is 3°C lower than the result (a drop of 7°C) deduced from sporo-pollen data. The deduced value is greater than that for South China as well as the northern part of Northeast China, since the variation value of the middle latitude area (especially between 40—50°N) was the highest so long as the magnitude of the Quaternary cold-warm variations is concerned.

---

* Same as the deduction based on data of sporo-pollen analysis.

These values are 4—5°C lower[4] than the values deduced from the paleoperi-glacial phenomena in Europe and North America. They also reflect a rather stable situation in climatic changes. This stability is presented by the fluctua-tion amplitudes of the southern boundaries of northern coniferous forest and permafrost.

REFERENCES

[1] Shi Yan-shi, 1978, ¹⁴C date from woody specimens near Zhalainuoer, and its stratigra-phical significance, Vertebrata Palasiatica, vol. 16, no. 2. (English abstract)
[2] Xie You-yu, 1982, The formation and involution of valley meadow lowland in south-eastern part of Hulun Buir League, Collected papers of surveys in Hulun Buir virgin land, Science Press (in press). (in Chinese)
[3] Pu Qing-yu, 1980, A preliminary study on climatic variations in China in the past 30,000, Nature Journal, vol. 3, no. 3 (in Chinese).
[4] Washburn, A. L., 1979, Geocryology, p. 279—320, Edward Arnold.
[5] Ono Y., 1980, Glacial and periglacial geomorphology in Japan, Progress in physical geography, vol. 4, no. 2, p. 149—160, Edward Arnold.
[6] Washburn, A. L., 1980, Permafrost features, as evidence of climatic change, Earth-sci. rev., 15, p. 327—402.
[7] Tang Ying-jun, 1980, An evidence of climatic fluctuation in Quaternary deposit—involution layers, Glaciology and Cryopedology, vol. 2, no. 1, p. 41—43. (in Chinese)
[8] Xu Ren, 1980, Pleistocene flora of *Pices* and *Abies* in China, and its implication on Quaternary research, Quaternaria Sinica, vol. 5, no. 1. (in Chinese)
[9] Kong Zhao-chen, Du Nai-qiu, 1980, Vegetational and climatic changes in the past 30,000—10,000 years in Beijing, Acta Botanica Sinica, vol. 22, no. 4. (English abstract)
[10] Chow Kun-su, 1978, The pollen analysis on late Quaternary of Beijing Plain and its significance, Scientia Geologia Sinica, (1), 57. (English abstract)
[11] Chow Ben-shun, 1978, The distribution of woolly *Rhinoceros* and woolly mammoth. Vertebrata Palasiatica, vol. 16, no. 1. (English abstract)
[12] Jiang Peng, 1977, Distribution of mammlian fossils of Late Pleistocene in Jilin, Vertebrata Palasiatica, vol. 15, no. 4. (English abstract)
[13] Heintz, A. E., Garult, V. E., 1965, Determination of the absolute age of the fossil remains of mammoth and woolly *Rhinoceros* from the permafrost in Siberia by the help of radio-carbon (¹⁴C), Norsk Geologisk Tidsskrift, vol. 45, part. 1.
[14] Brown, R. J. E. 1967. Factors influencing discontinuous permafrost in Canada. The periglacial environment, past and present.
[15] Wang Shao-wu, 1981, What is the trend of the climatic change, cooling or warming? Nature Journal, vol. 4, no. 7. (in Chinese)

232

THE CLIMATE IN SANJIANG PLAIN DURING THE PAST 36000 YEARS

Kong Zhao–chen and Du Nai–qiu

(Institute of Botany, Academia Sinica)

ABSTRACT

This paper is based on the macrofossil plants of *Larix gmelini* Rupr. and *L. olgensis* Henry and abundant pollen, spores, and green algae of 2 peat samlpes obtained from digging a irrigating canal at Jiansanjing farm (6.5 m and 3.5m deep below the surface) in Sanjiang plain. About 36,000 years ago Sanjiang Plain was covered by cryophilic needleleaf deciduous forest, prodominant of *Larix* spp., *Betula*, herb and green algae. The climate of Sanjiang Plain was rather humid and cold, with an annual mean precipitation higher than that of the present. The lakes and bogs were well developed. According to the information of macrofossil and palynoflora, we hold that in the late-glacial, *Larix, Abies, Picea, Betula* forests became predominant in the Sanjiang plain. Therefore, the development of swamp in Sanjiang Plain has been rapid since Late Pleistocene.

ON THE ENVIRONMENTAL EVOLUTION OF XIZANG (TIBET) IN HOLOCENE

Li Bing-yuan, Yang Yi-chou, Zhang Qing-song
(Institute of Geography, Academia Sinica)

and Wang Fu-bao
(Department of Geography, Nanjing University)

Abstract—The evolution of the natural environment of Xizang in Holocene can be divided into the favourable in Early Holocene (10,000—7,500 yrs BP), the optimum in Middle Holocene (7,500—3,000 yrs BP) and the deterioration in Late Holocene (since 3,000 yrs BP). The features of the changes in environment show evident regional differentiation. And all the changes were caused by global climatic changes and the intense uplift of the Qinghai-Xizang Plateau.

Xizang is the main part of the "world roof"—Qinghai-Xizang Plateau, with an average elevation of over 4,500 m above sea level and an area of about 1.2 million km², which occupies half the total area of Qinghai-Xizang Plateau. Since Holocene no obvious change has occurred in the general geomorphological framework of Xizang, but the evolution of paleogeographical environment is rather complicated. On the basis of the information obtained from the scientific expeditions in the last few years, this paper will inquire the evolution of the Holocene paleogeographical environment in Xizang along the following lines.

DIMINISHING OF LAKES

Xizang Plateau is one of the areas in China, where there are a great number of lakes. Since the water levels of lakes on the plateau were rather high during Early Holocene, a number of lakes in South Xizang and the eastern, western and southern parts of North Xizang existed as exterior fresh lakes. However, since the water level of lakes in that period showed a general tendency of decreasing, some lakes in the interior of North Xizang were enveloped and turned into interior lakes, and a salt-lake stage began since 9,000 yrs BP[1].

The age of third terrace around the present lakes has been determined to be 7,500—3,000 yrs BP by ^{14}C dating in peat and fossil snails in the terrace deposits. The finer grain size of the terrace deposits than that of the Early Holocene suggests the rising of lake water level in this period. In comparison between the terrace heights and the heights of saddles, which previously appeared as outlets of former exterior lakes, it has been found that almost all lakes in South Xizang and some in North Xizang belonged to exterior lakes or were drained into neighbouring lakes. It is evident that the extent of lake was large in the Middle Holocene than that at present. Therefore, Middle Holocene represents a period of high lake-water level in Holocene. On the other hand, some smaller lakes within the North Xizang Plateau without the supply of thawing water were already in enclosure and water began to concentrate. In Chagcam Lake, for instance, mirabilite and gypsum were precipitated due to the further concentration of water.

Since 3,000 yrs BP drastic diminishing of lakes took place throughout Xizang Plateau and water level of lakes fell rapidly. According to the analysis of some paleo-shore lines of lakes in South Xizang and the southern part of North Xizang, falling of the lake water level during this period was estimated to range between 10—20 m, causing some lakes to be seperated from a large integral lake into several small lakes, such as Yamzho Lake, Chem Lake and Bajiu Lake were formed in this way at that time. Moreover, some exterior lakes turned into interior lakes, due to the same reason. For example, Baiku Lake and Yamzho Lake were enclosed and became interior ones in Late Holocene. In the central part of North Xizang, among the interior lakes which were enclosed earlier, not a few of saline lakes reached the last stage of development, e.g. Chagcam Lake, where the precipitation of mirabilite and rocky salt was predominant during that period. Thus it can be said that Late Holocene was the main salification period in Holocene.

THE DEVELOPMENT OF PEAT AND MARSHES

Peat deposits are widely distributed in the wide valleys and basins on Xizang Plateau, except in the inner part of North Xizang. The most extensive distribution of peat deposits occurs in the plateau wide valleys at about 4,500 m above the sea level on the southern sides of Gandise Range and Nyainqentanglha Range. According to ^{14}C dating it can be judged that all the peat deposits, which have been found so far in the plateau, were formed since 10,000 yrs BP. On the basis of the analysis of peat deposits and the results of the ^{14}C dating, three stages of the peat and marsh development can be divided in the plateau.

The Early Holocene was the initial stage of peat and marsh development. According to the available information, peat and marsh deposits of this period are only discovered in two regions within the Yangbajain-Damxung Basin, that is, Wumaqu peat formation (9,970±135 yrs BP) at Damxung and Qilongduo peat formation (8,175±200 yrs BP) at Yangbajain. The deposits are characterized by peat layers of small thickness interbedded with sand and pebbles.

The Middle Holocene was a stage when peat and marshes developed widely. The peat deposits of this period extended throughout the wide valley and basin

235

areas on the plateau. The ages of ^{14}C dating show that most of the peats on the plateau were formed during this period, e.g. the peat formation in the lacustrine deposits of Nara lake, the Ngalalu peat formation at Ngamring and the peat formations along Dagu River near Baiku Lake and along the western side of Doqen Lake in south Xizang; the Maindong Lake peat formation, the Bangdag Lake peat formation, the upper Qilongduo peat formation at Yangbajain and the Wumaqu peat formation at Damxung. The ^{14}C ages of the above-mentioned nine peat formations are within the range between 7,670± 250—3,050±120 yrs BP. The peat deposits of this period are characterized by extensive distribution, multiple layers (e.g. Nara lake) and huge thickness. Late Holocene was a stage of peat and marsh declining when the peat deposits reduced in the extent compared with the previous periods; the development of peat deposits ceased in some areas (e.g. Wumaqu peat formation at Damxung) and the proportion of silt and pebble mixing with peat deposits markedly increased in other areas. For instance, in the Qilongduo peat formation at Yangbajain, the continuous peat deposits before 3,000 yrs BP was overlain by sand and pebble layers interbedded with peat.

EVOLUTION OF GLACIERS AND PERMAFROST

Since Early Holocene the climate on the plateau turned warm slowly, and the extensive retreat of glaciers from terminal moraines of the latest Ice Age occurred and several strips of retreat-type terminal moraines were left.

Middle Holocene was the warmest stage during the postglacial period when drastic retreats of glaciers occurred throughout Xizang Plateau. For Rongpu glacier on the northern slope of Mt. Qomolangma and Ruoguo glacier on the southern slope of East Nyainqentanglha Range, the positions of their ends at that times were 700 m or 400 m higher than those at the present, respectively (Zheng Ben-xing et al., 1978). Some larger glaciers have retreated back by more than 5km since the latest Ice Age.

During Late Holocene the worldwide climate became cold again and extensive advance of glaciers occurred throughout Xizang. The ^{14}C age of the buried wood preserved in the higher lateral moraine (with a relative height of 100 m) of Azha glacier at Zayu is determined to be 2,900±150 yrs BP[2]. This advance was much smaller in scale than that of the latest Ice Age and the glacial extension was less than 2 km. At present, all the large glaciers in South Xizang have 2—4 strips of terminal or lateral moraines in perfect shapes, and their ages are about 3,000 yrs BP, 1,920±110—1,540±85 yrs BP and in 17—19 centuries, respectively, among which the last advance is called the Little Ice Age. After that the climate turned warm and glaciers retreated back to the present positions.

The existing permafrost on the Qinghai-Xizang Plateau is the remnant of the latest Ice Age. According to the calculation by Ding De-wen, Early—Middle

Holocene was a period of permafrost deteriorating and thawing. Thawing zones occurred on the plateau surfaces of lower elevations, e.g. in most part of South Xizang and some areas of the southern part of North Xizang. During the New Ice Age permafrost expanded again in North Xizang and some higher regions, but it was not very extensive. After the Little Ice Age climate became warmer and the deterioration of permafrost returned. Thus, in the isolated permafrost zones along the southern boundary of combined permafrost zone in North Xizang, some lakes and ponds emerged due to the thawing subsidence[3].

BIOLOGICAL EVOLUTION

Vegetation evolution can be deduced from the pollen analysis of Holocene lacustrine deposits distributed on the plateau surface within 4,300—4,760 m above the sea level at Nara Lake, Chem Lake, Ngalalu Lake, Coqen Lake, Maindong Lake and Chagcam Lake[4] (Li Wen-yi et al, 1979). After entering Early Holocene, vegetation expanded and the pollen content gradually increased in which the proportion of woody plants grew slightly, such as *Picea, Pinus* and so on, but herbaceous pollen, such as *Compositae* and *Artemisia*, was still predominant e.g. in Maindong Lake and Chagcam Lake. In sediments dated 7,500—3,000 yrs BP on the plateau the most abundant content of pollen, the greatest variety of pollen and the largest proportion of woody pollen are found. Among them the content of woody pollen in those sediments of Maindong Lake, Coqen Lake, Nara Lake and Chem Lake are as high as 50% of the total of pollen. They are mainly composed of *Pinus*, Betulaceae (such as *Betula, Alnus, Carpinus, Corylus* and so on), but the temperate broadleaf component (e.g. *Quercus*) and thermophilous component (e.g. *Tsuga*) increased in the percentage and more shrubby species such as Rosaceae, Ericaceae and so on occurred. Nevertheless, herbaceous pollen was still predominant, with mesophilous and hydrophilous plant pollen such as *Artemisia*. Cyperaceae, Compositae and Gramineae being abundant species. Vegetation types were equivalent to those near the upper limit of the present forest. It can be inferred from this that the upper limit of forest rose up to about 4,500 m in these areas during Middle Holocene. However, the herbaceous pollen was still absolutely predominant in some areas, e.g. in Chagcam Lake, occupying 80—90% of the total of pollen. Since 3,000 yrs BP the pollen species and the proportion of woody pollen in deposits decreased markedly, or even disappeared (e.g. in Ngalalu). Comparing with those of the earlier the proportion of herbaceous pollen reached as high as above 80%, in which more xerophilous components occurred. The progressive transition of vegetation type to the existing type took place.

To adapt itself to the environmental evolution, certain changes took place also in the Holocene animal community. Eleven kinds of mammal fossils are included at Karub Neolithic site (4,690±135 yrs BP) near Qamdo. These animals are all the existing species. Among them *Hydropotes inermis* and *Capricornis* sp. only live in South China or areas of low elevation above the sea level and *Capreolus linnaeus* and *Cervus elaphus* are limited to the areas south to the Changjiang River at present. There are almost no trace of them in the present Qamdo District, or even throughout the whole Xizang Plateau[5]. From the variation of fauna, it can be assumed that the climate 5,000 years ago was warmer than that at present.

EXPANSION AND REDUCTION OF ANCIENT MAN ACTIVITIES

Traces of ancient man activities are widespread throughout the whole Xizang Plateau. They are verified mainly by various kinds of artifacts which can be grouped into three types:

Chip-type artifacts: altogether 110 pieces of them have been found at three sites. Among the three sites, one is located at the northern slope of the Himalayas and the other two are in the Bangong Lake-Serling Lake depression zone. They are all located at about 4,500 m above sea level[6].

Microliths: altogether 270 piceces of them exposed at 31 sites in Xizang. Except two sites located in South Xizang, all others were discovered on the North Xizang Plateau, where most of them were within Xainza County south to Serling Lake and nine were in the depopulated area of the northern part of North Xizang. Fine stone-core is of the largest number among the total 101 pieces of microliths[6,7].

Neoliths: distributed at three sites: Karub, Nyingchi and Medog (with highest elevation of 3,100 m), situated on river terraces in Southeast Xizang. Among them the ^{14}C age of Karub Cultural layer is 4,690±150 yrs BP[5,8,9].

Exsistence and distribution of Chip-type artifacts, microliths and neoliths provide a basis for studies of environmental changes. Belonging to the late Paleolithic Age, the period of chip-type artifact was equivalent to the end of Late Pleistocene. Chip artifacts are mainly distributed in further southern lower parts of Xizang. Since the environment was less suitable in these areas during the period, man activities though occurred, were not common. Microlithic Age, which belongs to the Midlithic Age or the early Neolithic Age, is equivalent to Early and Middle Holocene. Microliths are widely distributed and can be seen even in the northern part of the North Xizang Plateau (with highest elevation up to 5,200 m. above sea level), where no people inhabit at present, and near by dried rivers or streams and saline lakes, where man activities are rare at present. It shows that the extent of microliths distribution is beyond that of present man activities. Therefore, it can be inferred that the climatic condition in the northern part of North Xizang was better in that period than that at present and the natural environment was probably suitable

238

for ancient man to go hunting. Since the cultural relics of the Neolithic Age rarely coexist with microliths and mainly distribute in the valley areas of lower elevation in Southeast Xizang, it can be inferred that since Late Holocene the deterioration of environment occurred on the plateau, forcing ancient man to migrate toward lower areas.

SUMMARY AND DISCUSSION

1. During Holocene various components of natural environment in Xizang evolved progressively within the same track. Three evolutionary stages can be distinguished.

(A) The stage of improved environment in Early Holocene. From the dates of initial development of peat formation at Yangbajain and Damxung and the analyses of lacustrine deposits in some lakes of North Xizang, it can be inferred that Early Holocene was about 10,000—7,500 yrs BP.

(B) The stage of optimum environment in Middle Holocene. From the results of ^{14}C dating for snail fossils and peats preserved in lacustrine deposits, it is inferred that this period was approximately between 7,500—3,000 yrs BP.

(C) The stage of deteriorated environment in Late Holocene. From the age of the first advance of glaciers, it can be inferred that the deterioration of environment in Late Holocene began about 3,000 yrs BP.

2. The remarkable regional differentiation of environmental changes in Xizang occurred during Holocene. In the lake basin areas of the North Xizang Plateau, a dry and cold climate generally exsisted during the whole Holocene. The advance and retreat of glaciers were of smaller scope and peats and marshes developed relatively poor. The environmental changes were primarily characterized by the evolution from steppe towards the desert steppe and increase or decrease in salinity of lake water. In Southeast Xizang situated in the high mountain and gorge zone in the southeastern margin of the plateau, the climate varied from humid—warm to subhumid—warm and the vegetation from forest to shrubby steppe. In the high mountain—wide valley—lake basin zone of the central and South Xizang situated between the two above-mentioned zones, the environment varied from subhumid—warm to dry—cold and the vegetation varied from savanna through shrubby steppe to steppe. The exterior lakes and drainage systems changed into the interior ones. Peats and marshes were mostly distributed in this zone, therefore, it can be called the zone showing the most remarkable changes in the Holocene environmental evolution.

3. The effect that promoted the Holocene environmental evolution in Xizang is ascribed to the changes of global climate and intensive upheaval of the Qinghai-Xizang Plateau. The environmental evolution of the Holocene in Xizang had some common aspects as compared with other regions over the world. For instance, the climatic changes in Xizang, such as the optimum climate stage, the New Ice Age and the Little Ice Age[10] coincided with the global climatic

changes, and were undoubtedly resulted from the influence of global climate. However, some differentiations still exist in Xizang Plateau. For example, the interior of Xizang has increasingly become drier and colder since Holocene, but at the southern margin of the plateau the humid climate is still prevailing. This situation is obviously a result of the plateau upheaval forming a topographic barrier. Since the successive upheaval of Xizang Plateau caused the decrease of the temperature and blocked the warm air current from the south at the southern margin of the plateau which makes the precipatation abundant there, the interior has become drier and drier.

REFERENCES

[1] Chen Ke-zao, 1981, Acta Geographica Sinica, vol. 34, no. 1, p. 13—21.
[2] Li Ji-jun et al., 1979, Scientia Sinica, vol. 22, no. 11, p. 1314—1328.
[3] Wang Jia-cheng, 1979, Acta Geographica Sinica vol. 34, no. 1, p. 31.
[4] Huang Ci-xuan et al., 1981, Geological and ecological studies of Qinghai—Xizang Plateau, Proceedings of symposium on Qinghai—Xizang (Tibet) Plateau Science Press, Beijing, vol. 1, p. 215—224.
[5] Huang Wan-po, 1980, Vertebrata PalAsiatica, vol. 18, no. 2, p. 163—168.
[6] An Zhi-min et al., 1979, Archaeology, no. 6, p. 481—491.
[7] Dai Er-jian, 1972, Archaeology, no. 1, p. 43—44.
[8] Wang Heng-chieh, 1975, Archaeology, no. 5, p. 311.
[9] Shang Jian et al., 1978, Archaeology, no. 2, p. 136.
[10] Zhu Ke-zhen, 1973, Scientia Sinica, no. 2, p. 168.

VARIATION OF HOLOCENE CLIMATE IN CHINA: TEMPORAL AND SPACIAL DISTRIBUTION OF *ELAPHURUS DAVIDIANUS* SUBFOSSILS

Li Xing-guo

(Institute of Vertebrate Paleontology and Paleoanthropology, Academia Sinica)

ABSTRACT

Elaphurus davidianus is a kind of living deer, first known only in a half domes-
ticated state in Chinese Royal Zoological Garden. It is believed that *Elaphurus davidianus* living in wild condition disappeared about 2,000 years BP.
Elaphurus davidianus fossil was first discovered in the Middle Pleistocene in
Qizui County, Anhui Province. In Late Pleistocene or Holocene, it was widely
distributed in East and North China including: Zhejiang, Shanghai, Jiangsu,
Anhui, Henan, Hebei, and Heilongjiang.

In the famous archaeological site "Hemudu" in Yuyao of Zhejiang, antlers and
bones of the animal were found in a large amount. In recent years the Holocene beds in which *Elaphurus davidianus* bones were discovered are at Wujiang,
Wuxian, Peixian, Changzhou, Haian, and Rugao, in Jiangsu Province, and
Xiuxian, Lingbi in Anhui Province, and Beijing.

Middle Holocene *Elaphurus davidianus* was numerous in China, and 56 sites
have been reported, among which 17 have been dated by radiocarbon[14]. So
Elaphurus davidianus should distribute widely 7,000—2,000 years ago. The
temperature in Middle Holocene (7,000—4,000 years) in China was about
2—3 °C warmer than today. There is no doubt that the disappearence of wild
Elaphurus davidianus in wild condition was evidently correlated with the intensity of human impact.

GEOLOGICAL STUDY OF THE CHARACTERISTICS OF EARTHQUAKE HAZARDS AND PREHISTORIC EARTHQUAKES IN CHINA

ZHU HAI–ZHI

(Institute of Geology, State Bureau of Seismology)

Abstract—Geological study on earthquake hazards is to investigate earthquake-generated effects on the fault rupture, ground shaking and failure of unstable ground surface, such as the failure of slopes, avalanches, landslides, rockfalls, collapses, etc. including ground failure in plain areas such as liquefaction of saturated sand and mud softening during earthquake traces of historical and recent great earthquakes which obviously shows that various characteristics of earthquake hazards appear in different geological and geomorphological units. In this paper these characteristics have been discussed, and some categories of earthquake hazards are distinguished in some areas and showed in maps of vulnerability analysis in several sites.

The residual destruction by prehistoric earthquakes usually are preserved in Quaternary loose sediments and on recent or buried topographic unit. Extensive study on prehistoric earthquakes and earthquake hazards can provide a more reliable basis for estimation of seismic risk and for microzonation of a given area.

INTRODUCTION

According to statistics of incomplete data on 91 major earthquakes from 856 A.D. –1976 A.D. over the world, the casualties of killed or injured by earthquakes of China are more prominent than those of other countries. Earthquakes in China mainly occur within the continent, where some seismic zones and belts are sited. According to the analysis of conditions of genetic-seismic, earthquakes are mainly controlled by tectonics, which are characterized by fault block of the earth crust. But the wave caused by strong earthquake spread to the ground surface and give rise to destructive effects, which are controlled by various geological, geomorphological and ground water factors. The lithology of Quaternary loose sediments, thickness of covered layer, development on topography of gullys, valleys and gradient of slope particularly play the direct role. Attention has been paid to some areas that earthquake frequently occurred in prehistoric times are precisely the areas sensitive to earthquake destruction and will become a potential area of earthquake hazard in the future. These characteristics are obviously reflected on loess plateau. In order to lighten the menace of earthquake hazards, special study on their characteristics is

242

necessary.

THE MAIN CHARACTERISTICS OF EARTHQUAKE HAZARDS IN CHINA

Based on the analysis of Quaternary sediments and geomorphology, China is characterized by regionalization, there are many high mountains in the western part of China, with eluvium and diluvium mainly distributed on the surface. Thick loess deposits are found in the central part of China on loess plateau and in the eastern part where sedimentary plains lie. There are different characteristics of earthquake hazards in various areas of different geological and geomorphological features.

Earthquake Hazards in High Mountainous Region

Three categories can be distinguished i.e. earthquake-gravitational, earthquake-tectonic and earthquake-gravitational-tectonic destructions. The category of earthquake-gravitational destruction usually occurred at gradient of slope more than 30 degrees in topography. There are usually numerous avalanches, rock-falls and landslides distributed in seismic area after the earthquake. Rivers are stopped in some canyons. Ground surface is subsided and debris widly spread everywhere and a destruction landscape appears in the area. Diexi earthquake (Aug. 25, 1933, M = 7½) and Zhaotong earthquake (Feb. 6, 1973, M=7.1) belong to this category[1].

The earthquake-tectonic destruction in high mountainous area is directly controlled by seismic fractures, including collapse of slope surface, avalanches, deformation on the alluvial terraces, which all along the seismogenic faults characterized by linear or zonal deformation and destruction. The area of Luhuo earthquake (Feb. 6, 1973, M=7.9) in Sichuan Province is one of the examples. The earthquake-gravitational-tectonic destruction is a complex result of both tectonic and gravitational factors.

Earthquake Hazards in Loess Plateau Region

Loess is the main Quaternary sediments in the northern part of China. Due to the loose feature, it mainly consists of silts with a considerable thickness of 30—100 m, and in loess deposits, numerous joints and fissures are developed. The loess is distributed in an arid area, so it is scarcely covered by plants. Owing to the shortage of woods, more population lived in cave dwellings which has poor resistivity for earthquake. All these formed internal factors of easy destruction and serious earthquake hazards. Along the slope of loess Yuan

243

(loess upland), loess cliffs and ravines between loess ridges are easily destroyed during earthquake. As a result, the avalanches, collapses and landslides in large areas occur. Destructions of these categories occurred in Haiyuan, Xiji and Guyuan counties during Haiyuan earthquake (1920 M=8.5) in Ningxia. Especially, in Xiji County, where upper Pleistocene loess are developed and 41 seismic barrier lakes occurred, among which 27 lakes still remain now. The largest barrier lake has a volume $18 \times 10^6 m^3$ of water. From Xinying Village north of Xiji southwards to Jingning, a large belt of avalanches-landslides in a length of 60 km and a width of 20—30 km occurs which is consisted of 657 avalanches and landslides slipped in a long distance about 1.5—2 km. The gradients of the surfaces of avalanches are 47°—78°. The mechanism of the destruction in loess is the result of action both of seismic force and gravitation. Mainly the ground is usually broken, with collapses and slides along the slope, which was disrupted in dry condition. The structure of destroyed loess is characterized by prism and schistosity.

Earthquake Hazards in Sedimentary Plains

The sedimentary plains in East China is an industry and agriculture concentrated and densely populated area. In the delta plains of Changjiang and Huanghe Rivers, thick Quaternary sediments were deposited. These plains had suffered different kinds of earthquake hazards, such as ground cracks, landslides, sandy liquefaction, collapses are found in the littoral plain and buried river beds.

Large liquefaction areas were formed in the eastern plain that strong earthquakes frequently occurred in the historic times. Haicheng (Feb. 4, 1975 M=7.3) and Tangshan (July 28, 1976, M=7.8) caused terrible disasters[2], both of which resulted in liquefaction layer of large areas. The depth of the liquefaction layer is about 15 m below surface. The size of sand and silts spurted from the ground mainly is within the range of 0.1—0.01 mm. The ^{14}C age of some liquified layers is 5,240±125 yrs BP in Haicheng seismic area. The liquified layer is mainly consisted of Holocene silts and fine sand presented as single grains, of moderate sphericity. Based on the comparison between liquefaction layers and loess, though, both belong to silt texture, but their behaviours during earthquakes are obviously different, because they are of different degrees of concretion, ages, water contents, structure of soil layer, etc. Analyses both in field and laboratory show that the mechanisms of these two destruction categories are fundamentally different. Sand liquefaction is the result of the destruction of water film between grains during the shock, because the pore pressure increases, the water is squeezed out from interstices of sand, and when the displacement of grains occurs disruptures begin. This is a category of "wet damage". But loess destruction appears in dry condition, so it is a "dry damage".

244

Generally, the characteristics of destruction during strong earthquake can be shown in seismogenic fault, ground shaking and failure in unstable ground. But different features exist in different areas from high mountains to plains.

AN EXAMPLE OF APPLYING THE METHOD OF SEISMIC VULNERABILITY ANALYSIS

Based on the analysis of destruction of recent strong earthquake it can be seen that some geological elements have long played a direct role, which are closely related to the ground media, such as the different assemblage of foundation soil, saturated sand which influences liquefaction potentially, the fault related to earthquake hazard, the thickness of ground soil and so on. The analysis of these elements has prediction significance in estimating earthquake hazard. The office of the United Nations Disaster Relief Coordination (UNDRO) has named it seismic vulnerability analysis[3]. According to the actural condition in China the factors of vulnerability vary with different conditions in areas. Here a region with well developed Quaternary deposits is found situated in the southern plain of Changjiang River—Liyang Area, Jiangsu Province where two moderate-strong earthquakes occurred in recent years. The epicenters of the two earthquakes were almost at the same location. The results obtained by this method are as follows:

General Situation of Liyang Area

Liyang area is located to the west of Taihu Lake plain. In this area hills, highlands, lower paddy, lowlands and Quaternary deposits are distributed correspondingly to geomorphological units. Earthquake occurred in Liyang Area (M=5.5) on April 22, 1974. The intensity of the epicenter was 8 and then earthquake (M=6) occurred again on July 9, 1979. Although these earthquakes were not so large, but they resulted in serious damage. There were 41 persons died and more than 3,000 persons injured in M=6 earthquake. The characteristics of the two earthquakes in Liyan Area are: (1) The two epicenters were in the same location near Shangpei Village; (2) intensities of the two earthquakes all decreased rapidly towards the west than the east; (3) except the effect of burial seismicogenic fault, the foundation soil played on obvious role in destruction; (4) no seismogenic fault appeared on ground surface. For this area following factors should be taken into consideration in seismic vulnerability analysis: index of vulnerability analysis of foundation soil, sand liquefaction, soil thickness, fault, etc.

Index of Vulnerability Analysis of Foundation Soil (i)

Foundation soils in Liyang District can be distinguished to five categories:

(a) Bedrocks (Paleozoic and Mesozoic sand-shales, limestones, volcanic rocks); (b) Xiashu clay layer (Q_3); (c) loess-like deposits (Q_4); (d) soft clay (Q_4) and (e) saturated sand, silts and clay. There exists zonation from NW to SE. The results of the two earthquakes show that the damage effects on buildings of different foundation soil are various. If we consider the index of destroyed buildings of a section in NW–SE direction in a length of 23 km, the two earthquakes showed that the indexes are higher in d (soft clay) and e (saturated silty sand and clay). Although they were far from epicenter, higher index tendency is obvious. So the estimation on earthquake vulnerability analysis can be made. On the basis of the different resistance to earthquake of foundation soil, different data are given as follows:

Thickness of soil layer (m)	Categories of soils				
	a	b	c	d	e
	Vulnerability index i				
0—10	0.5	1	2	2	3
10—30	0.5	1	3	3	4
30—50	0.5	1	4	4	5
50	0.5	2	5	5	5

$$i = n_1 a_t + n_2 b_t + n_3 c_t + n_4 d_t$$

Where $n_1 + n_2 + n_3 + n_4$ indicates proportions of soil categories a, b, c, d, or a, c, d, e, ..., and t represents the soil thickness.

Sand Liquefaction (s)

Sand liquefactions appear in a few places to the east of Liyang (such as Nandu, Daxi villages) during the two earthquakes. The liquified layer is composed of saturated sand, silt and little clay (Q_4). According to repeated occurrence regularity of vibration hazards in this area, the potential area of earthquake liquefaction is recognized and still located at Nandu and Daxi Villages, and the sand liquefaction (s) $\leqslant 2$.

Soil Thickness (R)

When the soft soil is thicker, the period of resonance is longer, and the amplitude of earthquake is greater. According to the data from drill hole and natural profiles the thickness of loose soil is one of the elements of vulnerability analysis in estimating earthquake hazard.

246

Fault (f)

The striking earthquake hazard, in general, appears near the seismogenic faults. But the two earthquakes gave evidence that no seismogenic fault appeared on surface. So the seismogenic fault must be located underground, and the direction of seismogenic fault is a problem awaiting solution. But, comparing the indices of buildings destroyed in I–I' and II–II' profiles, the seismic hazards in NNE–SSW direction in front of Maoshan Hill increase, which shows the approximate orientation of a fault in depth. The index of vulnerability is $\leqslant 2$ along this belt.

Map for Earthquake Vulnerability Analysis

1. Index value of vulnerability analysis. The map of earthquake vulnerability analysis is presented by many grids. In the upper left of every grid indexes of G. S, R, i, f are shown. The total vulnerability analysis index is the sum of indexes from various seismic analyses expressed by a formula: $I = i + s + f + \ldots$

To obtain index (i), it is necessary to know the ground soil composition (G). But the types of the ground soil in each unit of vulnerability analysis are inadvisable to exceed four categories, so the ground soil composition (G) may be expressed as follows:

$$G = 1/4a + 1/4b + 1/4c + 1/4d$$

$$\text{or} \quad G = \frac{3}{4}b + \frac{0.5}{4}c + \frac{0.5}{4}e \quad \text{or} \quad G = \frac{4}{4}c \ldots$$

From the average thickness of Quaternary sediments in each unit for vulnerability analysis value i can be obtained, and for each unit we obtain (s) (f)...

2. Mapping. 168 units of vulnerability analysis are distinguished in Liyang Area, and the index value I is from 2 to 21. According to value I the mapped area is divided into four sections A, B, C, D as follows:

Section A: Weakest seismic ground-response (I=2—5)
Section B: Weaker seismic ground-response (I=6—9)
Section C: Stronger seismic ground-response (I=10—14)
Section D: Strongest seismic ground-response (I=15—22)

3. Estimation of resonance. The soil thickness is divided into three categories 0—5, 5—30, 30 meters and the corresponding resonance cycles (R) are also divided into 3 intervals (short, medium, long). The ratio (I/R) indicates the relationship between the two functions.

Ratio (I/R) is placed in the center of analysed unit. Those mentioned above are only an example of vulnerability analysis in hill, highland and plain area. The vulnerability elements will vary with different geological and geomorphological conditions, such as in high mountain, at the different gradients of

247

	$I=2$	$I=5$	$I=10$	$I=15$	$I=22$
Thickness of soil (m)	Resonance R I	Weakest A	Weaker B	Stronger C	Strongest D
0—5	Short 1	A1	B1	C1	D1
5—30	Medium 2	A2	B2	C2	D2
30	Long 3	A3	B3	C3	D3

slope, in the loess plateau, etc. Geomorphological elements of Yuan, Liang, Mao... should be considered. And potential damages caused by prehistoric earthquakes should also be considered as an element of vulnerability analysis.

PREHISTORIC EARTHQUAKE AND EARTHQUAKE HAZARDS

In estimation of future earthquake hazards, impacts of historic earthquake are easy to be paid attention by people, while those of prehistoric earthquake are easy to be neglected. Study on prehistoric earthquake has important significance both for earthquake prediction and the prediction of earthquake hazards. Potential disaster phenomena that the prehistoric earthquake left cannot be underestimated. There is a striking example in Xiji, Ningxia. During the earthquake in 1920, the distance from Xiji Area to the epicenter-Haiyuan County was 70 km, the intensity of Xiji was 9—10. As a result, more than 650 avalanches and landslides occurred. Later, in 1971, there occurred M=5.5 earthquake. Few avalanches occurred over a large area, which caused the death of 113. As compared with other areas of similar category the disaster of this earthquake was surprising.

It can be easily seen that Xiji Area is a sensitive Area of earthquake destruction. Why the sensitive area of earthquake destruction are formed here? By investigations in the field three facts are found:

1. Fissures are distributed in the interior of ground soil, buried fissures cut the loess of a thickness of 20 m into segments.

2. Some deposits of barrier lakes of multi-periods older than 1920 are found. The genesis of these deposits of barrier lakes in different periods, could also be earthquake, which belong to deposits of prehistoric seismogenic barrier lakes.

3. Surfaces of avalanches-landslides distributed on slopes of ridges older than 1920 are found. Older surfaces of avalanches and landslides are traces of prehistoric earthquakes, too.

According to the above-mentioned characteristics, it can be deduced that Xiji has been a frequently seismic active area since Late Pleistocene to recent. The result of frequent activity of earthquakes would cause "potential bruises" in the interior of ground soil, and when earthquakes occur in the future, it could become sensitive area of earthquake hazards. Four stages of deposits

248

of seimogenic barrier lakes during the 61 years prior to 1920 are preliminarily found as follows:

Stage E: 29,280±820 yrs BP
Stage D: 23,405±510 yrs BP
Stage C: 22,405±370 yrs BP
Stage B: 10,455±120 yrs BP
Stage A: 61 yrs BP (1920)

The interval of deposits of barrier lake represents that of great earthquakes. By ^{14}C dating the ages of the intervals are determined as from about 1,000 to 11,950 years, and the average is 7,305 years. Intensity, that caused seismogenic barrier lake, is judged preliminarily to be >7 (Because this area had subjected to earthquake of intensity 7 but no barrier lake was formed). Due to the result of multiplicity of prehistoric earthquakes, the Malan Loess was loosen and fissured. In the interior of the ground soil a lot of potential fissured surfaces are buried, and large structures lost their original stability which increase the seriousness, of destruction in future earthquake and cause a sensitive area for earthquake hazards.

REFERENCES

[1] Zhu Hai–zhi, Wang Ke-lu, Zao Qi-qiang, 1975, A discussion on the seismic ground failure in mountainous region as from the Zhao Tong earthquake damage. Scientia Geologica Sinica, no. 3.

[2] Zhu Hai-zhi, Wang Li-gong et al., 1976, The geological aspect of earthquake hazards caused by liquefaction of sand in the Lower Liaohe River region, Seismology and Geology, vol. 1, no. 2.

[3] Office of the United Nation Disaster Relief Co-ordination, Composite vulnerability analysis (revised technical report), Geneva.

MORPHOTECTONIC FEATURES OF THREE EARTHQUAKE REGIONS IN CHINA AND THEIR RELATIONS TO EARTHQUAKE

HAN MU–KANG

(Department of Geography, Beijing University)

Abstract—Taking the Linfen Basin, Shanxi Province where two earthquakes of M=8 occurred, the Tonghai earthquake region of M=7.7, Yunnan Province, and the southern section of piedmont seismotectonic zone with high seismicities in Taihang Mts. as examples, the author has found that there are conspicuous morphotectonic features there, which reflect the young crustal deformation and faulting under the effect of neotectonic stress field, indicating their close relations to seismotectonics and seismicities, and can be considered as the criteria for assessing the earthquake-risk regions.

China is a country with high seismicities, where a series of great earthquakes occurred in recent 20 years. It is of great significance for structural geomorphology to study the relationships between earthquakes and morphotectonics. From this viewpoint the author in recent ten years has studied some earthquake regions in China and found that there are conspicuous morphotectonic features. Here, three examples are briefly given.

1. The earthquake region of Linfen Basin in Shanxi Province, where the 1303 Hongtong earthquake and the 1695 Linfen earthquake took place[1]. Both were of magnitude 8. The Linfen Basin is one of a series of NE–NNE trending dextral tensional shearing "en echelon" depressions within the so-called Shanxi Graben system, which are formed under the action of NE–SW principal compressional stress since Neogene[2,3]. The author and his co-workers have found that the 1303 Hongtong earthquake occurred on the intersection of the eastern boundary fault of a NEE trending young deep downwarp with the northern boundary fault of a sub-east trending young transverse graben in the Linfen Basin, whereas the 1695 Linfen earthquake took place in the center of above-mentioned deep downwarp to the south of the transverse graben.

It is meaningful to note that the foresaid young deep downwarp and transverse graben with their boundary faults are clearly demonstrated by the deformation and displacement of river terraces and the Pliocene-Pleistocene strata, by changes in elevation of the sedimentary boundary between channel deposits and flood plain deposits of the Holocene alluvium, the distribution of archeological evidences and the results of 1955—1970 repeated levelling[1].

2. The region of 1970 Tonghai earthquake with M 7.7 in Yunnan Province.

250

The seismic fault of Tonghai earthquake is a NW trending Qujiang fault, along which the Qujiang River runs. It is revealed by the author's investigation that the Qujiang fault is a pivotal one with complicated dextral compressional shearing and scissorslike movement under the action of NNW trending principal stress since Quaternary. Along the western section of the fault, the fault plane dips to SW with the southern block thrusting over the northern wall, where the Neogene planation surface and the Early Pleistocene river terrace on the southern wall are located higher than those on the northern wall, while along the eastern section the situation is reversed, the fault plane dips to NE, with northern block thrusting over the southern one, where the Neogene planation surface on the northern wall and the Early Pleistocene river terrace along the main fault stand higher than those on the southern wall. (Fig. 1)

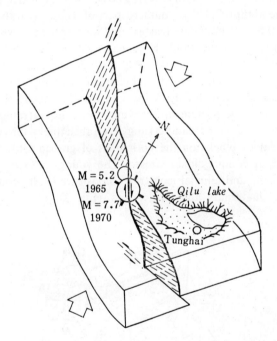

Fig. 1. The morphotectonics of seismic fault in Tunghai earthquake region.

Three (western, central and eastern) Quaternary transverse upwarps are formed along the Qujiang fault on its intersections with faults of other directions. They are indicated by the dome-like deformation of the Neogene planation surface and the Pleistocene terraces of Qujiang River. The western and central upwarpings caused interruptions, inversions and deviations of the Qujiang River. In addition, evidences of dextral lateral displacement of the Qujiang fault such as off-sets of tributaries of the Qujiang River running across the fault and off-sets of their alluvial fans are found to occur largely in the region of central upwarp, which indicates that the Qujiang fault in this region has dex-

251

trally displaced 1 km since Mid-Pleistocene. Moreover, the ground fractures with the maximum off-set of 2.7 m generated during the 1970 Tonghai earthquake also can be found here.

Among the foresaid transverse upwarps the central one is situated right near the pivot region of the Qujiang fault, where the dip of the fault plane is converted from southwest to northeast, and both the vertical upwarping and the lateral displacement along the fault are most conspicuous. It is suggested that this region is a "locked area" due to intense crustal deformation and susceptible concentration of tectonic stress, where exactly the 1970 great earthquake and its so-called early foreshock, the 1965 earthquake of M=5.2 took place.

3. The Tangyin Graben (Henan Province) in the southern section of NNE trending piedmont fault zone of Taihang Mts. It is an earthquake-risk area, assessed by the author[6]. The southern part of Tangyin Graben is the graben proper confined by two boundary (eastern and western) faults, while the northern part is a halfgraben, bounded only by an eastern boundary fault. The western boundary fault of the graben is not only a deep-seated fault along which the Pliocene basalts were ejected, but also a hinge fault with progressively decreasing vertical displacement and finally dying out toward the north. In addition, two series of NW trending young upwarps and downwarps arranged "en echelon" are formed within the NNE trending Tangyin Graben due to the dextral compressional movement of blocks on the both sides of graben under the action of NE–SW principal compressional stress. The young upwarps consists of the Pliocene lacustrine and alluvial lacustrine deposits, and the young downwarps are filled with Quaternary alluvial deposits. (Fig. 2)

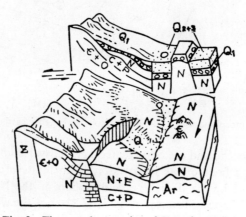

Fig. 2. The morphotectonics of Tangyin Graben.

The young activities of above-mentioned boundary faults of the graben are indicated by the displacements of river terraces running across them and the Pliocene-Pleistocene strata, and the young activities of upwarps and downwarps arranged "en echelon" within the graben are verified by the drainage features, distribution of isopachs of the Quaternary and results of 1959— 1973—1980—1981 repeated levellings.

The author made a prediction after a survey in the area studied that the northern end of western boundary (hinge) fault of Tangyin Graben would be prone to be the site of future earthquake since the fault extended northward under the action of regional tectonic stress. Later, the prediction was verified by an earthquake of M=3.9 just occurred in this region[6].

Besides, the piedmont fault zone of Taihang Mts. to the north of Tangyin Graben consists of three faults, which have been highly mobile since Quaternary and have formed a complicated flexure-graben-horst system. This section of the piedmont fault zone is separated from its southern section (Tangyin Graben) by a NWW trending pivotal fault, which passes through the Anyang City and moves in sinistral sense. Moreover, the further northward extending of the western boundary fault of Tangyin Graben will lead to its intersection with the NWW trending pivotal fault. All these will result in formation of a "locked area", where the tectonic stress is easy to concentrate and probably will be a medium-strong earthquake-risk region.

The above-mentioned three examples clearly illustrate that the morphotectonic study has been found to be effective for identifying seismotectonics and evaluating the earthquake regions. Chinese outstanding geologist' Prof. Li Si-guang (J. S. Lee) in his monograph "Seismogeology" summarized that earthquakes largely occurred in structurally specific locations such as at the end, the sharp curvature, and intersection parts of active faults, where the tectonic stress is easy to concentrate[7]. It is obvious that such structurally specific locations are well indicated by the morphotectonic features, which, therefore, must be taken into full account in the study of an earthquake-risk region.

REFERENCES

[1] Er Ke, Li Zhi-zheng*, 1976, Characteristics of neotectonic movements in Linfen Basin, Shanxi Province, and their relations to seismicities (in Chinese). 'Geological Sciences and Technology", no. 4. Geol. Pub. House. (*The pen name of the present author and his co-workers).

[2] Chang Wen-you et al., 1975, Preliminary note on the origin and development of rock-fracture and its bearing on earthquakes (in Chinese with an English abstract). "Acta Geologica Sinica", no. 1.

[3] Deng Qi-dong et al, 1979, On the tectonic stress field in China and its relation to plate movement (in Chinese with an Engish abstract). "Seismology and Geology", no. 1.

[4] Han Mu-kang, 1980, A discussion on the location of 1970 Tonghai earthquake of M= 7.7 in Yunnan Province (in Chinese). "Seismology and Geology" no. 2.

[5] Han Mu-kang et al., 1981, The Holocene migration of the Qilu Lake in southern Yunnan and its relations to the neotectonic movements in the 1970 Tonghai earthquake area (in Chinese with an Engish abstract). "Geological Review", no. 6.

[6] Han Mukang, Zao Jing-zheng, 1980, Seismotectonic characteristics of Tangyin Graben, Henan Province, and its earthquake risk (in Chinese with an Engish abstract). "Seismology and Geology", no. 4.

[7] Li Si-guang (J. S. Lee), 1973, "Seismogeology", (in Chinese). Science Press.

DISCUSSION ON THE NEOTECTONIC MOVEMENT IN YANSHAN MOUNTAIN THROUGH FREQUENT CHANGES OF LUANHE RIVER CHANNELS

Li Feng–Ling

(Tianjin Institute of Geology and Mineral Resources, Chinese Academy of
Geological Sciences)

ABSTRACT

Based on the reconnaissance survey in the area from Jixian to Changli Counties, the explanation of the satellite images and some drilling records, it is shown that the present Luanhe River gap—the entrance into the piedmont plain was captured from Qinglong River in recent period. But the river entered into the sea through the plain in Yutian and Fengrun Counties before Late Pleistocene. Yet it changed its river channel directions considerably from SE to SW from time to time in the past. Furthermore, eight right-angle turns along river valleys, thought to be caused by tectonic movements, have been found in the course of a distance of 1,200 km. But from Panjiakou Village downward, seldom branches enter into the river valley, because of the Yanshan Mt. uplifted rapidly leading the river incise its valley in a straight-line form and, in addition, several faults of NE direction occur, along which some river valleys lie and directly enter into the sea.

But there is also evidence that the mountain is still uprising, while the piedmont plain subsiding, which results in forming quite a number of depressions, arranged successively along the river course Jiyun River in the piedmont. It is evidenced that the mentioned piedmont belt has become the lowest belt of Hebei Plain, where the hot springs spread out from place to place, and earthquakes occur from time to time.

The above-mentioned facts show that the frequent changes of Luanhe River valleys reflect the frequent occurrence of neotectonic movements in Holocene, and this should be taken into consideration for the future construction in the mentioned region.

THE RELATIVE MOTIONS AMONG INTRAPLATE BLOCKS AND QUATERNARY FAULT-BASINS AROUND ORDOS BLOCK NORTH CHINA

Lu Yan–chou

(Institute of Geology, State Bureau of Seismology)

and Ding Gou–yu

(State Bureau of Seismology)

Abstract—This paper attempts to inquire the relative motions between Ordos block and its adjacent blocks, and the effect of the motion on the distribution and the internal structure of Quaternary fault-basins around Ordos block in North China.

INTRODUCTION

In North China block fault region, there is a striking phenomenon that a series of the active faults and Quaternary fault-basins are distributed on the boundary-zones between Ordos block and its adjacent ones, and along the zones strong earthquakes frequently occur. Within these blocks, however, neotectonic activity was so weak that no large earthquakes (Ms$\geqslant$6) or rare happened during the past 2,000 years. It is suggested that the intraplate blocks appear to be internally rigid, as stable regions, but their margins are mobile, as macro-seismic zones. This may be similar to the pattern of the plate tectonics suggested by Wilson (1965) and Le Pichon (1968)[1,2]. Therefore, we intend, with the aid of the geometrical means of plate tectonics[3,4], to inquire the relative motions between Ordos block and its adjacent ones, and to discuss the effects of the motions on the formation and development of Quaternary fault-basins around Ordos block.

ROTATION ABOUT A POLE: THE RELATIVE MOTIONS BETWEEN ORDOS BLOCK AND ITS ADJACENT ONES

Large shakes occurred around Ordos block have been identified as shallow structural earthquakes, and the depths of foci range from 10 km to 40 km. Moreover, the components of strike-slipping commonly were larger than those of dip-slipping for those earthquakes, which suggests that the slip vectors of the

255

earthquake faults may imply some information of relative movements of these blocks.

According to the fault plane solutions for the earthquakes around Ordos block[5,6],* the directions of horizontal projection of slip vectors vary in a systematic manner over the entire region. Along the eastern boundary of the block, for example, the direction of slip vector is NE at the southern end and gradually turnning to NNE or nearly SN towards the northern end. On the other hand, along the western boundary the directions of the slip vectors run progressively from SSW to SW.

It is recognized that the major-axes trending of the isoseismic line may represent the striking of earthquake fault. Based upon the isoseimic maps of the shake events around Ordos block for the last 1,000 years, the directions of earthquake faults changed in accordance with the directions of slip vectors. Within the boundary-zones between Ordos block and its adjacent blocks, most (80%) of the normal horizontal projections of the slipping directions and/or strikes of about 40 earthquake faults converge in a small area of 41°—44°N and 91°—95°E. Using the spherical trigonometrical method, relative rotational pole of plates can be determined[7,8], and a center in the small area is fixed at 42° N and 94° E near Hami Basin in Xinjiang, China. It is shown that the relative motion between Ordos block and its adjacent ones is a rotation about a pole at 42° N and 94° E.

Considering these intraplate blocks as rigid internally, it is reasonable to suggest that this initial result is consistent with the idea that any relative motion of two plates on the surface of a sphere is a rotation about some axes, a theory deduced from Euler's theorem[4]. If all adjacent blocks were relatively fixed, Ordos block would bring itself to rotate or torque anticlockwise. And its eastern part should move northeastwards or north-northeastwards relative to Taihang Mts. block, and its western part south-southwestwards and south-westwards relative to Alxa block and Qinghai-Xizang block, respectively.

AN EFFECT OF THE RELATIVE MOTION: THE DISTRIBUTION AND INTERNAL STRUCTURE OF QUATERNARY FAULT-BASINS IN THE BOUNDARY-ZONES

Since late Tertairy Ordos block with an area of about 250,000 km² has been slowly uplifted as a whole, and overlain by thin *Hipprion* Red Clay, Quaternary eolian loess and eolian sand. The total thickness of the loose deposits is generally not more than 200 m. On the contrary, Quaternary alluvio-lacustrine sediments of considerable thickness are developed in many fault–basins or intermontane troughs circumfused Ordos block. Each of these basins, in

---

* Part of the data was given by the Bureaus of Seismology of Shanxi Provi̇nce, and Shaanxi Province.

general, is not large in area, but has a specific shape and internal structure, and the basins are arranged in an extraordinary pattern. How did it happen? A possible answer could be found from the different structural features of various boundary types resulted from the relative motion among the intraplate blocks.

Shanxi Fault-Basin Zone

This zone is located on the boundary between Ordos block and Taihang Mts. block. It is mainly composed of seven fault-basins from north to south such as Datong—Shuoxian, Fanshi—Yuanping, Dingxiang—Xinxian, Taiyuan—Jiexiu, Linfen, Quwo—Hejin and Yuncheng Basin. They are arranged dextrally en echelon with the trending about NE30°. Each individual basin is very narrow in its width, giving the ratio of width/length 1:3—1:7. And all their boundaries are controlled by two groups of active faults striking NEE–NE or NWW–NW. On the basis of the isopach map of Quaternary sediments, it is obvious that most of the basins incline towards their margins where the active faults are situated[9]. The sedimentary center is generally situated near the southeastern or northwestern margin of the basin, in which Quaternary deposits with a thickness of about 400—700 m are accumulated.

All characteristics mentioned above show that each basin in the zone seems to be the tension-shearing one developed within shear-fracturing belt. Similar pattern can be observed in the surface shear-fracturing zone resulted from the large shakes, for example, from Tangshan earthquake (Ms=7.8) in 1976.

It is significant that the major-axes' strikes of the basins vary systematically from south to north along the Shanxi Zone. They are shown in the following table.

Names of fault-basin	Strikes of the major axes	Motion direction of Ordos block vs. Taihang Mt. block
Yuncheng	∼68°	38°
Quwo—Hejin	∼70°	36°
Linfen	∼35°	29°
Taiyuan—Jiexiu	∼52°	24°
Dingxiang—Xinxian	∼45°	21°
Fanshi—Yuanping	∼40°	20°
Datong—Shouxian	∼38°	16°

According to the table, some synchronous variations took place between the strikes of major axes and the directions of relative motion, maintaining an angle of about 25°—35°. It we take it into consideration the angle of 25°—30° being precisely equal to the angle between the earthquake fracturing plane and the maximal shearing plane 10—30 km under surface[10], it is evident that Shanxi fault-basin zone would be considered as a result of the relative motion between Ordos block and Taihang Mts. block.

Weihe Graben and Hetao Graben

These two grabens or fault-troughs are located on the southern and northern boundaries of Ordos block, respectively, near Qingling and the Yinshan block. However, they show some common characteristics to compare with each other. For instance, from plan the narrow trapezium form with nearly EW trending is exhibited. The Weihe Graben is much more wider in the east than in the west but the Hetao is just opposite. Similarly, the Quaternary deposits in the eastern part are much more thicker than those in the western part of the Weihe Graben. In the contrary, the thickest deposits occur in the western part of Hetao Graben. In addition, Weihe Graben is separated into the eastern Gushi fault-Basin and the western Xian Basin by Lintong—Gaoling transversal horst striking in NW, while Hetao Graben is separated into the western Houtao fault-Basin and the eastern Hohhot Basin by the Dalaqi—Baotou horst trending in NE. Both Gushi and Houtao Fault-Basins are situated along the south-eastest and northwestestern margins of Ordos block respectively, and the two basins were not only subjected to sediments of thickness over a thousand meters during the Quaternary Period, but have also been continually spreading and subsiding till the present day. As a result, the wide valley plains have been formed in the two fault-basins.

The pattern mentioned above may be explained by the relative motion of the blocks. Because Ordos block underwent a movement towards NE in the eastern Weihe Valley, and towards SSW in the western part of Hetao Valley, and tensional deformation would obviously occur along the two boundaries to form the grabens or fault-troughs. On the contrary, the western part of the Weihe Valley could become a convergent boundary in certain degree, as Ordos block moved towards SW relatively to Qinling. The similar condition also exists at the eastern end of Hetao Graben.

Yinchuan Fault-Basin

The basin is located on the boundary between Ordos block and Alxa block. It shows a narrow trigon which extends in a direction of about 35° N. There is a sedimentary centre at Pingluo near the northwestern margin of the basin, where the maximal thickness of Quaternary deposits reaches 1,605 m. Because Ordos block, relative to Alxa block, moved towards SSW, Yinchuan Basin, especially its northern part, has subjected to tension-shearing. A good evidence shows that the Great Wall built along the northwestern margin of the basin during the Ming Dynasty was dextrally displaced by 1.45 m. along NNE direction, and together with a vertical displacement of 0.9 m.

258

Liupan Mt. Compress-Uplifted Zone

On the eastern side of Liupan Mt. there are active faults striking NNW and NW, which confine the southwestern boundary of Ordos block. When the block moved southwestwards as above mentioned, the boundary is subjected to compression, because there is an angle of 55°—80° between the block movement direction and the strike of the fault zone. This could be the reason that the Tertairy sedimentary beds were uplifted, and folds and thrust faults occurred. Within the zone, no typical Quaternary sedimentary basin has been found. Since 1954, along the zone, from Guyuan to Baoji, the earth's surface has uplifted at an average rate of 1—2 mm per year from the vertical deformation measurement. In addition, there exist the Holocene thrust faults, for example, one of them is observed in the northwest of Guyuan County. So the fault zone at the eastern side of Liupan Mt. is a certain kind of convergent boundary resulted from the relative motion of the intraplate blocks.

In summary, the distribution and the internal structures of the Quaternary sedimentary basins around-Ordos block correspond with the different types of the boundary-zones resulted from the relative motion between Ordos block and its adjacent blocks. In other words, the state of relative motion of the blocks may give an explanation for the formation and evolution of the fault-basin within the boundaries of the blocks.

CONCLUSION

It is suggested that intraplate blocks, in which there are no or rare strong earthquakes during the recent 1,000—2,000 years, may be internally rigid. The relative motion among them which could be carried out on the surface of a sphere, in turn, would be a rotation about a certain axis. Here, we offer an example of the relative motion between Ordos block and its adjacent blocks, it is a rotation about a pole at 42° N and 94° E. Supposing the adjacent blocks were relatively fixed, Ordos block would bring itself to rotate or torque anticlockwise. And the eastern and western boundaries would belong to the type of shearing boundary, while the eastern part of the southern margin and the western part of the northern margin would belong to the type of tension boundary, but the southwestern boundary would be the compressing one.

The distribution and internal structure of the Quaternary fault-Basins around Ordos block correspond with a variety of boundary types, which suggests that the relative motion among the blocks probably controls the structural evolution of the fault-basins within the boundaries of blocks.

REFERENCES

[1] Wilson, J. T., 1965, A new class of faults and their bearing on continental drift, Nature,

v. 207, p. 343—347.

[2] Le Pichon, X., 1968, Sea–floor spreading and continental drift, J. Geoph. Res., v. 73, p. 3661—3697.

[3] Bullard, E. C. et al., 1965, Fit of continents around Atlantic, in Blacktt, P.M. S., et al. (eds.), A symposium on continental drift: Roy. Soc. London, Phil. Trans., ser. A, v. 258, p. 41—75.

[4] McKenzie, D. P. et al., 1967, The North Pacific: An example of tectonics on a sphere, Nature, v. 216, p. 1276—1280.

[5] Li Qin-zu, 1980, General feature of the stress field in the crust of North China, Acta Geophysica Sinica, v. 23, p. 388—394.

[6] Sun Jia-ling, 1979, The regional stress field, and the focal mechanism for earthquakes in the Inner Mongolia Autonomous Region, China. Seismological Research, no. 1, p. 28—35.

[7] Morgan, W. J., 1968, Rises, trenches, great faults, and crustal blocks. J. Geoph. Res., v. 73, p. 1959—1982.

[8] Savostin, L. A. et al., 1981, Recent plate tectonics of the Arctic basin and of northeastern Asia, Tectonophysics, v. 74, p. 111—145.

[9] Den Chi-tung, et al., 1973, On the tendency of seismicity and their geological set-up of the seismic belt of Shanxi Graben, Scientia Geologica Sinica, no. 2, p. 37—47.

[10] Go Zeng-jien et al., 1979, The physics on the earthquake foci, Seismology Press, p. 104—123.

CHARACTERISTICS AND EVOLUTION OF THE LONGITUDINAL PROFILES IN THE MIDDLE AND LOWER REACHES OF THE CHANGJIANG (YANGTZE) RIVER SINCE LATE PLEISTOCENE

You Lian–yuan

(Institute of Geography, Academia Sinica)

Abstract—Based on the features of terraces and fluvial deposits, and palaeoclimatic and paleohydrological changes as well as characteristics of the neotectonic movement, this paper expounds the general characteristics of the longitudinal profiles of Changjiang River and its development since Late Pleistocene.

In the view of traditional geomorphology, the problem of evolution of river longitudinal profile of permanent river is considered that under the interaction between both internal and external agents, longitudinal profile will tend to be equilibrium profile gradually, showing a concave curve upward with gentler gradient from upstream to downstream, and the upward concave degree becomes greater with the passage of time. Undoubtedly, this concept is true. However, first of all, since the process stated above carries on so slowly, it is often difficult to analyse the development of longitudinal profiles of natural rivers over a long period of time. Secondly, because this process does not carry on in a straight line, it adds more complex to analyze the development of longitudinal profile. Based on the features of the terraces and the fluvial deposits shown in the section from Wuhan to Jiangyin of the middle and lower reaches of the Changjiang River, and palaeoclimatic and palaeohydrological changes as well as characteristics of the neotectonic movement since Pleistocene, this paper tries to expound the general characteristics of the longitudinal profile and its development of different stages since that time.

THE RIVER TERRACES AND CHARACTERISTICS OF FLUVIAL DEPOSITS

One of the important characteristics of the river valley geomorphology in the middle and lower reaches of the Changjiang River is the widely developed river terraces of four grades. The features are as follows[1]:

T_1: 10—20 m. usually above water level. It is an accumulated terrace consisting of Late Pleistocene brown–yellow sandy clay. This terrace was formed in late Late Pleistocene.

T_2: 20—40 m. usually above water level. It is an accumulated terrace consisting of Middle Pleistocene red earth with white networks, and Late Pleistocene brownyellow sandy clay, sand and gravels. This terrace was formed in early Late-Pleistocene and is distributed most widely.

T_3 and T_4 having no direct relation to this paper, so are omitted here. Below the terraces is the modern flood plain of Changjiang River, most of which consists of fluvial deposits of Holocene and is characterized by dual texture in vertical direction. The top bed is deposits of flood plain phase consisting of finer grains with an average thickness of 12—13 m. The basal bed is deposits of river bed phase consisting of coarser grains. Moreover, reaching the depth, grains become more coarser. Further downward, the basal bed directly contacts with the surface of the bed rock, but the thickness of the deposits presents quite a large undulating.

Owing to the variation between scouring and depositing, and the different river patterns along the course, the modern river bed has considerable undulation. The water depth, in average, is between 13 and 22 m. The top bed of the river bed deposits is mainly composed of fine sand, with medium grain size of 0.15—0.20 mm; slightly fining along the course down. Besides, similar to the change of the flood plain, the grain size becomes coarser, when reaching the depth: from medium sand to coarse sand, to gravel, and to pebble, and finally the bed rock is met. The whole sequence of deposits belongs to Holocene. According to drilling data, the thickness inclines from upper to lower reaches and gradually thickens. The average thickness is 23.5 m near Wuhan 35 m at Wuxue, 37 m at Jiujiang, 40 m at Anqing, 41 m at Yuxikou and 41 m at Nanjing.

THE RIVER LONGITUDINAL PROFILES OF DIFFERENT STAGES

Determination of the Longitudinal Profiles of Different Stages

According to the above-mentioned topographic surfaces and boundary surfaces which have been referred to different stages, joining all the sites mentioned successively, a series of undulate curves along the course can be obtained. In the order from older to younger, they are as follows:

  1. Joining line of the tops of the second terraces.
  2. Joining line of the tops of the first terraces.
  3. Joining line of the bed rock surfaces of the modern river beds.
  4. Joining line of the boundary surfaces between the deposits of flood plain phase and the deposits of river bed phase in the flood plain.
  5. Joining line of average water depths in the modern river bed.

262

6. Joining line of the surfaces of the modern flood plains.

Since these lines are undulate along the course, they do not represent the river longitudinal profile of different stages.   One of the important causes is resulted from the differentiation of neotectonic movement in later period.   According to earlier studies, the general characteristic of neotectonic movement in the middle and lower reaches of the Changjiang River is periodic uplift and subsidence in large area, but since the course goes through different tectonic units and is influenced by the tectonic inheritance, and by the difference between the magnitude of uplift and subsidence[1][2], which is of secondary importance, the lines mentioned above actually represent the accumulative deformation value of the uplift and the subsidence since longitudinal profiles of different stages have been formed.

Furthermore, the influence of the variation in scour and deposition plays also an important part in forming the undulation of the longitudinal profile of modern river bed.

Therefore, in order to represent its original features, the best way should be to eliminate the undulation formed from later unequal magnitudes of uplifting and subsiding and instantaneous change of scouring and depositing.   But, in fact, it is very difficult to make this, because the magnitude of the neotectonic movement has never been constant and is not allowed to use modern measurements as the calculating basis.   Thus, other means must be applied.

The writer maintains that this requirement can be basically fulfilled by means of averaging and outlining these surveyed sites along the course by least square method to determine the profile lines.   Comparing with the method of eliminating the uplift and subsidence, we can see that although they have different physical sense, yet practically both aims are to eliminate undulation and to represent their original features, so the same results can be obtained.   As for those joining lines of 4—6 stages mentioned above though their ages are younger and they have undergone less influence of the tectonic movement, yet, to be averaged by least square method is also a better means to eliminate small difference between scouring and depositing and to obtain the time-mean values. According to actural situation, it is reasonable to choose an equation with parabolic form to determine the form of longitudinal profiles, i.e.

$$H = H_o e^{-bL}$$

where H, altitude at each site along longitudinal profiles; H_0, altitude at the first site; L, distance at each site from the first site; b, exponent, repesenting the gradient of longitudinal profile curves along the course.   The greater the b value is, the steeper the longitudinal gradient presents, and vice versa.

The Characteristics of Longitudinal Profiles of Different Stages

The following data can be obtained with the above-mentioned means, i.e. value of the exponent b, average general longitudinal gradient of longitudinal

263

profiles of different stages and some local average longitudinal gradients of some of the stages, as shown in Table 1.

Table 1 Characteristics of different stages of the longitudinal profile

Time	Years BP	Topographic surface or depositional interface	b value of parabola equation
Q^1_3	70,000	The top of the second terrace	0.001
Q^2_3	25,000	The top of the first terrace	0.001
Q^1_4	12,000	The surface of rock of the river bed	0.003
Q^2_4	5,000*	The boundary surface between the flood plain deposits and the river bed deposits in the flood plain	0.003
Q^3_4	Recent	The surface of the flood plain	0.002
Q^3_4	Recent	Recent river bed (mean water depth)	0.0025

Time	General	Longitudinal gradient ‰. Upper Wuhan-Hukou	Middle Hukou-Digang	Lower Digang-Jiangyin	Coefficient of correlation	Form of the longitudinal profile
Q^1_3	0.24				0.78	concave upward
Q^2_3	0.26	0.28	0.20	0.23	0.82	concave upward-convex upward
Q^1_4	0.84	0.45	0.68	2.4	0.99	convex upward
Q^2_4	0.36				0.83	
Q^3_4	0.25				0.99	
Q^3_4	0.28	0.38	0.20	0.24	0.75	concave upward-convex upward

* Based on: Ancient trees, 4,890±100 years of age was discovered in the sand strata of the river bed of the Changjiang River 30 m below water level in 1975. The position of the sand strata corresponds with the river phase sand strata in the river bank, therefore, it is deduced that the latter should be formed in Middle Holocene.

Table 1 shows that there is considerable difference among the characteristics of different stages of longitudinal profiles. Generally speaking, the longitudinal gradient of early Late Pleistocene approximately equaled to that of late Late Pleistocene. Thenceforth, it gradually increased and did not reach the maximum value untill the beginning of Holocene. Later it decreased gradually and now basically equals to the Pleistocene gradient. From changes along the course, we learn that the longitudinal gradient normally decreases gradually, but in some cases the maximum longitudinal gradient is displayed in the upper reach, the next is in the lower and the least in the middle one. So apart from a general concave upward curve, that of the middle and lower reaches is slightly convex upward. However, the first stage of Holocene is an exception. The longitudinal gradient at this time gradually increases from the upper to the lower and the longitudinal profile has convex upward form.

THE DEVELOPMENT OF THE LONGITUDINAL PROFILE SINCE LATE PLEISTOCENE

Climatic changes and tectonic movement of the earth crust as well as changes of the base level resulting from them are the basic elements, affecting on the development of the longitudinal profile. In below we will roughly analyse and explain this process from general variation and interaction of these elements in the area of the middle and lower reaches of the Changjiang River since Late Pleistocene. But details of the evolution or local variation is omitted. Recent studies indicate that during early Late Pleistocene, i.e. about 70,000 yrs BP (the interglacial stage between Lushan and Dali), it was a continually warm and humid climate interglacial of an age in the middle and lower reaches of the Changjiang River. The annual average temperature was 2—5°C higher than that of the present time, the precipitation was quite abundant and the vegetation lush[3]. Thus the hydrological and sedimentary characteristics are: large discharge, less and finer sediments coming from the upstream. Especially in the area belonged to monsoon climate, the annual mean temperature is lower than those of the same latitude, so the formed runoff was relatively greater and the sediment concentration was lower under the same precipitation. The general characteristic of the neotectonic movement was the tendency of slight subsidence at that time. As corresponding with the continuous and stable warm and humid climate, the sea level was also about 5—7 m higher than that of the modern sea. The sea water reached far into the river valley and kept in relatively stability, only with slightly falling or fluctuating tendency. Thus, a natural environment is formed for the Changjiang River where it can steadily widen and swing in lateral directions. The large discharge corresponds with the river bed of a wider section leading to form broad flood plains in both sides of the river bed. Red finer sediments, which is a feature of warm and humid climate, mainly deposited on the floodplains and the longitudinal profile was gentler. Besides, since the sea level slightly fell down and the regressive erosion was still going on in a smaller degree and a smaller range, the gradient of the lower reach was inversely a little steeper than that of the middle reach. In middle Late Pleistocene, i.e. about after 70,000 years BP. Early Dali glacial stage came. The maximum annual mean temperature was 5°C lower than that of the recent, the precipitation decreased and the Changjiang River basin belonged to semi-arid climate in certain aspects. The vegetation failed. The inflow water to the Changjiang River decreased and the sediment content increased, but since the sea level fell down rapidly to about 90—100 m below modern sea level, its effect on increasing activity of the river could enough compensate or even had a surplus decreasing due to for the less discharge (the

265

explanation of this phenomenon see below). In addition, the feature of the tectonic movement during this period was slightly uplifting[1], all these made the river bed incised and the second terrace was formed from the original flood plains. As uplifted magnitude of the tectonic movement was varied along the course, different heights of terraces occurred.

From middle to late Late Pleistocene, i.e. about 50,000—25,000 years BP, the sub-interglacial stage in the middle of Dali glacial stage came. The temperature rose again, the precipitation increased and the sea level correspondingly rose too, yet, was still lower than the modern sea level. Therefore, the inflow water and sediments did not show much increase and decrease. As the regressive deposition resulted from the rising of the sea level which kept relative stability for a long period of time same as the preceding interglacial stage, again occurred a depositional environment that led Changjiang River to maintain its relative stability, in association with widening river bed and slight incising. The flood plain expanded, too. But, this period was not as long as the preceding one, and so no well development of the flood plain occurred. In late Late Pleistocene, i.e. about after 25,000 years BP, late Dali glacial stage arrived. The reduction of the temperature, precipitation and sea level was far more greater than that of early Dali glacial stage and the sea level reached the lowest level of Dali glacial 130—140 m below the modern sea level. The large-scale sea regression pushed the river mouth of Changjiang River forward 600 km from today's shore, and gave rise to intense regressive erosion. The magnitude of the regressive erosion compensated enough and had a surplus for the decreasing activity resulted from the decreasing discharge: This can be explained by the following brief calculation. In consideration of the fact that the mean longitudinal gradient of the river bed of Changjiang River increased from 0.24‰ to 0.86‰ in this period, i.e. 3.5 times as much as before, according to Manning's formula $v = \frac{1}{n} R^{2/3} J^{1/2}$ the velocity (v) should correspondingly increase to $3.5^{1/2}$, i.e. 1.88 times as much as before. But, in fact, as the volume of the inflow water decreased, and so did the water depth, if the velocity is to remain unchanged, the hydraulic radius (R), (the same to the water depth) should decrease 2.5 times. Assuming the lateral section of the river bed was in trapezium, its water area passed would decrease approximately 7.3 times as the water depth decreased 2.5 times (It would decrease 10.3 times when the lateral section was in triangle). Therefore, the corresponding discharge would increase 7.3 times (or 10.3 times) when the velocity maintains constant. However, according to the relation among the discharge, runoff and precipitation, obviously the discharge is not possible to decrease to so many times. So, the fact should be that the water depth did not decrease much, but the velocity increased a great deal. Because the erosive capacity was in direct ratio with the higher power of the velocity, a slight increase of velosity would cause the great increment of erosive capacity. In addition, the neotectonic movement generally will cause a rapid increase of the erosive capacity. The whole Changjiang River valley uplifted except around Boyang Lake, all this made the following situation possible:

1. The river bed was incised and the original flood plain became the first terrace.

2. Owing to the tremendous activity of the river, not only no or little sediment was deposited, but also almost all the original bed deposits, formed in early and middle Late Pleistocene were carried away leaving only the surface of the bed rock.

3. The falling of the base level and the rejuvenation of the erosion made the longitudinal gradient steeper rapidly, reaching the maximum value of $0.84\%_{000}$. at the beginning of Holocene. Besides, because of the intense regressive erosion, the gradient of the lower reach was much larger than that in the middle reach, and the latter was larger than the upper one. So the whole longitudinal profile had a convex from upward. Probably, the boundary between the lower reach and the middle reach was the upper limit of the regressive erosion. After the end of late Dali glacial stage, the temperature began to rise again. A new geology period, Holocene, arrived. The temperature and precipitation rose rapidly so did the sea level. At last it reached the maximum at 6,000 years BP or so. At that time the river mouth of Changjiang River was situated near Yangzhou City and the annual mean temperature was a few degrees higher than now. In this climatic condition the volume of inflow water of Changjiang River increased again. The bankfull discharge could reach 58,000 cu.m per second (now is 47,000 cu. m per second), and the inflow sediments decreased again. However, since the sea level rose very rapidly causing the regressive deposits, and the universal subsidence of the neotectonic movement resulting in the deposits along the course, the longitudinal profile of the river bed became gentler gradually and deposits on the river beds were thicker along the course down. Since the activity of the river was still relatively strong, only coarser grains, such as pebbles and gravels deposited. As the gradient continually became gentler, deposits also finer.

In Middle Holocene, i.e. after about 6,000 years BP, temperature, precipitation and sea level began to fluctuate and fell, but in a small magnitude. The variation of the annual mean temperature was between 1—2°C, and the sea level was not more than 3—5 m. And this led to the slow decrease of the volume of inflow water and the slow increase of the inflow sediments. The earth crust still subsided slowly at that time. All mentioned above led to the gradient of the longitudinal profile flattened steadily and constantly, and the flood plains were widely developed during the same period. Sand finer than the preceding period was continously deposited on the river bed. Since the neotectonic movement had a tendency of greater subsidence toward the lower reach of the river and the sea level had a slight subsidence, the lower reach of the longitudinal profile was a little steeper than the middle reach. The later stage of this period entered mankind historic epoch.

From 1,700 years BP on, the natural environment was similar to that of the preceding period. Only owing to mankind's greatly strengthened activities, such as felling the forest, developing of agriculture, the inflow sediments increased from slow to rapid and the fine grains increased greatly, which directly speeded up expanding of the delta of the Changjiang River. The regressive deposition intensified, which accelerated the flatting of the gradient. Although it is only 10,000 years and more since the beginning of postglacial, which is a much shorter period than the two preceding interglacial stages, yet, the flattened magnitudes of the longitudinal profiles were nearly the same.

REFERENCES

[1] Yang Huai-ren et al., 1962, Geomorphology and Quaternary geology in the middle and lower reaches of the Changjiang River (from Yichang to Nanjing) Proceedings of 1960 National Symposium of Geography pp. 6—44.

[2] Huang Di-fan, et al., 1965, Geological studies of the formation and development of the three large freshwater lakes in the Lower Yangtze valley, Oceanologia et Limnologia Sinica, vol. 7, no. 4, pp. 396—424.

[3] Wang Jing-tai, Wang Pinh-sien, 1980, Relationship between sea-level changes and climatic fluctuations in East China since Late Pleistocene. Acta Geographica Sinica, vol. 35, no. 4, pp. 299—312.

PRELIMINARY STUDY ON PALEOLITHIC BONE ENGRAVING IN SHIYU SHANXI PROVINCE

YOU YU-ZHU

(Institute of Vertebrate Paleontology and Paleoanthropology, Academia Sinica)

Abstract—This article presents the analysis of carving bone fragments found in Shiyu Site. The traces on the surface of bone fragments recorded hunting behavior and mode of thinking of Shiyu Man.

Since 1911, many Ice Age artifacts and engravings of bone have been excavated in limestone caves in southwestern France and northern Spain. The earliest carving notations and symbols in Europe found so far were made by Cro-magnon hunters who lived 30,000 years ago during the great glacial age. The Magdalenian arts in Europe, for instance, arose at 10,000—20,000 yrs BP. The engraved materials appear in all the culture of Upper Paleolithic of Europe, though they are not present at every site. Nevertheless, early carving notations and symbols are still difficult to recognize, and there may be some different explanations about them.

Dr. Alexander Marshack has examined thousands of Upper Paleolithic artifacts and engravings—a large proportion of unearthed artistic and symbolic remains similar to those of Cro-magnon man. Dr. Alexander Marshack held that Upper Paleolithic notations and symbols indicated some of the probable origins of later formed systems, such as writing, and arithmetic. He believed that the modern characters, mathematics and arts were all derived from notations, symbols and images made by ancient man who lived in Late Paleolithic age. And carving notations, symbols and images in Paleolithic provided a kind of evidence which could preserve and record our ancestor's knowledge about their environment and their own activities. Therefore, they become extremely precious materials for us to study the development of human intelligence.

A few materials were found in China before. But there has been no report except five tubular bones with one to five cutting-marks from the Upper Cave Man, so nobody has made any study on Paleolithic carving notations, symbols and images.

The present author and his colleagues discovered a large number of carving bone fragments, excavated in Shiyu Site, Shanxi Province in the summer of 1963. By the microscopic observation and comparison, the traces on the surface of bone fragments are identified to be carved artificially, rather than to be corroded or formed by plant traces and bacteroidal traces.

269

Before describing the traces on the surface of bone fragments, we have to introduce briefly Shiyu Site.

Shiyu Site is located at the northern bank of Shiyu River near Shiyu Village, and 15 km northwest away from Shuoxian County. It is situated right at the boundary between northwestern Shanxi Plateau and Shanggan River Plain. The stratigraphical section at Shiyu Site are rather simple, which can be divided into 4 layers:

Upper Pleistocene

    4. Loess, 18 m.

    3. Grey sand of median grain size, 9.8 m.

    2. Brown sandy clay (or cultural layer), 1.5 m.

    1. Greyish brown conglomerate, 1 m.

This Site was excavated in a large scale. About 20,000 pieces of stone artifacts such as core, flake blade, scraper, point, burin and chopper-chopping tool were excavated. In addition, several dozens of burnt stones and bones, a piece of ornament and a piece of occipital fragment of *Homo sapiens* were collected. In respect of stone artifact, this site is characterized by small flakes and small tools. So Shiyu Culture belongs to the Small Tool Tradition of North China.

The fauna of this site is rather rich, more than 6,000 teeth of various animals were unearthed. Up to now 14 forms have been recorded as follows: *Strathio* sp., *Erinaces* sp., *Crocuta* sp., *Panthera tigri*, *Myospalax* sp., *Cervus elaphus*, *Megaloceros ordosianus*, *Procapra picticaudata przewalskii*, *Gazella* cf. *subgutturosa*, *Bubalus* cf. *wansijocki*, *Bos sp.*, *Ceolodonta antiquitatis*, *Equus przewalskii*, *Equus hemionus*. The fauna reflects an environment of grassland and the climate was colder than the present.

The age of Shiyu Culture is determined to be 28, 135±1,330 BP by radiocarbon method.

There are two kinds of bone artifacts with different length: 8—10 cm and 15—18 cm.. Obviously, they were artificial products with special purpose. From these bone fragments, about a thousand pieces with carving traces were found. Microscopic observation indicates that these traces on the surface of bone fragments were carved by stone tools as burin or point.

There are some kinds of carving traces: pits (round pits and triangular pits), straight lines, netted lines and images. The economical activity of Shiyu Man was hunting. The round pit and triangular pit traces are inferred to have the meaning of kinds and numbers of large animals (e.g. rhinoceros, tiger) hunted by them. Straight or netted lines were supposed to be the animals of ordinary size (e.g. horse, gazelle).

Among these bone fragments the most interesting thing is a horse humerus with quite complicated images. The bone fragment is 8 cm long; 3.1 cm wide. The images consist of the following two groups which recorded hunting behaviour of Shiyu Man. (Fig. 1)

A (left) Group: A *Procapra* (one kind of gazelle) killed by Shiyu Man lies on

270

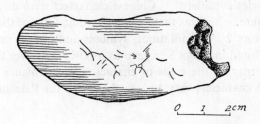

Fig. 1. Bone engraving of Shiyu, an Upper Paleolithic Site.

the grassland, and two hunters run towards the died *Procapra* from different sides.

B (right) Group: A *Struthio* (one kind of ostrich) is being surrounded by hunters from three directions.

From the numerical statistical analyses of the animal fossils in Shiyu Site, we can see that main hunting subjects are horse, gazelle, ostrich and deer. It is likely that an image of horse is too complicated for Shiyu Man to express with burin, whereas an image of gazelle is much easier to be carved. In addition to this, as subjects of human hunting, gazelle and ostrich were very abundant in northern China during Late Pleistocene. Therefore, it is natural that Shiyu Man carved their images on the surfaces of fragments.

The following points about the carving traces and images from Shiyu Site are discussed here:

1. Shiyu Man had simple numerical conception. One of the evidences is that there were notations different in number, and there were many fragments with notations less than five. Another evidence is that in the images mentioned above, the gazelle has two antlers instead of one or three. Three narrow traces were added below its belly. It shows that Shiyu Man knew gazelle had four legs instead of one.

2. The images are simple and rough, but the carving traces are very deep. We have tried to make image on surface of bone fragment with steel knives or stone artifacts, yet unexpectedly, the depth of the traces is less than half of that made by Shiyu Man. According to the directions of strokes observed under microscope, the maker held the bone fragment with his left hand and held the stone tool with his right hand.

3. The surface of the bone engraving is very smooth. Even some narrow traces also become smooth after being worn. The bone engraving is inferred to be kept in the hand of its owner, probably the head of primitive tribe.

4. It is not accidental that marks, symbols and notations carved on bone fragments appeared in Late Pleistocene, neither earlier nor later. It is the result of earlier development of human ability and intellegence. And it was the beginning of later writing characters, mathematics and arts. Therefore, the appearance of the notations, symbols and images was an important event in the

271

history of human cultural development.

5.   Academic circles consider that Chinese characters were derived from Neolithic Banpo Culture.   But was there any embryonic form of those before Neolithic Banpo Culture ?   The notations, symbols and images on the surface of bone fragments found in Shiyu Site gave us a positive answer to this question. Undoubtly, they may be the seeds of characters, mathematics and arts.   The origin of Chinese characters may be traced back to the Paleolithic.

REFERENCES

[1]   Alexander Marshack, 1972, Cognitive Aspects of Upper Paleolithic Engraving. Current Anthropology, vol. 13, no. 3—4.
[2]   Alexander Marshack, 1975, Exploring the mind of Ice Age Man. National Geographic, (1).
[3]   Chia Lan-po, Gai Pei, You Yu-zhu, 1972, Report on the excavation of Shiyu Site, Shanxi. "Kaogu Xiebao", (1).
[4]   Peng Xi, 1981, Preliminary research on the early mathematics.   "Kaogu Yu Wenwu", (2).

A FOSSIL HOMINID SKULL AND MAMMALIAN FAUNA FOUND IN HEXIAN COUNTY, ANHUI

HUANG WAN-PO

(Institute of Vertebrate Paleontology and Paleoanthropology, Academia Sinica)

ABSTRACT

A fossil hominid skull was found in Longtan Cave, Hexian County, Anhui Province, on November 4, 1980, the first discovery of this kind in South China. By November 1, fossils of nine human teeth and the left part of a mandible were found. In view of that the main features of the skull make himself relate more closely to Java Man than to Peking Man, it is reasonable to attribute the Longtan Cave skull to a new subspecies and assign it a tentative name *Home erectus hexianensis*.

From the vertebrate assemblage of Longtan Cave it can be seen that: these fossils *Sinomegaceros pachyosteus*, *Hyaena sinensis*, *Trogontherium*, *Pseudaxis grayi*, etc. are so far known mainly from the Peking Man fauna of Zhoukoudian. There are, however, other genera such as *Stegodon*, *Tapirus* and *Megatapirus*. etc. mainly from the *Ailuropoda—stegodon* fauna in South China. It seems to be a new fauna. On this account, we call it the Hexian Man Fauna, which is considered to be of Middle Pleistocene in age.

Palaeoecologically, the mammalian assemblge of Longtan Cave shows as a whole, a dominance of animals of forest and woodland. The grassland types are comparatively few. The climate then was warm.

273

SOME GENERAL REMARKS ABOUT THE LAND IN THE LOESS PLATEAU

ZHU XIAN-MO*

(Institute of Soil and Water Conservation, Academia Sinica)

Abstract—The basic characteristics of the loess plateau are described, and some new views on the soil erosion, type and evolution of the land on the loess plateau is proposed by the author.

The loess plateau in China is the most developed region of loess in the world where this substance, being with a great thickness and a complete sequence, is widely distributed. The loess plateau is just located in the middle reaches of Huanghe River and loess spreads mainly between 33 and 41°N (latitude) and 102—114°E (longitude), controlled by the alignment of mountain systems. To the south, the loess region is bordered by Qinling and Funiu Mts. The true loess in this region is also related to climatic belt, being distributed mainly in the northern arid or semiarid regions, where the temperature in the cold months is generally below 0°C, annual precipitation averages 250—500 mm, and evaporation is over 1,000 mm per year.

In regard to the area of coverage in the middle reaches of Huanghe River, there exist different views. The total area of the middle reach is about 610,000 km²; the area of loessial coverage about 530,000 km², the area seriously eroded is more than 430,000 km²; and the area of typical plateau with true loess deposit is only 275,000 km².

The loess region in the middle reaches of Huanghe River is mainly distributed at the altitude 400—2,400 m above sea level. In the plateau it is always at the altitude of over 1,000 m, which to the east at the altitude between 1,000 and 2,000 m and in the areas lower than 1,000 m loess are deposited in some of the basins and plains in the east and scattered along some of the piedmonts in the west.

The thickest loess deposits lie in the middle reaches of the Huanghe River. To the west of Liupan Mt., spreading northwards from the terrain between Huajialing and Mahanshan to the vicinity of Lanzhou and to the west of the Baiyu Ranges, it attains a thickness up to 200—300 m. To the east of the Liupan Mt. and the west Luling Mt., the thickness is between 100 and 200 m. But in the hilly region west of Liupan Mt., the thickness of loess deposits is always less than 50 m. In the basins and large valleys, the loess deposits, being interbedded with alluvial sediments, are not very thick.

---

* S. M. Chu

The underlying landform of the loess plateau, which was formed before the loess deposition, can be divided into: 1. The Wutai, Luliang and Zhongtiao shield — two parallel folded mountain series of Taihang and Luliang and a series of basins; 2. The high plain of the ups and downs of the tableland were formed due to the continuously cut and denudation of the Ordos platform; 3. The Longzhong basin situated on the east by Liupan Mt. — a series of long highland, sharp-top mountain, lowergentle hill and front-mountain (piedmont) plain, etc.

The loess at various times was deposited on the paleogeomorphological basis mentioned above. In various stages of loess deposit owing to the variability of natural conditions, the different distance of deposit nativity, the difference of geomorphological basis and the different forms of deposition in distributed area, the tendency of loess deposition varied in stages, however, accelerated the silt accumulation in original gullies and enabled the gullies nearby approaching the watershed to change their own wages.

The surficial features of loess deposits are closely connected with the palaeo-topography of the underlying bed rocks. In terrains of ancient rolling hills, it is generally represented by "Liang" (elongated loess mound) and "Da" (round loess mound). In case the bed rock is flat and broad, the landform appears as "Yuan" (a high table-like plain with abruptly descending edges). On the terraces of different heights produced by faults, the loess deposit forms a flat "Daiyan" (a loess-covering on terraced landform) extending along the direction of the faults, whereas the loess on the terraces along the river valleys is usually flat and less thick stretching along the valleys. On the piedmonts of great mountains trending E–W or NW–SE, loess is often characterized by belt, distribution with a little thickness and width variation in properties. The geomorphology in loess terrain is thus not only controlled by paleo-topography, but also deeply affected by the denudation process after the deposition of loess. Although a dry palaeo-climatic conditions may account, generally, for the formation of loess in China, the multilayered fossil soil (paleosol) intercalated in loess deposits indicates that during the formation of loess alternate changes from dry and cold to moist and warm should have happened many times.

The loess in the middle reaches of the Huanghe River is mostly rather strongly calcareous, and where it occurs in large bodies and has been eroded, it stands in vertical cliffs and exhibits a strongly developed columnar structure. Loess columns are frequently sharply angular and form rough pentagons and hexagons exposed on cliffs and road cuts. This columnar structure developed immediately after reworked or rewashed loess becomes dry, and in places where the people have built walls of it, the columnar structure develops immediately. This is true not only in the loess plateau but also in the dry parts of China as well. Even moderately heavy clay can develop also a loess-like columnar structure which contain a high percentage of finely-divided and separated particles of lime. Another point of interest along this line is that strongly leached loess deposits which have lost their lime tend to lose their columnar structure. It is true that they frequently stand in vertical cliffs, but the strongly developed and sharply angular pentagons and hexagons are lacking.

The loess deposits contain horizons in which there are many hard lime concretions or "loess puppets". Many of these horizons occur directly under bands of reddish brown clayey materials. In such cases the lime concretions usually represent old Bca-horizons, the reddish bands, old A-horizon of a pedocal soil (as paleosol mentioned above). In many places, however, it is quite evident that some of the lime concretions were gradually formed by precipitation from under-ground water and here is no question but that the old Bca-horizons have received increments of lime from the same source since their original formation as a part of a soil profile. Many of the deeper loess deposits contain a large number of superimposed soil profiles, indicating that the loess accumulation was sporadic and not a continuous process.

The loess, except for the bands of reddish clay and lime concretions already mentioned, is of a uniform silty loam throughout the whole plateau and the loessial region in the middle reaches of Huanghe River. Mechanical analyses indicate in itself that only a very low percentage of particles of more than 0.1 mm occur in soils of loess region, while there is a high percentage of particles less than 0.005 mm. in diameter, some times exceeding 30%. These clay sized particles, however, cannot remain as separate individuals in the presence of a high percentage of lime. They form aggregates which almost entirely of sizes varying between the limits are well-known, and most of them are probably at least 0.01 mm in diameter.

Loess plateau is the birth place of Chinese agriculture, in which abundant and varied soil resource is distributed. The soil stratum is rather thick and fertile, which makes up an excellent condition for growing plants and developing of agriculture. On this area, irrational uses of the land and plunder management were being carried out for over thousands of year, the whole land almost having been cut into fragments, and the soil erosion being very serious, however, it still has a great producing potentiality. In fact, on the basis of the same land, we have to carry on a series of investigations of accumulated experiences in order to draw up a plan of rational uses, the soils on the loess plateau can be cultivated and turned to be further fertile.

At the present day, on the Fen-Wei Plain, "loutu" soil (a very well developed agric soil with dualistically genetic profile) is widely spread, which is well known in producing cotton and wheat. The "loutu" is formed on the basis of loam-clay soil cultivated by continuously applying earth manures, as a result of that measure throughout long period of year, it has become a fertile soil as seen today, "Heilutu" (an argillified chernozem-like soil) distributed on the lands of loess plateau has a thick and mellow humic horizon, as such in the Dongzhi-yuan of Gansu Province, where it is called the natural granary for its fame in producing wheat. The slope-land covered with "Huangmiantu" (a mellow loessial entisol), on which soil erosion is rather serious, and under the prevailing normal climatic conditions, usually a satisfactory yield can be also obtained. As for

276

the slope of waste-land where if even a little care is to be taken of, it surely would have lush growth of forage grass, green manuring crop and economic forest. In fact measures are being used to change the "3–running field" into the "3–conserving field", so as to improve manuring and culture technics, the poor soil in this field would be yielding high and stable with bumper harvests. Besides, this "heimatu", the blackish meadow steppe soil, chestnut soil, and chernozem wide spread in the loess plateau are also fertile soils.

The main types of the water erosion may be introduced as follows:

Splash erosion (splashing)–soil particles are splashed and transported by the direct attack of raindrops. Therefore, splashing is an unnegligible soil erosion, which often happens without any runoff or before the runoff occurs in the field. The nature and properties of soil are the internal cause of soil splash, while the nature of rainfall may be considered as the external causes. The whole phenomena of soil erosion are the unity of the two opposites, i.e. erosibility and erosion-resistibility, and the two, either opposites or unites with each other within the same soil body. If we had made a detailed study of the soil properties, we could have easily found out that there would be at least two pairs of contradictions within the process of soil erosion. The first pair of contradiction is between denudability and denuda-resistibility or anti-denudability, while the second is between washability and wash-resistibility or anti-washability.

The corrosion may be divided into two different forms in general. One is the leaching phenomenon, which includes the chemical leaching and mechanical leaching in the soil forming process. Only in the case of loose and bare surface lacked of enough plant cover during the heavy rainfall can the mechanical leaching occur. This phenomenon is apparent only in some regions, but, in general, the mechanical leaching can reach only to the under-ploughed layers. The other is the so-called karst erosion process developed only in limestone or strongly calcareous regions. The loess contains much $CaCO_3$. And the caves and sink holes can be seen everywhere in the loess plateau. So, this kind of soil erosion is often considered as corrosion. But, in fact, there is only a very little relation with corrosion to be found.

Sheet erosion is a soil erosion that the soil particles mainly in suspended state are carried away by the sheet flow or laminar current along the slope of ground surface. Regardless of rainfall, snow-melted water, irrigation water, etc. the amount of each is more than the soil filtration capacity, the runoff on the land surface will soon be formed. When directly attacked by the raindrops, or dissolved by water, or disturbed by the runoff flow, the part of the soil body will be disrupted into the separated soil particles, and micro-aggregates. If it is only at this time when the runoff formed can carry away these separated soil particles, and aggregates, they will be migrated along with the directions of the runoff flow. At that time, the runoff is more or less equally distributed and the kinetic energy is so little that it can only carry away the dissolved substances and suspensible soil particles.

As the result of sheet erosion, the surface layer of a soil becomes sandy, thin, unfertilizable and barren at last. The strength of sheet erosion is generally defined (divided) by checking the eroded thickness and remaining layers of soil profiles. Because the layers of loess is very thick, loose and abundant in mineral nutrients and even the soil profiles in the vast areas are shifted or eroded wholly, farms are cultivated continuously, the effect of erosion is still seriously occuring, this kind of erosion has been named as the parent material sheet erosion by the author. Naturally, the sheet erosion of a soil profile has been named as the profile erosion. In a certain case, the practical strength of parent material sheet erosion is related positively with the slope of field (land) surface at less than $35°$, and so the gradient of slope is considered as the standard of strength of parent material sheet erosion. The surface of high flat plain and high terrace are more levelled and the soil profiles are retained completely in greater parts. At the same time, since the peasants often apply various earth manure every year, the top of the soil profile has been accumulated in different thickness. With or without soil erosion, its strength should be studied and examined by means of geographical contrast method. That is the distinguished accumulative thickness on the more level part near the divided line as a standard. Gully erosion when the runoff occurs on a slope is concentrated to such a degree that its kinetic energy is strong enough to cut the soil surface with some clear line traces. This shows that the gully erosion has already begun. It is evident that the role of gully erosion is not limited in suspension, but it has much enhanced with the pushing role caused by the current flow, too. Therefore, the pushing phenomenon has become more apparent. It will be seen that the role of water-denudation is the leading factor of gully erosion, which is entirely determined by the two opposites, the denudation-resistance and the runoff kinetic energy pushing against soil surface. Formerly we used to divide the gully erosion in rills (fine gullies), shallow gullies and gullies, and on the basis of the gully density appeared in per unit area or the gully volume for the classification of the gully erosion intensity.

The vertical sections of gullies disregarding their slope and sizes on the slope often assorts decline with the decline of the slope. But their horizontal sections often vary with some different conditions, such as the utilization of the slopes, denudation-resistance of the gully cutting soil horizons, different gully types and their developing stages. However, it is slmost developed as follows: shallow and narrow chute-shaped or V-shape→deep and rather wide chute shape→ladder shape. The cut-in gullies at the edge (marginal space) of plateau, terrace, high plain or any other upperlands have hanging gullies and sink holes at their ends, there are also waterfall dipping holes, earth water fall and even natural bridge at gully bottoms. In the late (mature) stage because of silting up and cutting down, the vertical section at the place near the ditch mouth does not coincide with the slope, then it changes into valley gradually. Water course or water system erosion from the beginning of the valley, under the slope, it belongs to the natural water system, namely, the main gulch, branch-

278

gulches and fine gulches or even streams, rivers and great river are included. As far as the influence of present erosion is concerned, the main kinds of erosions are the backward erosion of gulch end, expansion of gulch cliff and gulch bottom brushing and dipping and the lateral brushing of the river banks, etc. Its forms and shapes include the hanging gully, brush rock cave, sink holes, collapses, landslide, natural bridge, pond-lets, debris cone and so on. There are waterfall of earth and stone, waterfall holes, natural bridges and trough lines or gullies at the gulch or stream bottom and collapse of embankment, land slip, earth debris rolling at the ditch cliff. It will be seen that the erosion of water course is not merely the water erosion.

The wind erosion is apparently carrying out on the loess plateau especially at its northern and northwestern parts inside and outside of the Great Wall. The blowing barren land, monadnock, sand mounds, sand dunes and sand ridges may be seen every where.

As far as the outward appearance and the status quo in production are concerned, the land in the loess plateau mainly falls into four categories, namely: Shan – mountains, Qiu – hills, Yuan – high–flat–land and Chuan – stream plain or – valley to river valley.

With mountains, there are high and big (deep) mountains in addition to small ones, the former being mainly stony while the latter earthy blended with stone. The land in the mountainous areas according to its location, shape and nature of earth, can be divided into the following:

Ling – ridges, referring to lands in the vicinity of watersheds.

Liang–on the tops of mountains there are sometimes stretches of flat land.

Places below the ridges are called slopes.

On the slopes terraced land is built.

The land on the loess hills is generally termed slope land or mountainous land. The large tracts of land on the top are called whole land. The top area of ridges is also called covers. While the land adjacent to the covers is called slope land. The land on both sides of water-sheds is named with reference to its shape, size and location. They are respectively called Da – knob or top-like knob, Geda – knoll or kop–knoll (koppie), Gejian – pinnacle, hogback ridge, narrow hogback ridge, flat ridge, donkey tail liang (ridge) and so on. The slope land below the tops of ridges according to its shapes and different degrees of slope (rather gentle), is called Po – slope, Mao – projection slope, Ze – steep straight slope, Wa – recess–shaped. What is noteworthy is that all these names are given in reference to their appearance and not to their geomorphological formation. As for Yuan – high tablelike plain – is essentially an open and flat tableland at the top of a hill or flat upland. The slope near watershed is small, usually 1—3°. They all have the tendency of slanting towards the edges of the gullies and the whole tableland usually bevels towards the lower reaches of rivershed.

The slopes on the edges of the tableland are usually more slanting than others, sometimes as much as 5°. These places are called Pan – edge. At the places where two big gulches, streams or rivers are joining each other, the tableland would become rather narrow and more sloped and it is called Yuanzui-Cape, promontory or hill of the tablelands.

Chuan is the term for stretches of land on river banks. The local people use different terms to differentiate them from each other. They are called in the order of their size and length: jiang, he, chuan, gou – river, stream, streamlet, gulch.

Water courses smaller than gulches are divided into three gradations: the main, the tributary and the small. The main usually has an area of 20—30 km² or even 100 km². On both sides of it there are terraced lands.

The tributary or the small is sometimes called qu gully or cao gullet. Cultivated lands are mainly concentrated on these places.

The evolution of the land in the loess plateau is the direct result of water erosion and its functions. It is well-known that the economic activities of human being also have a direct impact on the evolution of land, and what is more, they even play a decisive role on it. Generally speaking, the role of human efforts can only be played under the specific conditions and in accordance with the law of natural development.

As a result of the difference in the properties of the earth, all the places that are covered with loess have their own peculiarities either in the course or in the rate of evolution. This can be explained as follows:

The evolution of the land in the loess plateau chiefly depends on the original types of geomorphology, the properties of the loess, the characteristics of rainfall and the impact of the economic activities of mankind on the earth's surface.

The geomorphological outlook of this area, as has been explained above, has mainly inherited from that before the deposition of loess. Therefore, the geomorphological outlook is divided into four different types:

1. Low hill and wide valley in the form of depressions formed by the not very thick loess deposits in the karst region.
2. Knob-like loess hill formed by thick loess deposits in the hilly region.
3. Typical loess plateau-high plain.
4. The sediment of loess along the bank of river cover terraced land of ancient times and adjacent slightly undulating slopes, thus, forming strips of plain-like land.

Erosion by water and impact of human activities all play their role on the basis of the above-mentioned characteristics. The general trend is that the earth's surface is changed from big to small and from whole to fragmented, while gullies are from shallow to deep, small to big, short to long and then from erosion to sedimentation: as for slopes they are changed from gentle to steep (after broken) and then in reversed order, from recess to straight and to projecting and then backward. This process is going on and on infinitely.

After long years of erosion, even a high flat plain will change into the geomorphological outlook just mentioned above.

280

A ridge may become fragmented and the ridge top become a series of knoll-shaped covers as a result of flushes by rivers and the deepening effects by rivulets. Judging from the evolution of the geomorphology of the loess plateau, we can say for certain that it will be changing in the following pattern:

Yuan (high table-like plain)→Liang (flat ridge)→Da or knob→Jian (pinnacle) This can be easily visible by remote sensing through images taken by plane and satellite. But it should be pointed out that the majority of ridges and gullies did not result from erosion by water but were formed on the basis of ancient geomorphology coupled with loess sediment and erosion of various times.

The whole loess plateau bears remarks of erosion of third degree and upward. Wherever, erosion is not strong at present time, there are still comparatively flat ridges of three gradations, that is, the main, the lateral, and the sublateral elongate liangs and loess mounds: where there is erosion being serious, there the main mound will become knoll-shaped, i.e. the lateral ridges becoming fragmented.

We must take into consideration the concrete conditions, learning advanced measures in a scientific way so as to wage production rationally, then, and only by way of such doing can we make the best use of the land.

THE ENGINEERING GEOLOGICAL
PROPERTIES OF VALLEY LOESS IN CHINA

ZHAI LI-SHENG

(Chinese Academy of Building Research)

Abstract—This paper expounds the engineering geological properties of valley loess in the main loess area of China. By the investigations of geology-geographical environments and the experiments of the material composition and physicomechanical properties and the subsidence features in four major industrial cities, it is recognized that the engineering geological properties of valley loess in China have regional difference with the variation of physical geology-geographical environment where alluvial loess originated and developed, and as a rule trends of variation from the west to the east and from the north to the south occur.

INTRODUCTION

Industrial and civil buildings are extensively gathered on the valley plains which are the focal point of building research. We have made some investigations and experiments in the major industrial cities which are distributed in different physical, geology-geographical environments. The data given here are only for referential use.

A BRIEF INTRODUCTION OF EXPERIMENTAL SITES

The general situation of geology and geomorphology at the experimental sites is showed in Table 1.

Table 1   The general situation of geology and geomorphology at the experimental sites

Place	River basin	Geomorphological site	Type of genesis	Depth of ground water (m)
Taiyuan	Fenhe River	the second terrace in Taiyuan Basin	alluvial	40
Luoyang	Luohe River	the second terrace in Luoyang Basin	alluvial	20
Sian	Weihe River	the second terrace in Sian Basin	alluvial	26
Lanzhou	Huanghe (Yellow) River	the second terrace in Lanzhou Basin	alluvial	20

MATERIAL COMPOSITION

Particle Composition

Using hydrometer method to analyze the particle composition, the results are shown in Table 2.

From Table 2, we learn that generally speaking, in the extensive regions the distribution of clay content of the loess on the second terrace in each valley has a general trend with gradual increase from the west to the east and from the north to the south which is similar to that of the loess on highland. It is understandable because the valley loess is the regenerated materials from the highland loess which were transported not very far from the original locality.

Table 2 Particle composition of valley loess

Place	Total amount of samples	0.05 mm (%)	0.05 0.005 mm (%)	0.005 mm (%)
Taiyuan	39	15	68	17
Luoyang	80	14	47	29
Sian	55	7	68	25
Lanzhou	41	18	69	13

Clay Mineral Composition

Judging from the analytical data (Table 3) of clay mineral composition, the clay mineral in the valley loess collected from three sites are mainly illite, but the content from each place is different. The difference was also reflected in the colloid chemical analysis for clay mineral identification.

Table 3 Clay mineral composition of valley loess

Place	Differential thermal analysis	X rays analysis	Clay colloid, ($<1\mu$) SiO_2/R_2O_4	
Taiyuan	illite	illite	2.7	2.8
Sian	illite	illite	2.5	2.6
Lanzhou	illite	illite	2.9	3.1

Chemical Composition

By the analytical results of chemical composition (Table 4) several features

283

are found:

1. The chemical compositions from the three places are different, especially the highly soluble salt, moderately soluble salt and exchange capacity, etc. which have significant effect to engineering geological properties of loess show large difference.

2. The maximum content of highly soluble salt is found at Lanzhou, the next is at Sian, then is at Taiyuan. The loess in Lanzhou has a large amount of easy dissolving sulphuric acid ion, but at Sian and Taiyuan have only little. As far as pH index is concerned, Lanzhou loess is less than Sian and Taiyuan. So, Lanzhou loess is of sulphate type, and Sian and Taiyuan are of carbonate type.

3. The content of moderately soluble salt (gypsum) is found higher in Lanzhou loess, but have not been discovered in Sian and Taiyuan loess.

Table 4 Chemical composition of valley loess

Place	Bulk chemical analysis (%)		Highly soluble salt (%)		Moderately soluble salt (%)	Low soluble salt (%)
	SiO_2	Al_2O_3	HCO_3	SO_4	$Ca\ SO_4$	$Ca\ CO_3$
Taiyuan	56.97	11.90	46.35	trace		9.85
	61.38	12.30	57.95			14.48
Sian	53.39	14.31	43.81	trace		8.00
	55.60	17.55	69.54			16.90
Lanzhou	57.20	12.13	25.32	8.40	58.22	8.10
	60.20	13.60	31.72	194.40	158.60	11.40

Place	pH value	Organic matter	Exchange capacity (m.e./100g. soil)	Exchange calcium (m.e./100g. soil)
Taiyuan	6.00	0.15	12.62	10.43
	7.57	0.43	17.29	12.47
Sian	7.46	0.26	18.52	12.69
	8.30	0.46	23.08	20.24
Lanzhou	6.48	0.36	8.16	7.82
	7.06	0.58	13.31	8.95

PHYSICAL AND MECHANICAL PROPERTIES

The physical and mechanical properties are shown in (Table 5 and 6).

1. Among the physical indexes, the natural moisture content, etc. show fairly large regional difference, and have a general trend of gradual increase from the west to the east and from the north to the south, but in some places the porosities are basically similar.

2. Among the mechanical indexes, the regional variation of cohesion and internal friction angle is small, but the difference of modulus of deformation in field is quite large.

Table 5 Physical properties of valley loess

Place	Total amount Index of samples	Moisture content (%)	Unit weight (g/cm³)	Dry unit weight (g/cm³)	Specific gravity	Void ratio	Plastic index (%)
Taiyuan	39	15	1.58	1.37	2.71	1.00	10
Luoyang	50	20	1.70	1.43	2.72	0.82	15
Sian	55	19	1.61	1.35	2.72	1.10	13
Lanzhou	41	13	1.52	1.34	2.70	1.05	10

Table 6 Mechanical properties of valley loess

Place	Indoor experiment			Outdoor experiment
	Compressibility (cm²/kg)	Cohesion (kg/cm²)	Angle of internal friction (degree)	Modulus of deformation (kg/cm²)
Taiyuan				20
Luoyang	0.038	0.027	18.0°	120
Sian	0.047	0.027	21.5°	80
Lanzhou	0.035	0.025	20.0°	50

THE FEATURES OF SUBSIDENCE

In the past few years, the knowledge on subsidence loess was gained, besides the coefficient of relative subsidence, the index of initial pressure of subsidence, was taken into account.

Coefficient of Relative Subsidence

1. In horizontal distribution (Table 7) the coefficient of relative subsidence of valley loess have a regional variation and a general trend with gradual decrease from the west to the esat and from the north to the south as that of the highland loess.
2. In vertical distribution (Table 7 and Fig. 1) the coefficient of valley loess as that of the highland loess also have a decreasing regularity with depth. Generally, the coefficient of subsidence near the surface is the largest, wavily de-

creases downward, and disappears at certain depth. The depth of subsidence
disappearance is regional, and has a trend of deeper in the west and shallower
in the east, and generally the subsidence all disappears at a depth of 10—15 m
below the surface.

Table 7 The relative subsidence coefficient of valley loess

Place	Thickness of loess (m)	Thickness of the subsidence loess (m)	Coefficient of the relative subsidence
Taiyuan	20	9	0.040
Luoyang	19	8.5	0.030
Sian	27	12	0.055
Lanzhou	11	14	0.062

Remerks: The coefficient of relative subsidence is the values under 3 kg/cm² pressure.

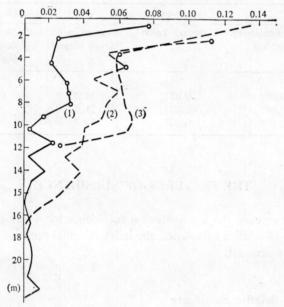

Fig. 1. Changes coefficient of relative subsidence with depth.
1. Taiyuan 2. Lanzhou 3. Sian

Initial Pressure of Subsidence

In our country, the value of initial pressure of subsidence has developed into
a mechanical index with practical significance. It also bears evident regional

286

features (Table 8) and a general trend with gradual increase from the west to the east and from the north to the south.

Table 8 The value of initial pressure of subsidence on valley loess

Place	Value (kg/cm^2)
Taiyuan	0.5
Luoyang	>1 or no subsidence
Sian	1.0
Lanzhou	0.35

CONCLUSIONS

1. The variation of material composition is regional in which the particle composition and the composition of soluble salt are even more notable and they generally have the variation trend from the west to the east and from the north to the south.

2. The variation of physical and mechanical properties is regional, and some indexes among the physical properties generally also have the variation trend from the west to the east and from the north to the south, but some physical and mechanical indexes such as porosity, etc. usually have little change.

3. The feature of subsidence is regional, in which the initial pressure of subsidence manifests considerable difference, while the variation of coefficient of relative subsidence is usually not evident, but they all have a general trend of variation from the west to the east and from the north to the south.

4. According to the synthetical analysis of material composition, physical and mechanical properties, and the features of subsidence, we have gained an overall knowledge about the engineering geological property of valley loess bearing regional difference with the differentiation of natural geology-geographical environment where the alluvial loess originated and developed and also have a general trend with orientations varied from the west to the east and from the north to the south.

PRINCIPLE AND METHOD OF BUILDING ENGINEERING GEOLOGIC REGIONALIZATION OF SUBSIDENCE LOESS AREAS IN CHINA

ZHAI LI–SHENG

(Chinese Academy of Building Research)

ABSTRACT

Loess is widely distributed over China. It has an area of 440,680 square kilometres, and subsidence loess occupies about three — fourths. At present it is an important task to study the principle and method of building engineering geologic regionalization of subsidence loess areas in China in a small-scale. Many natural factors have influenced on the regionalization, but under different conditions these influences of each factor are different. The practical work of regionalization requires: from large territory to small scope these factors should be analysed and summarized step by step, to find out each leading factor in different scope and used as the dividing basis of different levels. This is the principle of regionalization. According to these systematic principle, the same or similar principal aspect of building engineering geological conditions in all levels, can be maintained, and the practical effect of prediction can be reached.

We have accumulated the available data from 163 cities, among which 18,000 test data of undisturbed samples, including about 14,000 test data of undisturbed samples from 79 cities dispersed over the main loess area. According to these data the regionalization is made out as follows: Firstly we put the characteristic data of building engineering geological condition of investigated sites, on the base map of a scale of 1:1,000,000, then, we put on the corresponding stratigraphic and geomorphologic data is regarded as the relevant characteristic data of building engineering geological conditions. Then we draw all the stratigraphic and geomorphologic boundaries of natural conditions so that the work spread from sites to the whole area is completed. After that, by the similarity in building conditions, in a large area, combining with the scientific knowledge of distribution regularity of geological and geographical conditions, all the available data are generalized and divided into the following three levels i.e. 5 regions, 10 districts and 10 zones.

THE UTILIZATION AND PROBLEMS OF THE QUATERNARY DEPOSITS RELATING TO THE CONSTRUCTION OF WATER CONSERVANCY

PAN YAN–LING, YU YUNG–LIANG, AND CHEN ZU–AN

(Planning and Designing Administration, Ministry of Water Conservancy)

Abstract—The utilization of Quaternary residual lateristic soils, rudaceous clayey soils and various weathered rock waste (or fragments) as materials for dam construction are briefly described in the paper. The major problems of possibility of piping, liquefaction under shock, sliding resistance and settlement of alluvium in dam and lock foundations are also described with construction engineering cases.

INTRODUCTION

Quaternary deposits of alluvium, diluvium, eluvium and slope wash are distributed in vast areas in China. These deposits are important as relating to human activities and engineering constructions. We can do better in construction of water conservancy if we study the erosion, transportation and deposition of Quaternary deposits and learn their distribution regularity and engineering properties well.

THE UTILIZATION AND PROBLEMS OF QUATERNARY DEPOSITS IN THE CONSTRUCTION OF WATER CONSERVANCY

Either rebuilding river levees, constructing locks and dams, or digging channels are all related to the Quaternary deposits. Selecting dam (or lock) sites and investigating materials suitable for dam construction are the two items of utilization of Quaternary deposits relating to the construction of water conservancy

Materials for Dam Construction

A variety of soil and rock materials have been used for dam construction, such as loess, clay, and sand-gravel. As the construction of water conservancy

289

has been developed, these materials, however, are short of in many areas where dams are being built. In these cases, a lot of farmland could be destroyed if only the clayey soil is used for dam construction. In addition, the thickness of rolled—earth fill is limited. As a result, this will prolong the construction period and cost high expense in construction. Recently, since the technique of designing and building have been improved, constructional materials available for dams tend to be diversified. Even materials used for impervious core are also various in kinds. For example, weathered rudaceous materials, red clay, silt, fine sand, excavated waste formerly considered as "bad materials" are now used widely for dam construction. From the trend of development, almost all the soil and rock are available except materials containing organism more than 5%. So, the utilization of Quaternary deposits are expanded.

Weathered residual lateritic soil is widely used for dam construction in humid areas in South China. This clay is also distributed in other low latitudes areas over the world. It is of high moisture content, low density after rolled filling, and with stable granular structure. Based on the structural characteristics, its engineering properties are: moisture content is as high as 30—60%, its density is as low as 0.95—1.40 g/cm^3, liquid limit 50—70%, plastic limit 30—50%, the content of clay and colloidal particles is as high as 60%, friction angle $27°$—$30°$, cohesion 0.4—0.8 kg/cm^2. For instance, the dam of Hengshan Mt. reservoir in Zhejiang Province has a height of 48 m. The material of its impervious core is weathered residual clay (from tuff), when clay content is 20—35%, the relevant optimum moisture content and the dry density after rolled filling were 23—25% and 1.50 g/cm^3 and when clay content was 35—50%, the relevant values were 25—28% and 1.48 g/cm^3. The dam shell is consisted of sand and gravel, within which less than 5 mm occupies 40%, 5—20 mm 15%, 20—150 mm 45%. The filled dry density adopted was the dry density of the material with a relative density of 0.67.

Rudaceous soil is a good material for dam construction. It can be used either as impervious material or for dam shell embankment. The advantages of this material are easier to compact and with a high shear strength after rolled filling. However, the property of this material varies due to the difference of its weathered degree and gravel (or fragment) content. The dam of Xikeng Reservoir in Guangdong Province is 42 m in height, and the gravel clay (gravel 7%, sand, silt and clay 31% each was used for impervious core. Also, the dam of Zhaopingtai Reservoir in Hunan Province, has a height of 34 m. The rock fill toe at downstream and in the central part of the dam were filled with weathered rudaceous soil, containing coarse grains (>5 mm) 3—50%, clay 12—19%. In this case, the construction period was shortened and construction costs were decreased. But, the gravel content is to be cautious not to exceed 50% when using the gravel – bearing soil for impervious core. In general, when using gravel – bearing soil as dam shell materials, the gravel content is not limited. e.g. in Maojiacun earth dam, the gravel content reached as high as 60—80%. Using weathered rock waste (or fragments) for dam construction has the obvious advantages of using local materials, destroying less farmland, suffering less from rainy season during constructing, low construction costs and short construction period. A lot of dams consisted of weathered rock waste have been built successfully in some provinces in China. For example, the Zhelin Reservoir in Jiangxi Province, the dam is 62 m in height with a impervious core consisted of weathered Rock waste. Both upstream and downstream shells were filled individually with rock wastes of quartzose sandstone and slate. Also, the dam of Tangyu Reservoir in Shaanxi Province is a sloping core dam, 39.3 m in height, and both slopes were filled with muscovite—biotite quartzose schist fragments, which occupies about 40% of the total dam embankment. Also, the dam of Bikou reservoir in Gansu Province is 101 m in height. The lower parts of both slopes were filled with phyllite excavated waste. According to the experiences from embankment dams, in principle, every kind of hard or soft rock and even excavated wastes from tunnels, underground powerhouses or spillways can be used for rolled embankment dams. When fine content is excess in materials, the effect and variation of the coefficient of permeability should be taken into consideration. Rolling compactly the filling materials is the most important thing in using rock fragments or excavated waste for dam construction. Keeping the rock fragments lasting long is also very important to the stability of dam.

Recently, experiences in using various kinds of very fine sand for dam construction have been accumulated in some areas in China. For instance, Sangzhu Reservoir in Xingjiang Autonomous Region, even aeolian fine sand were used for dam construction, the dam is 53 m in height. Another example is the Jinjisha Reservoir in Shaanxi Province, a central core dam was built with fine sand shells, 46 m in height. Also, the Jingjiang River levee of Changjiang (Yangtze) River is 10 m in height, consists of very fine sand and fine sand, containing silt 1—11%, very fine sand 5—21%, fine sand 31—90%, medium sand 1—4%.

Foundations for Dams/Locks

The properaties of Quaternary deposits are the major factors influencing the selection of dam/lock site. In general, foundations can be divided into two types:

A. Sand-gravel foundation: It contains four groups—boulder, gravel, debris and sand foundations. Among them, sand foundation can be further divided into beach, aeolian, and alluvial deposits from their origins. Dry aeolian sand is even in size and loose, and is one of the "bad ones" among sand foundations. Generally speaking, sand foundation has certain bearing capacity, higher shearing strength with a coefficient of internal friction of 0.45—0.60, rather stable under water surface.

Building a dam/lock on sand-gravel foundation, the main engineering problems are seepage and piping. As a rule, seepage prevention should be made in sand-

291

gravel foundation when a dam is being built. For instance, Shitouhe Reservoir in Shaanxi Province, the dam is 104 m in height. There is a trench on each side of the bottom of the river, the sand-gravel bed is about 26 m in thickness and of high permeability. Subsident loess and extensively distributed sand-gravel deposits occur below the normal top water level at both slopes of the river bank. Owing to the serious seepage and settlement after filling, cut-off and curtain wall were performed according to the different thickness of the sand-gravel deposits. Also, the Bikou Reservoir in Gansu Province, the dam has a height of 101 m, the sand-gravel in foundation is 34 m thick, and was treated by using a curtain wall in depth. The dam of Guanting Reservior, is a central core dam with a height of 45 m. The foundation consisted of fine sand and sand-gravel, about 22 m in depth was treated by using upstream blanket and cut-off. The concrete gravity dam of Shangmaling Hydropower Station is 32 m in height. The sand-gravel in foundation is 34 m in thickness and a grouting curtain of 50 m deep was used. As to medium and small reservoirs, problems in different seriousness occurred due to the improper treatments or no treatment to dam foundations, or deficiencies in constructing. For example, Dongyuling Reservoir in Shanxi Province, the auxiliary dam is 8 m in height, its foundation failed in a consequence of seepage piping. Huangbizhuang Reservoir, the foundation of auxiliary dam was treated by using upstream blanket, downstream drainage and relief well. After it had been put into operation for twenty years, more than 230 holes and some ten cracks were observed in the upstream blanket. Some measures were taken again and the reservoir now is under normal operation. The Yuqiao Reservoir, where sand, sand-gravel and gravel clayey soils are widely distributed on the river bed and terraces I, II with an average thickness of 20—30 m and a maximum thickness of 40 m. Seepages were observed at the left bank slope after operation and the position of seepage spring rose along with the rising of water levels of the reservoir. Marshland and piping were also found at downstream.

Liquefaction under earthquake shock is another serious problem in sand foundations and sand-gravel foundations. In the previous ten years, eight $M>6$ earth-quakes occurred in China. Macroscopic investigation on these earthquakes indicated that loose saturated fine sand was easy to be liquidated, and its shock resistance was very poor. In VI° influencing area, water pumping, sand boil and cracks were widely observed in high/low flood plains, old channels and low lands, e.g. during Xingtai earthquake in 1966, cracks and sand boil occurred at some places in downstream flood plain of the dam of Huangbizhuang Reservoir and the old channel of Fuyang River. Another example was the 1968 Bohai Gulf earthquake in Shandong Province, it was in the VII° influencing area near by the river levees along both sides of Huanghe (Yellow) River, where water pumping rose to a height of 1—2 m during earthquake. Investigation information also indicated that not only the fine sand was easy to be liquidated but also the sand-gravel consisted of certain proportion of different grain size (coarse content is less than 60%) as well. Besides, the sand shell of dams can also slide by liquefaction, if it is not well filled. Also, during Bohai Gulf earthquake, slides occurred at upstream slopes of thick impervious core dams with sand shells at Yeyuan, Huangwu and Huangshan Reservoirs in Shandong Province. At Baihe dam of Miyun Reservoir, the sand-gravel protection of upstream slope slided into the reservoir during Tangshan earthquake in 1976, where the influencing scale of intensity was VI° and the volume of the slided materials reached 150,000m³. For this reason, attention must be paid to protect the sand foundation from liquefaction in constructing of water conservancy. The other type is clayey soil foundations which are distributed both in mountain valleys and plains. The origin, stratification and thickness of this type are various due to the difference of concrete sedimentary environments, sources of deposits and seasonal water flows. Among them, clay, loam and silt soil are mainly related to the construction of water conservancy. The characteristic of clay foundation is its low coefficient of internal friction (0.2—0.35 in general) higher cohesion, poor permeability, difficulty in self-drainage, and easy to have pore water pressure, which results in decreasing the foundation strength, or even losing its strength. There were cases of slide in building dams/locks on clay foundations, e.g. a large slide of 170,000 m³ took place at the 1:5 slope on the left side of the diversion channel downstream of the Haihe River tide lock. In order to enlarge the base of the structure and improve the shear-resistance between the structure and the foundation, silt soil of 2 m in thickness was taken away below the base surface and replaced by sands. Meanwhile, sheet piles were drived to 8 m in depth to enclose the foundation materials by which the foundation was protected from deformation.

The plastic flow and settlement of dam foundation are other critical problems in clay foundation, e.g. the dam of Bajiazui Reservoir at Puhe River, a secondary tributary of Huanghe River, is a homogenous dam with a height of 66 m. It was heightened 8 m in height on the base of the silt soil at the upstream of the dam in 1975. The silt deposited at the upstream of the dam in recent years was fine in size, which contains silt 50—70%, clay 15—20%, and shows a characteristic similar to loess with high moisture content, less permeability and poor strength. Using this deposits as a dam foundation in Bajiazui Reservoir more than 35% earthfill of the total embankment can be saved. But the pressure of the heightened dam would cause plastic flow and settlement in its foundation. So, the constructing speed was controlled in order to prevent the failure of the foundation in limiting the plastic zone and to accelerate the dissipation of pore water pressure in improving the property of the soil. At the same time, at the vicinity of the foundation and on the foundation itself earth-fill loading were covered, and a good result was achieved.

CONCLUSION

Origins of Quaternary deposits are complicated. They are mainly the products of endogenous and epigenous geodynamics. The utilization of Quaternary deposits is one of major studies in the earlier stage of the construction of water conservancy. To study the origin, the engineering properties and to find out the general regularity of Quaternary deposits are the prerequisite to evaluate the engineering geological conditions of a site. Only after the distribution regularity of Quaternary deposits is learned, can the dam/lock sites be selected properly. Only if we have learned the engineering properties of Quaternary deposits, can we utilize the deposits properly and transform the Quaternary deposits effectively.

ON THE SHALLOW FRESH WATER OF
OLD CHANNEL ZONE IN HEBEI PLAIN

Wu Chen, Wang Zi-hui, and Zhao Ming-xuan
(Institute of Geography of Hebei Province)

Abstract — In Hebei Plain, the fresh water of the shallow aquifer buried 30 metres and more in depth belongs to the fresh water of the shallow aquifer formed by the old channels. The old channels of three stages formed separately three shallow fresh water aquifers, and three shallow fresh water aquifers formed the shallow fresh water aquifer group in the old channels zone.

Several shallow fresh water belts with a base plate about 30 m deep occur on the heavy saline-water body of Hebei Plain* in SW-NE direction. The belts occupy more than one third of the total area of Hebei Plain and have an annual exploitation capacity of 23 hundred million cum, providing 50 percent of the water for the irrigated land of that area. Comparing with saline-water area, the crop yield in the shallow fresh-water belts is higher and the advantage is obvious, therefore, the shallow fresh-water belts are reputed by the peasants as "Treasured Land" or "Rich Land".

How were the shallow fresh-water belts formed ? What features do they bear On the basis of a preliminary analysis on the Map of Distribution of Shallow Fresh Water in Hebei Plain**, We have chosen the shallow fresh-water belt of Daming-Qing-he-Jingxian (coded as II 2 (1) B), which is the largest and most typical one for exploration profiling***, and come to a conclusion that the shallow fresh water belt well conforms to the old channel zone.

FEATURES OF RIVER-FACIES SAND BODY AND
SUBDIVISION OF OLD CHANNEL STAGES

The reason why we say the shallow fresh-water belts are actually the old chan-

---

* The Hebei Plain related here is the alluvial plain and littoral plain lying to the east of the fan-shapped piedmont alluvial plain of Taihang Mts. correspondent to the area of Heilonggang of Hebei Plain and with an area of more than 36,500 square km.

** The map was compiled by the Hydrogeological Section of the Geological Bureau of Hebei Province in 1975.

*** Along the 340 km-long shallow fresh water belts, there are 2,100 surface physical prospecting sites with the exploration depth of 50–80 m, 144 loggings and 163 prospecting holes with the depth of 35–60 m. Also 1,000 samples for analyses in granularity, mineral, clay mineral and sporo-pollen composition, micro-structure of quartz sand and isotopic age determination were made.

nel zone is that the water-bearing dene has the following sedimentary features of river-facies sand body:

1. Washed surface and river-bed holdup-facies sediments. The washed surfaces in depths of about 30, 20 or 8 m are widely found within the exploration depth. On the 30, 25 and 20 m–deep washed surfaces, sediments of river-bed holdup-facies are seen composed mainly of flat ellipse-shaped calcium pits and brown clay balls. Among the surfaces, the 30 m ones have been cemented by calcareous cement.

2. Sedimentary cycle fining upwards. Within the exploration depth, the wet-lithology of the shallow fresh water is a large sedimentary cycle from medium-fine sand, fine sand, silty fine sand to silt in ascending order which consists of 3 medium sedimentary cycles taking 20 and 8 m washed surfaces as their boundaries. Each of the medium cycles can be further divided into 2 small cycles, each of which consists of a binary structure of sand in the lower part and clayey sand and sandy clay in the upper.

3. Thinning of the deposits with fining of the sand grains, decreasing of the content of the bed-load transported matters with increasing of the suspended-load transported matter occur along the current direction caused by one-way current.

4. The plane shape of the water-bearing dene appears to be a belt with ropy margins.   The entire shallow fresh-water zone is a belt from southwest to north-east of 340 km long within the boundaries of Hebei Province and, generally speaking, the upper reaches is wide (7—12 km) while the lower narrow (1—4.5 km).   If a detailed division is made, it is in wide and narrow alternations and the outer margins of the dene appear ropy.

5. There are a considerable amount of shell remains of *Lamprotula aneigu-aoahar* as well as carbonized timber and tree branches in the dene. The clay and the sandy clay contain a large number of debris of fresh-water shells and a few ostracod fossils. On the surface of the quartz there exist conchoidal fractures and V-shaped traces as well as traces of pits and shallow channels dissolved by river water.

Connecting from the shallow fresh-water belt of Xunxian-Neihuang (in Henan Province), the dene joins another dene coming from the southwest nearby Daming and Guantao, and then extends to Tianjin.

Comparing the heavy minerals of the dene with those of the denes of the current channels, the similarity between heavy minerals of the dene to those of the Huanghe (Yellow) River and Zhanghe River can be observed. It was recorded, in many historic documents, that the Huanghe River, Qinghe River and Zhanghe River once flew northwards to the sea (1,2) in the past. Therefore, it can be determined that the dene is a river-facies one, which is the old channel of the Huanghe River-Qinghe River-Zhanghe River in a southwest to northeast flow direction.

The old channels usually remain on the terrene or is buried underground with the form of dene lens (in cross section) and belt (in vertical view), therefore, any dene which appears as a belt on the plane and a lens in the cross section and have the marks of the alluvial sediment is considered as an old channel. Since the rivers are different in property and composed matter and the dene lens are different in composition, we consider the dene lens of the lowest grade consisting of coarse sand as the main stream facies; the larger dene lens composed of several continuously-connected main stream facies with roughly the same altitude and washed surfaces at the bottom as an old channel; and the largest composite dene lens composed of overlapping channels of several stages as an old channel zone.

In accordance with the above-mentioned principles, we have made a comprehensive analysis on the data of 163 drilling holes and the vertical and cross sections constructed from the data. The figures demonstrate that the old channel zone is composed of large old channels of three major stages which are constituted separately by of six stages minor small old channels.

PALEORIVER PATTERN AND PALEOHYDROLOGY

It is mentioned above that the shallow fresh-water belt was precisely the old channel zone, which is composed of overlapping channels of three stages. But, why do the capacity and quality of water in the same old channel zone differ from each other? For the purpose of answering this question, it is neccessary to make further studies on the paleoriver pattern and paleohydrology of the old channel of each stage.

We have chosen four representative sites i.e. Daming, Qinghe, Jingxian and Qingxian from the Old Channel Zone II 2(1) B and, according to some on-the-spot surveying data, we have made a preliminary estimation of the paleoriver pattern and paleohydrology of the sites with certain formulas.

The data demonstrate clearly that the discharge of the old river of the 1st stage was the largest with a mean total annual runoff of 7,000 hundred million cum, a little more than two thirds of that of the current Changjiang (Yangtze) River. The amount of the bed load transported sand in the channel upstream of Qinghe was large (with an average of 32.44%) while that of the suspended-load transported sand was small (with an average of 4.16%) and the vertical gradient of the runway was large (with an average of 0.29 mm/m). So the channel upstream of Qinghe belongs to braided channel taking the bed load and saltation load as its main sand-transported form. Though the total annual runoff, vertical gradient of the runway and the amount of bed-load transported sand of the channel downstream of Qinghe all decreased while the amount of the suspended-load transported sand increased, it still belongs to braided-consequent channel with hybrid-load transported sand. Therefore, the dene is thick and wide with large-granular sand, so that the groundwater stored in the channel of this stage is large in capacity and good in quality and, what is more, the conditions of the runoff were favourable. The discharge of the old river of the 2nd stage decreased (with a mean total annual runoff of 2,500 hundred million cum, equal to one fourth of that of the current Changjiang River) and the content of the bed-load transported matter decreased (with an average of 5.92%) while that of the suspended-load transported (with an average of 11.7%) increased. The runway vertical gradient decreased (with an average of 0.085 mm/m). The channel upstream of Jingxian belongs to braided-consequent channel taking the hybridload as its main sand-transported form, so the accumulated dene is narrow and thin with small-granular sand. Owing to the facts mentioned above and the dene being mixed up with oozy, lacustrine accumulation in particular, the capacity of the groundwater of the channel of this stage is smaller and the quality is worse than that of the channel of the 1st stage. There was a further decrease in discharge for the river of the 3rd stage (with a mean total annual runoff of 500 hundred million cum, equal to one twentieths of that of the current Changjiang River). The content of the bedload transported matter in it was less than 1% while that of the suspended-load matter more than 40.71%, therefore, the channel of the 3rd stage belongs to consequent-curved channel taking the suspended-load as its main sand-transported form. In this stage, the river often breached and changed its course, resulting in an abundance of flood-plain facies deposits. Thus the dene deposits are wide and thin with small-granular sand, which cause the groundwater of channel of this stage to be less in capacity. However, the water is good in quality owing to fine circulation between the channel and aerial precipitation as well as open water.

DIVISION AND CHARACTERISTICS OF AQUIFERS

Since the shallow fresh-water belt is controlled by the old channel zone and both the water quality and capacity are dependent on the paleoriver pattern and paleohydrology, the division of the aquifers agrees well with that of the stages of the old channels, that is to say, the composite dene lenses of the old channel zone constitute the shallow fresh-water-bearing group and the dene lenses of the old channel of each stage a shallow fresh-water aquifer, while each main-stream facies dene acts as the smallest water-bearing body.

The 1st aquifer formed by the old channel of 3rd stage, located at a depth of about 8 m below the surface, is composed of two smallest water-bearing bodies (three bodies locally) of small old channels of two stages. The lithology is mainly of silt and, next the clayey sand with a thickness ranging from 2 to 5 m and reaching up to 8 m individually. The aquifer is referred to epiflag fresh-water aquifer.

The 2nd aquifer formed by the old channel of 2nd stage at a depth of about 8—20 m is composed of four to six smallest water-bearing bodies of small old channels of two stages. The lithology is of fine silt and silt with a thickness of 2—10 m. The 2nd aquifer is an aquifer of shallow fresh water which is second only to the 3rd.

The 3rd aquifer formed by the old channel of the 1st stage at a depth of 20—

30 or 35 m consists of six to ten smallest water-bearing bodies of small old channels of two stages. The lithology is of fine silt medium-fine sand and a little coarse-medium sand with a thickness of 10—15 m. It acts as the principal aquifer of shallow fresh-water-bearing group of the old channel zone in Hebei Plain. A part of the shallow fresh-water zone is slightly under pressure because a oozy water-resisting layer overlying this was an aquifer formed from the deposits of the old river of the 2nd stage.

It is worthy to point out that an old channel zone is not always the overlap of the old channels of three stages, therefore, the whole old channel zone is not definitely to be the shallow fresh-water belt. Only the old channel of the 1st stage, or, parts of the overlapping of the 1st and 2nd or the 1st, 2nd and 3nd stages channels can deserve to be called the shallow fresh-water belt. While existing alone, the 2nd stage belongs to an incompetent shallow fresh-water belt and the 3rd epiflag fresh-water belt.

MORPHOLOGIC TYPES OF THE OLD CHANNEL AND VARIATION OF WATER CAPACITY AND QUALITY

As described in the preceding sections, both water capacity and quality of the shallow fresh water are basically governed by the paleoriver pattern and paleohydralogy. Nevertheless, it is by no means that, in the same channel, the water capacity is similarly large and the quality identically fine. The water capacity and quality are also dependent on the different morphologic types of the old channel zone and, in particular, the variation of water quality caused by other factors in the later period.

The water capacity of the shallow fresh water of the old channel, generally speaking, ranges from 2.5—10 T/hr. m. Apart from depending on the above-mentioned overlapping relations among the old channels of three stages, the rich aquiferousness of the shallow fresh water has two distributive regulations as follows:

1. The water capacity is gradually reducing from upper reaches to the lower, e.g. in the area of Daming: 8—10 T/hr. m; in the district from Guantao-Zaoqiang: 5—8 T/hr. m and in the district to the north of Zaoqiang: 2.5—5.0 T/hr. m.

2. The water capacity is gradually reducing from the mainstream facies to the marginal facies, e.g. in the main-stream facies nearby Daming, 10 T/hr. m; the intermediate transitional facies, 5—10 T/hr. m and the marginal facies, 5 T/hr. m; in the area near by Cangzhou, the main-stream facies, 5 T/hr. m; the intermediate transitional facies, 2.5—5.0 T/hr. m and the marginal facies: less than 2.5 T/hr. m.

In this article, water quality is expressed in terms of the values of the apparent resistivity of the electrical logging. According to the unified stipulations of Hebei Province, when the value of PS is smaller than 15 ΩM, the water is of saline water and, while the value is larger than 15ΩM, it is of fresh water. In order to demonstrate the relations between the water quality and the old

299

channels more clearly, this article further stipulates that the water with a PS value more than $30\Omega M$ is called high quality fresh water and that with a PS value about $15\Omega M$ semisaline water. Making a comprehensive analysis on the surveying curves of 144 wells, the regularity is found as follows:

1. The base plate and borders of the shallow fresh water are basically consistent with those of the dene lens of the old channel. But somewhere abrupt contacts still remain between the saline and fresh waters.

2. The top plate, base plate and borders of the high-quality fresh water are extremely consistent with those of the main stream facies of the old channel. The contact relation is basically of gradual variation.

3. Most of the water stored in the center of the comparatively wider central beach and epibeach is semi-saline water.

4. When the dene of the old channel of the 3rd stage is absent, the water quality in the section presents a distributive regulation of being salt in the upper part and fresh in the lower. The dividing line is on the top of the dene lens of the old channel of the 2nd stage.

5. The water in the dene lens of Qingxian old channel is saline, which, perhaps, was caused by the transgression in Middle Holocene; nevertheless, the water in holes 5, 6 and 7 of the dene lens turns fresh, which is apparently caused by later-stage supply of the Grand Canal.

From the facts that the distribution of the shallow fresh water is well consistent with that of the dene of the old channel and, some of the boundaries between the saline and fresh water still keep abrupt contact with each other, the shallow fresh water of the old channel is the connate fresh water formed together from the old channel; as for the variation of the water quality occurring somewhere, it is caused by the pollution in the later stage.

To sum up, three main conclusions are drawn as follows: 1. The shallow fresh water belt of Hebei Plain owes its origin to old channels; 2. Both the capacity and quality of the shallow fresh water are dependent on the paleoriver pattern and paleohydrology and; 3. The variation of the fresh water somewhere in the old channel zone is caused by the pollution in later stage.

Acknowledgement — The departments that took part in the work of physical exploration, exploration drilling and sample analyses are: the Physical Exploration Team, the Drilling Team and the Sediment-Analysis Lab. of our Institute; Institute of Geology: Institute of Vertebrate Paleontology and Paleoanthropology, Academia Sinica; Institute of Geology, State Bureau of Seismology; Institute of Archaeology, Chinese Academy of Scocial Sciences; Beijing University; Tongji University; the Hydrological Team of the Geological Bureau of Hebei Province. The authors wish to express their thanks to all the above-mentioned departments.

REFERENCES

[1] Cen Zhong-mian, 1957, Changes of the Yellow River, published by People's Publishing House.

[2] Zou Yi-lin, 1980, An introduction to the changes of the channels in the lower reaches of the Yellow River and their influences. Fudan Journal, no. 8.

A GENERAL VIEW OF THE QUATERNARY
PLACERS IN CHINA

WANG ZHENG–TING
(Tianjin Institute of Geology and Mineral Resources,
Chinese Academy of Geological Sciences)

ABSTRACT

Quaternary placers in China are very developed. The Quaternary placers are
characterized by the perfect developement of genetic series. The eluvial,
talus, alluvial, beach, glacial and man-make placers consist of the genetic series
of Quaternary placers in China. The alluvial and beach placers are of the
most important chains among them. The glacial placer is less abundant
than the first two types, but plays an important role in formation. Prof. Li
Si-guang (J. S. Lee) considered that the gold and diamond placers in western
Hunan and eastern Guizhou Provinces were transported by glacial flows and
concentrated by the fluvioglacial deposition. The man-made accumulation
for example, the tails on a few old mineral mines may become a placer type
too.

Another characteristic of placer in our country is the good varieties of mineral
product series, including mainly the natural gold, cassiterite, dianite, tantaline,
zircon, monazite, ilmenite, xenotime, etc and next rutile, diamond and garnet,
and a small amount of platinum, piezo-electric quartz, etc., Some sapphire
placers are found in recent years. The alluvial tin, niobium, tantaline, diamond are predominant, but zirconium, cerium, titanium, mainly are beach
placers.

As to the stratigraphical position in the Quaternary sequence, the alluvial
placers can be found in many horizons, while beach placers are only found
in the upper part of Holocene deposits.